Springer Tracts in Civil Engineering

Springer Tracts in Civil Engineering (STCE) publishes the latest developments in Civil Engineering - quickly, informally and in top quality. The series scope includes monographs, professional books, graduate textbooks and edited volumes, as well as outstanding PhD theses. Its goal is to cover all the main branches of civil engineering, both theoretical and applied, including:

- Construction and Structural Mechanics
- Building Materials
- Concrete, Steel and Timber Structures
- Geotechnical Engineering
- Earthquake Engineering
- Coastal Engineering; Ocean and Offshore Engineering
- Hydraulics, Hydrology and Water Resources Engineering
- Environmental Engineering and Sustainability
- Structural Health and Monitoring
- Surveying and Geographical Information Systems
- Heating, Ventilation and Air Conditioning (HVAC)
- Transportation and Traffic
- Risk Analysis
- Safety and Security

Indexed by Scopus

To submit a proposal or request further information, please contact:
Pierpaolo Riva at Pierpaolo.Riva@springer.com (Europe and Americas) Wayne Hu at wayne.hu@springer.com (China)

More information about this series at http://www.springer.com/series/15088

Sai On Cheung
Editor

Construction Dispute Research Expanded

Springer

Editor
Sai On Cheung
Department of Architecture and Civil Engineering
City University of Hong Kong
Kowloon, Hong Kong

ISSN 2366-259X ISSN 2366-2603 (electronic)
Springer Tracts in Civil Engineering
ISBN 978-3-030-80255-4 ISBN 978-3-030-80256-1 (eBook)
https://doi.org/10.1007/978-3-030-80256-1

This Springer imprint is published by the registered company Springer Nature Switzerland AG
The registered company address is: Gewerbestrasse 11, 6330 Cham, Switzerland

To May, Veronica, and Bryan

With Everlasting Love

To Members and Friends of the

Construction Dispute Resolution Research Unit

With Deepest Gratitude

Foreword by Professor Chimay Anumba

Disputes have been an integral part of construction projects for many years. This is often attributed to the adversarial culture of the industry, which is rooted in the fragmented nature of the construction industry and the associated traditional procurement methods used for projects. These disputes sometimes result in legal proceedings that cost the industry considerable time and money. This makes an understanding of the nature of construction disputes and the means to their successful resolution a very important field of study.

Professor Sai On Cheung is one of the leading experts in this field, and his work has provided new insights and approaches that have advanced both research and practice. His 2014 book, *Construction Dispute Research: From Conceptualization to Resolution*, provided a comprehensive treatise of the subject and covered critical aspects of dispute conceptualisation, dispute avoidance, and dispute resolution through negotiation.

Following considerable additional work, Professor Cheung has built on the first book to produce *Construction Dispute Research Expanded*, which is destined to be a *vade mecum semper* of construction dispute researchers and professionals worldwide. This new book presents novel and leading-edge work on the following key issues: bias in construction dispute negotiation, impediments against settlement, and reality check of construction dispute negotiation conditions.

The book characterises and analyses bias in construction disputes and provides an innovative tool for detecting bias, and proffers strategies for mitigating the potential impact of bias on construction dispute outcomes. It also provides detailed insights into the impediments that prevent the settlement of construction disputes and discusses practical approaches to avoiding and/or overcoming these impediments. In the final section of the book, a reality check is provided on construction dispute negotiation conditions, and a new alternative dispute resolution (ADR) intervention model is proposed as a means of reducing the problems and costs associated with resolving construction disputes.

There is no doubt that this is a very timely book that will be invaluable to the construction industry as it seeks to transition to a less litigious culture that is based on more collaborative and relational procurement methods. It deals with real problems faced in the industry and offers a plethora of tips, guidance, suggestions, models, and mechanisms for success in addressing these problems. I commend this book to researchers working on aspects of construction disputes and to professionals devoted to fairly resolving these disputes. The construction industry stands to benefit in the long term.

June 2021

Professor Chimay Anumba
B.Sc., Ph.D., D.Sc., Dr. h.c.,
FREng, C.Eng, FICE, FIStructE, FASCE
Fellow of the Royal Academy of Engineering (UK)
Member of the National Academy of Construction (US)
University of Florida
Gainesville, FL, USA

Foreword by Professor Peter Brandon, OBE

At the heart of progress in any industry, there needs to be a strong element of investigation, an opportunity to gather and broadcast new information, and a desire to see the results being implemented in practice. Construction is no exception, and over several centuries, there has been a willingness to improve the choice of materials used, to develop new procedures, and to assess performance. It has been a relatively slow process of improvement, and the technological development has been focussed on the physical building or construction method. Other industries relating to the development of a product have found it much easier to engage with advances in science and engineering, but construction fell behind. It is only in the last century that construction began to see the value of seeking new improved solutions which would enhance its performance. Engineering played a leading part but with the advent of new specialisms and new professional disciplines, there was a demand for each to want to improve their contribution to the total construction process and product performance. Initially, the new disciplines did not have a strong university base of academic enquiry which would act as a springboard for advancement. This was true of construction management until postgraduate courses were developed. Subsequently, the subject was introduced as part of the curriculum in undergraduate degrees, and a profession of construction management was established. In the last 50 years, the development of the profession has been significant.

We are now entering a new phase where the techniques are developing fast, often driven by information technology and the complexity of large-scale construction, and we are now seeking to gain new insights into the way managers work and the tools they use. It is a time to gain a better understanding of what we do and to challenge our traditional processes. The authors of this book belong to a research group called 'The Construction Dispute Resolution Research Unit' at the City University Hong Kong. They are looking at the forces that are acting upon those engaged in the management and interpretation of construction information. They provide new insights into such aspects as negotiation, the management of disputes, and the behaviour of individuals and teams seeking to improve their understanding of human behaviour in an increasingly complex world. As we move into larger cities with high density, the question of scale is important. The techniques, we used for small developments, do no not always hold good. Hong Kong is an excellent laboratory for exploring such advancements. Tradition should be challenged, by not only the physical sciences, but also the social sciences. Indeed, we need to understand the mental processes whereby decisions are made and how this leads to the interaction between members of the construction team including the client. It is a very interesting and stimulating area of study and involves multi-disciplinary working across a very large swathe of skill and knowledge.

Research of this nature is not easy. One of the challenges of management research is the environment in which it attempts to provide knowledge upon which we can build. It tries to establish understanding, and a degree of certainty, in a world of business which is outside of its control. This world is transient. It is influenced by such issues as market forces, data which is difficult and expensive to collect, fashion and, of course, the problem that no two construction sites are the same. The range of variables is enormous, and communication is difficult to achieve for everyone concerned. Applied research has as its target something which is universal and repeatable. However, this is often difficult to achieve in a business climate which is subject to so much change and uncertainty. The studies of the Construction Dispute Resolution Research Unit are illustrating examples of how conscientious and enduring efforts can meet with these challenges.

The previous volume published by the authors from the Research Unit (in 2014) focussed on *Construction Dispute Resolution.* It provided suggestions for analysing disputes and suggested a framework for conceptualising the source and structure of such potential differences in opinion. This was based on their research at the time. Since then, they have developed their work to further explain the causes of adversity in disputes and the ways they may be resolved. They have chosen to explore the subject through important topics such as 'bias', 'intention to settle', 'power asymmetry', and 'inequity'. These terms will not be very familiar to many practitioners, and yet they summarise key aspects of behaviour and structure which create problems. In some cases, practitioners may not be aware of the way these issues affect their own behaviour. In addition, they may not recognise these issues in those with whom they work. This understanding can have a profound impact on the way we behave in the future and encourage us all to look for better ways of addressing the need to find better solutions. The authors have attempted to use this new knowledge in important government initiatives such as Hong Kong's 'Apology Ordinance' which came into effect on 1 December 2017. Such studies are useful in giving new insights because they reveal further understanding and suggest the limits to which a new concept, in this case legal, can be applied successfully in practice. These empirical studies are vital as the industry and its support documentation progress. Collectively, these studies have further integrated theories of dispute resolution with practice. This volume will be a key and useful reference for the Construction Dispute Resolution communities.

My own experience in judging construction research over many years suggests that management research is important in gaining an understanding but does not always develop long-lasting tools. If we take a typical research project, it can take a year to get the funding, two or three years to undertake the research work, and a further two years to begin to disseminate the information. It then takes a further one to any number of years to apply it in practice. Meanwhile, the industry is slowly being prepared for whatever might arise in the future. Unfortunately, the demand from industry is for an immediate piece of knowledge or a tool which the industry can pick up and use immediately. This book prepares the ground for new knowledge and techniques and is essential reading for all those engaged with disputes or the design of industry practice. I congratulate Professor Sai On Cheung in leading the Construction Dispute Resolution Research Unit and publishing this Construction Dispute Research Expanded. I highly recommend it to you.

June 2021

Professor Peter Brandon
OBE, D.Sc., D.Eng., D.Univ.,
M.Sc (Arch), FRICS, ASAQS
Professor Emeritus
University of Salford
Salford, UK

Foreword by Dr. Christopher To

Majority of legal textbooks on construction disputes focus on the hardcore aspects of the law. This book is a turning point, where Professor Sai On Cheung and leading experts in their respective fields have focussed on analysing the soft side of construction disputes, which is rare to say the least.

The book is split into three main parts with the first part on the *Bias in Construction Dispute Negotiation*. Under this part, there are five chapters which provide the reader with insights as to how bias occurs, what one deems to be bias, tools used to detect bias, and what steps one can take to minimise bias. The authors have used a lot of empirical data to back up their logic and reasoning, which makes reading the chapters enjoyable from a practical perspective. As a person who was to a certain degree involved in shaping the landscape of Hong Kong's construction environment, it gives me great pleasure to see in place, at last some meaningful pragmatic and verifiable information to assist those involved in the construction industry to understand the obstacles and how one can work towards reshaping the industry from a bias slant for the benefit of all mankind.

The second part anchors on the *Impediments Against Settlement*. Under this part, there are five chapters which centres on inequity, balance of power in negotiations, the use of mediation to the incentive of apologising during the settlement stages. Each chapter pivots on the important aspects of human interactions in settling differences during the negotiation stages within the construction context. Some of the writings go further in providing the reader with insights into situations where complex scenarios touching on individual relationships and behaviour within the wheeling and dealings stages of entering legal relations to the resolution of disputes. A fascinating combination of multiplexing styles of human interactions occurs within the construction context.

The third part concerns *Reality Check of Construction Dispute Negotiation Conditions* in which five chapters touch on the competitive environment of the construction industry and the manifestation of disputes, the interdependence of project team members that manage projects that have an impact towards disputes, the incentive towards minimising disputes, what one need to know about engaging third-party neutrals to the tight fist approach of using alternative dispute resolution to resolve one's disputes.

This book will be invaluable to local and international construction practitioners and parties and will also be a vital teaching tool to assist those in understanding the intricacies of operating under the fast-paced construction environment. Sai On and his team who have given so generously with their time, commitment, and knowledge are to be congratulated for their efforts in compiling a piece of work that blends theory into actual practice.

With the growth and focus of building smart cities of the future in Asia, construction practitioners will have the necessary awareness and the tools to effectively manage negotiations from the vantage point of knowing that human interaction plays a significant role in shaping how one prevents and resolves disputes efficiently and effectively.

The book will no doubt provide readers with indispensable guidance of the inseparable human nature which plays a crucial role within the dispute arena of the construction industry.

June 2021

Dr. Christopher To
D.Eng. (honoris causa), Ph.D.,
LL.M, M.A., LL.B. (Hons.), B.Eng. (Hons.), Chartered Arbitrator
Chartered Information Technology Professional and Chartered Engineer
Fellow of the Hong Kong Institute of Arbitrators
Fellow of the Chartered Institute of Arbitrators
Fellow of the Hong Kong Institute of Directors
Fellow of the Institution of Engineering and Technology
Fellow of the Hong Kong Institute of Engineers
Fellow of the Asian Institute of Alternative Dispute Resolution, Hong Kong
Barrister-at-law, Gilt Chambers, Hong Kong

Preface

In 2014, we published the research book *Construction Dispute Research* that presents a collection of studies conducted by the Construction Dispute Resolution Research Unit at the City University of Hong Kong. The 2014 volume focusses on conceptualisation of construction disputes and the use of relational techniques to manage them. The much-treasured theoretical anchors for construction dispute studies are thus articulated. Since then, our research direction expanded to looking into the particulars that affect the success or otherwise of dispute negotiations. We are pleased to include these studies in this volume titled as *Construction Dispute Research Expanded*, and there are 15 chapters arranged in three parts.

Bias in Construction Dispute Negotiation

Bias may well be fact of life and will also affect construction dispute negotiators. Our study first examines the happening of biases (Chapter 'The Happening of Bias in Construction Dispute Negotiation') and its conceptualisation in construction dispute negotiation (Chapter 'Conceptualising Bias in Construction Dispute Negotiation'). The conceptualisation is primarily developed from the effect of biases. Accordingly, four forms of bias were unveiled: preconception, self-affirmation, optimism, and interest oriented. The value of having a tool to detect these four forms of bias to facilitate bias study in construction dispute negotiations and beyond is self-evident (Chapter 'A Bias Detection Tool for Construction Dispute Negotiation'). The detection tool indeed had proved invaluable in the triangulated examination of bias manifestations that was repeated in the chapter 'Conceptualising Bias in Construction Dispute Negotiation'. The manifestations of bias in construction dispute negotiation have further been enriched by studying the application of two infamous special forms of bias: endowment effect and reactive devaluation (Chapter 'Special Forms of Bias: Endowment Effect and Reactive Devaluation'). Acknowledging the existence of bias in dispute decisions, the trendy use of multi-tiered dispute resolution process was put to test. It is argued that repeated evaluation of a dispute may bring unintended outcome of hardening of position due to judgmental biases of preconception and re-affirmation. We are indebted to a group of senior dispute resolution experts in providing invaluable advice on ways third-party neutrals can help in minimising the occurrence and effect of biases.

Impediments Against Settlement

Negotiation has been well recognised as the most efficient way to resolve dispute. Moreover, many negotiations do not result in settlement even the proposals on table are extremely attractive. This part of this volume discusses the impediments against settlement. Construction contracting is typically a buyer market. As a result, one-sided contracts, unilateral risk, and responsibility allocation are very commonly in construction industries across the world. Chapter 'Inequity and Dispute' discusses inequitable treatments would give rise to retaliatory responses ex post. Opportunistic exploitation therefore derails dispute settlement. As a continuation of this line of argument, Chapter 'Inter-organisational Relationship and Conflict Resolution' advocates the soothing effect of amicable interorganisational relationship in bridging the equity gap between the contracting parties. It is further suggested that incentive scheme can be used smartly to build interorganisational relationship. The dis-incentivisation arrangements used in a super mega project are used to provide the empirical support. After all, settlement through negotiation can only be possible when negotiating parties are having the intention to do so. Thus, insufficient intention is an impediment against settlement. Chapter 'A Note on Intention to Settle' summarises the signs and symptoms of settlement intention. Attention is then shifted to mediation—a form of assisted negotiation. Voluntary participation is considered a cornerstone of mediation. It is believed that mediated settlement will only be honoured by parties participating voluntarily. Moreover, the power asymmetry between contracting parties runs against underlying spirit of voluntary participation. Chapter 'The Paradox of Power Asymmetry and Voluntary Participation in Construction Dispute Mediation' therefore serves as a timely reminder for advocates of 'compulsory' mediation. Can emotion a stumbling block against settlement? The answer seems to be affirmative according to supporting reasons for the enactment of the Hong Kong Apology Ordinance in 2017. Chapter 'The Values of Apology in Incentivizing Construction Dispute Settlement' further suggests that offering an apology at the right time, in the right tone, and to the right person would release the tension between the parties whereby the chance of reaching a settlement will be greatly enhanced.

Reality Check of Construction Dispute Negotiation Conditions

Most negotiation theories assume that negotiators can negotiate at their own will and have adequate, if not perfect, information of their counterparts. In reality, these assumptions seldom fully hold. This part of the volume aims to provide reality check on several construction dispute negotiation conditions. Are construction markets really that competitive and hence so dispute prone as commented by many researchers and practitioners? Chapter 'Market Competition and Dispute Occurrence' introduces two commonly used indicators to measure market concentration. In examining the mega projects market in Hong Kong, cut-throat type of competition was not identified. Thus, disputes are being driven by severe competition may only be true when the market has no or low barrier to entry. Instead, self-interest maximising behaviours are more damaging as far as conflict is concerned. Chapter 'Recognising the Importance of Interdependence' is a wake-up call to the construction communities that we need to go back to the basics. The commercial reality is that project team members are interdependent. Dispute is counter-productive to the sustainability of interdependence. Chapter 'The Power of Incentivisation in Minimising Construction Disputes' offers some practical suggestions on how interdependence can be created. Typically, incentive arrangements include establishing project outcome targets that are agreed by the parties involved. These targets somehow would engender joint efforts of the stakeholders whereby certain level of interdependence can be resulted. Concerted effort would then be directed to accomplish the targets instead of practising self-interest taking behaviours. The last two chapters address issues arising from the use of alternative dispute method. Chapter 'Caveats for Using Third-Party Neutrals' presents a realty checking on the susceptibility of mediators being affected by the opportunistic moves of disputing parties. The result of the simulation used for the study indicated that continual training is necessary to remind mediators the need to stay impartial in providing mediation services. A parsimonious alternative dispute resolution intervention (ADR) model is proposed to raise the adoption of ADR to resolve construction dispute. Trusting the value of ADR was found to be the determining factor.

Construction Dispute Research Expanded includes pioneering works on biases, impediments against settlement and reality check. The problems addressed in these studies are real and detailed analyses shall contribute to the body of knowledge in construction dispute research. We are in debt to Professor Chimay Anumba, Professor Peter Brandon, and Dr. Christopher To for their enlightening forewords. This book cannot be possible without the excellent editorial assistance of Dr. Liuying Zhu and contributions of the members of the Construction Dispute Resolution Research Unit.

June 2021

Professor Sai On Cheung
D.Sc, Ph.D., M.Sc., MBA, LL.M., LL.B.
Director, Construction Dispute Resolution Research Unit
Department of Architecture and Civil Engineering
City University of Hong Kong
Kowloon, Hong Kong

Contents

Bias in Construction Dispute Negotiation

The Happening of Bias in Construction Dispute Negotiation

Sai On Cheung and Keyao Li

Introduction

As the world population grows, rapid urbanization has been the key form of development across the world. Naturally, infrastructural facilities are major parts of urbanization. In addition to the ever-growing in scale, construction projects have also become more and more complex and sophisticated. Close collaboration among technically and commercially specialized professionals is needed for the delivery of these projects. Very often, these professionals are coming from different organizations and even from different countries. The inevitable differences in organizational and personal interests add further managerial complexity for decision makers. In fact, it has been reported that many major construction projects did end with substantial disputes that took years to resolve [73, 116, 126]. Davis and Pharro [52] and Meng and Boyd [133] further found that human factor is one of the key barriers against dispute settlement. It has also been well documented that conflicts among project participants underpin disputes [42]. In construction dispute resolution, it is not uncommon to find parties involved overlooking notable chances of settlement, in extreme situations even win–win options are forsaken [124]. Therefore, addressing human factor is pivotal in settling construction dispute negotiation [43, 63].

Rational evaluation is considered to be the prerequisite of quality decisions. On this basis, negotiation studies are mostly conducted with this assumption whereby decision makers are rational and able to make sense of the available information and select the most appropriate options [15, 19, 26, 49]. As a matter of fact, human

S. O. Cheung (✉)
Construction Dispute Resolution Research Unit, City University of Hong Kong, Hong Kong, China
e-mail: Saion.cheung@cityu.edu.hk

K. Li
Future of Work Institute, Faculty of Business & Law, Curtin University, Perth, Australia
e-mail: Keyao.li@curtin.edu.au

S. O. Cheung (ed.), *Construction Dispute Research Expanded*, Springer Tracts in Civil Engineering, https://doi.org/10.1007/978-3-030-80256-1_1

decisions are not always made through deliberate analysis [4, 51]. The limitation of human decisions was pinpointed by Simon [174] whose concept of bounded rationality remains one of the key theories on cognitive human judgments. In this regard, Bromiley and Papenhausen [26] echoed that complete rational human decisions cannot be expected. In reality, many decisions are made by applying heuristics for the ease to achieve direct and quicker decisions. For this purpose, the problems are typically simplified with the application of heuristics [102, 175]. Moreover, Tversky and Kahneman [191] commented that systematic and predictable errors could arise if heuristics are used too grossly. It has been further reported that application of unconscious heuristics could lead to bias [10, 63, 106]. Specifically in construction, Stingl and Geraldi [180] reported that project success could be compromised by biased decisions [3, 193]. Other undesirable impacts include ineffective risk management [117, 118], suboptimal project planning [72, 158] and failure to respond to failing signals [85]. Notwithstanding the well documented influence of biases on decisions, study on bias in construction decisions is very limited and particularly in construction dispute negotiations (CDN hereafter) is almost uncharted. Notably, conceptualizing bias in CDN would be instrumental in detecting biased behaviors during dispute negotiations. Minimizing biased decisions is no doubt an effective way to improve the possibility of achieving successful dispute settlement. Efficiency will be enhanced when valuable resources are directed appropriately with more rational decisions. Against these backgrounds, the following questions are addressed in this study: (i) are construction disputing parties rational as assumed in most negotiation studies? (ii) do biases affect their decisions? (iii) if biases do exist, what are the underlying constructs of biased behaviours that would hinder amicable dispute negotiation? This study therefore fulfills the research gap of biases in CDN by proposing a bias conceptualization framework.

Admission of bias is not likely for dispute decision makers; however, their dispute negotiation behaviours might present evidence. Drawing from literatures on effects of bias in decision-making, a list of possible manifestations of bias in CDN was assembled. To conceptualize bias in CDN, the underlying constructs of biased behaviours are first explored. In these regards, an empirical survey was conducted in Hong Kong to study the propensity of construction project disputants of practicing biased manifestations. A conceptual framework of bias in CDN (the Framework hereafter) was proposed. With data collected from professional construction practitioners, the Framework was validated. The study is reported in seven parts: (i) Manifestations of bias in CDN; (ii) Data collection; (iii) Data analysis; (iv) Findings; (v) Implications on construction project dispute management; (vi) Limitations and future direction and (vii) Concluding remarks. Figure 1 shows the research plan of the study.

Manifestations of Bias in CDN

The artifacts of bias are drawn from literature. As indicated in Table 1, anchoring effect (AE), overconfidence effect (OE), self-serving effect (SE), hindsight effect

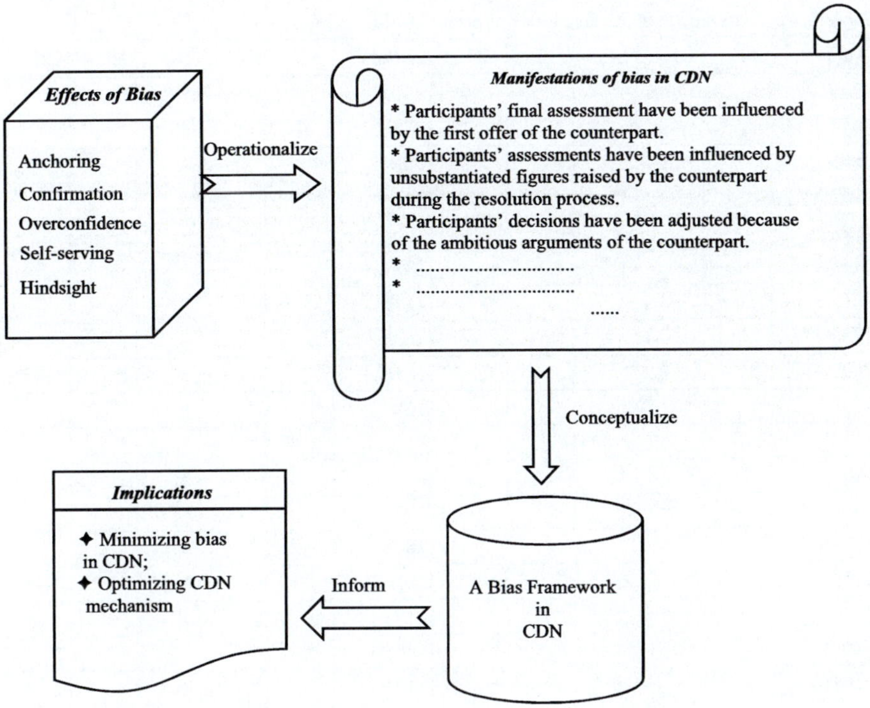

Fig. 1 Research plan of the study

(HE) and confirmation effect (CE) are found to be the mostly reported bias effects. The respective impacts on irrational human decisions have been summarized by Li and Cheung [121]. These five types of bias effect are discussed seriatim.

Anchoring

Anchoring effect describes the lasting impression derived from the first set of information uptake [135, 139, 145, 172]. It means that once an impression was made based on the first received information, it becomes anchor for any further decisions to be made. In construction dispute negotiation context, disputants under anchoring effect would rely on information received at the earlier negotiation stage without validating their applicability [33, 76, 191]. As a matter of fact, it is quite natural for people to analyze their problem at hand with reference to previously accepted information. However, Strack and Mussweiler [182] and Mussweiler et al. [145] found that influencing anchors derived from previously received information could surprisingly be irrelevant, uninformative, implausibly extreme or even self-generated. Tversky and Kahneman [191] demonstrated that even estimation about the percentage of African

Table 1 Key references of the five types of effect of bias

Authors	AE	OE	SE	HE	CE	Authors	AE	OE	SE	HE	CE
[88]	*	*				[20]				*	
[5]		*		*	*	[165]		*	*	*	
[89]	*	*		*		[86]	*			*	
[137]			*			[22]	*			*	
[92]	*	*				[160]	*			*	
[141]	*		*	*	*	[132]	*	*		*	
[61]	*		*		*	[111]				*	
[78]	*			*	*	[46]		*		*	
[60]	*		*			[161]	*		*	*	
[33]	*	*		*	*	[12]	*	*	*		
[101]	*		*			[7]		*	*	*	
[105]		*	*		*	[8]			*		
[99]					*	[193]		*			
[197]	*				*	[186]		*	*		
[150]		*	*		*	[80]			*		
[127]	*			*	*	[147]	*	*			
[107]		*		*	*	[182]	*				*
[108]					*	[113]		*	*		
[100]	*	*	*		*	[36]		*			
[109]		*			*	[54]			*		
[81]		*			*	[162]			*		
[173]		*				[191]	*				
[25]					*	[6]				*	
[85]		*				[28]				*	
[122]		*	*			[179]				*	

Notes Anchoring effect (AE); Overconfidence effect (OE); Self-serving effect (SE); Hindsight effect (HE); Confirmation effect (CE)

countries in the United Nations were affected by the anchors randomly selected by spinning a wheel of fortune in the subjects' presence. Russo and Shoemaker [166] also experimented that the estimates on the date when Attila the Hun was defeated in Europe had been impacted by anchors like irrelevant telephone numbers. Apparently in these experiments, subjects used the information provided by the researchers as anchors for their evaluations without challenging their relevancy and reliability. One possible explanation is that people start analyzing a problem with the first set of available information and subsequent decisions are then made through adjustments there-from Tversky and Kahneman [191]. These adjustments are typically not thorough and terminate prematurely when reaching a region of acceptable answers [34, 61, 145]. Therefore, with insufficient adjustments, the final estimation would be close to the anchors and likely to be suboptimal. Mussweiler [144] provided another explanation of anchoring effect with a selective accessibility process whereby decision makers estimate a target with the hypothesis that the target is similar or close to the anchor [37, 143, 145, 182, 181]. Further, Chapman and Johnson [33] proposed that under anchoring biases, factors that are common to the anchor are considered during decision making while the others are ignored [34]. Besides, Bergman et al. [16] demonstrated that anchoring effect could be reduced with greater cognitive skills. Furnham and Boo [76] and Furnham et al. [77] further argued that human attributes, such as mood, knowledge and personality would influence the strength of anchoring effect. Moreover, computer-based training programs have been found to be effective in mitigating anchoring effect [1, 47, 56, 142].

Overconfidence

Previous studies have found that decision makers with high confidence are likely to overestimate the accuracy of their judgment—a phenomenon called overconfidence [55, 67, 68, 88, 109, 123, 134, 188]. Moore and Healy [140] and Bazerman and Moore [13] described three principal forms of overconfidence: (i) over-estimation of one's actual performance, (ii) over-placement of one's performance compared to others'; and (iii) over-judgment of the accuracy of one's assessment. Klayman et al. [109] and Tsai [190] added that confidence level is idiosyncratic and varies with the way a problem is presented and the reference knowledge domain. Prior researchers also explored the relationship between confidence and difficulty [100, 123, 183]. The more difficult the problem is, the more confident one tends to be. Whereas easier problems are likely to acquire lower confidence [23, 163, 190]. Overlooking new information while sticking with previous knowledge is another manifestation of overconfidence [177, 190, 191]. Interestingly, Klayman et al. [109] proposed that overconfidence can stem from inherent judgmental error. To this end, one possible source of inherent judgmental error is inappropriate assessment of the validity of the collected information [62, 69, 109, 177]. This judgmental error is akin to the concept of cognitive limitation whereby people are not good at interpreting additional information to verify the accuracy of their judgment [190]. Furthermore, Radzevick and Moore [164] proposed that objective circumstances, like social pressure could also exacerbate overconfidence. Bazerman and Moore [13] demonstrated that decision makers may use overconfidence to relieve internal dissonance and tension when faced with tough decisions. More recently, Dunning [57] and Feld [66] found that lower-skilled subjects are showing higher level of overconfidence.

Self-serving

Self-serving effect is a kind of cognitive discourse whereby an individual is inclined to claim contribution for achievement of positive outcomes. However, if the outcome is negative, one would either blame the counterpart or take external factors as excuses [29, 53, 122, 137, 149]. Furthermore, self-serving tendency was found to be a type of self-protection mechanism to maintain self-esteem by denying the responsibility of negative outcomes [171, 200]. Campbell and Sedikides [29] described a self-threat model—when one encounters unfavorable feelings of self-threat such as being challenged, questioned, blamed or despised, one would leverage self-serving mode with the aim of shedding responsibility of the negative results and protecting self-image. Miller and Ross [137] examined the relationship between optimistic attitude and self-serving tendency, they claimed that self-serving behaviors are underpinned by over-optimistic predictions and expectations [184]. Similarly, self-serving behaviors are backed by the desire to protect and enhance positive self-image. Thus, it can be summarized that self-esteem is the prime motivator of self-serving acts [24, 48]. From another perspective, Lerner and Keltner [120] and Coleman [48] found that emotion also matters. For example, Baumgardner and Arkin [11] identified positive emotions and bright self-image made one more likely to bring about self-serving effect than negative emotion. In addition, pessimism would generate detrimental effects on one's self-image and self-concept. In negotiation context, decision makers under the impact of self-serving bias only take actions that are beneficial to themselves and believe that this is not unfair to the other stakeholders [114, 149, 152].

Hindsight

Hindsight effect describes that people over claim their estimation of happening likelihood of an event after knowing the outcome [18, 38, 70, 155, 157]. Hindsight explains the influence of outcome knowledge on judgment under uncertainty. The phenomenon of claiming known outcomes as being inevitable is known as "creeping determinism" [17, 94, 165, 196]. Hawkins and Hastie [89] further demonstrated that creeping determinism is the result of instantaneous integration of outcome information into the knowledge frame of an event. One explanation of the effect of outcome knowledge is assimilation [70]. Subjects assimilate the already known outcome into their memory about the issue. Assimilating outcome knowledge to match with the event background, thus making the outcome appears more likely to happen right at the outset. The effect of hindsight is also called knowledge updating [21, 89, 165].

Fischhoff [70] opined that one is typically not aware of the influence of hindsight effect [159]. It is because hindsight effect occurs as the cognitive activities of information integration, resulting in the simplification of cause and effect [50]. As a result, known happenings are then incorporated into memories that are taken as background information of the event [115, 125]. The better one makes sense of the stories by injecting meaning into the past, the stronger is the hindsight effect [21, 89, 165].

Confirmation

Confirmation effect describes one's tendency to search for and interpret information therefrom to confirm existing beliefs and assumptions [112, 150, 192]. Klayman [107] added that confirmation effect can be in the forms of inclination to retain or reluctance to abandon a favored position. When one takes on a position or forms an opinion on an issue, the subsequent information search becomes highly selective and aims mostly to defend the previous position [112, 150, 155]. Characteristics of confirmation effect include: (i) focusing only on favored assumptions while neglecting alternatives, (ii) giving greater weight to evidence that support existing beliefs and undermining those that suggest otherwise; (iii) testing only cases that fit the hypotheses; and (iv) interpreting information into the way that one is looking for [35, 71, 91, 138, 150]. With the wishes to confirm existing assumptions, one would render unwarranted weights to information that supports the same [112, 150, 170]. Millward and Spoehr [138] also found that decision makers were inclined to only test cases that they expected would tie with the hypotheses [64, 185]. Nickerson [150] demonstrated confirmation effect by the following example: suppose the concept to be discovered is small circle, if the subject's hypothesis is small red circle, he would then only test those cases with the hypothesized features of small, red, and circular. It is obvious that this approach would not unveil the small circle concept because the confirmation effect would exclude other qualifying examples like small yellow circle.

With the afore-mentioned deliberations on bias effects, a list of manifestations of bias in CDN was developed and summarized in Table 2.

Data Collection

The study requires data on the frequency of construction dispute negotiators practicing biased behaviours. Measurement statements were developed from bias manifestations summarized from the aforementioned literature. For example, "Participants' final assessments have been influenced by the first offer of the counterpart" was changed to "My final assessments have been influenced by the first offer of the counterpart". The data collection survey has two parts. The first part collects personal particulars, including professional background of the respondents. In the second part of the survey, the respondents were asked to evaluate their practice in CDN against a six-point Likert scale of frequency from 0 (not at all) to 6 (always) for each of the measurement statements. The Hong Kong construction professionals with dispute resolution experience are the target, including professionals working in contractor,

Table 2 Manifestations of bias in CDN

Manifestations	References
1. Participants' final assessments have been influenced by the first offer of the counterpart. (Anchoring)	[41, 78, 187]
2. Participants' assessments have been influenced by unsubstantiated figures raised by the counterpart during the resolution process. (Anchoring)	[97, 147, 191]
3. Participants' decisions have been adjusted because of the ambitious arguments of the counterpart. (Anchoring)	[32, 59, 128, 130]
4. Participants cannot get away with the assessments made at prior round of dispute negotiations. (Anchoring)	[58, 59, 194]
5. Participants become immune to alternative reasonable assessments after forming their first assessments about the dispute. (Confirmation)	[55, 71, 107, 150]
6. Participants have paid more attention to the information which is consistent with their prior knowledge of the dispute. (Confirmation)	[155, 156]
7. Participants incline to interpret further information as evidence to justify their assessments. (Confirmation)	[90, 95, 189]
8. Participants search for information that confirms their assessment. (Confirmation)	[30, 112, 150, 155]
9. Participants consider that their party has contributed more to the positive outcomes of the resolution. (Self-serving)	[29, 65, 82]
10. Participants endorse information that supports their assessments. (Confirmation)	[150, 155]
11. Participants are very optimistic about the likelihood of winning irrespective of the arguments of the counterpart. (Overconfidence)	[14, 147, 168]
12. Participants totally believe that the outcome of the resolution will be good for their party. (Overconfidence)	[14, 147]
13. Participants are very confident that their ambitious requests will succeed. (Overconfidence)	[14, 113, 186]
14. Participants believe that their party is able to avoid bias. (Overconfidence & Self-serving)	[136, 162]
15. At the conclusion of the dispute, participants feel "I know the outcome all along". (Overconfidence & Hindsight)	[94, 165, 196]
16. Participants think that the counterpart is having bias. (Self-serving)	[27, 162]
17. Participants think the counterpart should take greater responsibility to the negative outcomes of the resolution. (Self-serving)	[65, 82, 171]
18. Participants stick to the arguments that are beneficial to their party. (Self-serving)	[8, 65, 113, 186]
19. After knowing the negative outcome of the resolution, participants consider the demands of the counterpart during the dispute as unreasonable. (Self-serving & Hindsight)	[176, 179]
20. At the conclusion of the dispute, participants consider the failure to settle as inevitable because of the negative attitude of counterpart. (Self-serving & Hindsight)	[70, 87, 155]

client and consultant. Both on-line and paper-based questionnaires were used. The contacts of respondents were mainly collected from: research networks; websites of government departments including Hong Kong Housing Authority, Buildings Department, Civil Engineering and Development Department and Department of Justice; websites of professional institutes including The Hong Kong Institute of Engineers, The Hong Kong Institute of Architects, The Hong Kong Institute of Surveyors, Institution of Civil Engineers and Hong Kong Institute of Construction Managers. Hard copy of the questionnaire was also distributed at learned societies' seminars and workshops. 347 questionnaires were distributed, and 134 valid responses were obtained, representing a satisfactory response rate of 38.6% [2, 9]. The organization, professional background, years of experience of the respondents as well as the dispute types they were involved are presented in Table 3.

Table 3 Profile of respondents

Profession	Percentage (%)	Organization	Percentage (%)
Architect	3.0	Contractor	29.9
Builder	9.7	Client	24.6
Engineer	48.5	Consultant	45.5
Building surveyor	3.7	Total	100.0
Project manager	12.7		
Quantity surveyor	17.9		
Others	4.5		
Total	100.0		
Years of experience cons	Percentage (%)	Dispute types	Percentage (%)
Below 5 years	28.4%	Building services installations	16.4
5–10 years	27.6%	Building work	37.3
10–15 years	11.2%	Civil engineering work	36.6
15–20 years	7.4%	Maintenance work	6.0
Above 20 years	25.4%	Others	3.7
Total	100.0	Total	100.0

Data Analysis

Bias Framework Development

The respondents were grouped according to different professional background: Contractor group (N = 40), Client group (N = 33) and Consultant group (N = 61). The results are shown in Table 4. Item 1 to Item 20 are the twenty manifestations listed in Table 2. For each professional group, the mean value of the responses of each manifestation was calculated. Standard deviation (S.D.) was also calculated to indicate the dispersion of the responses. The manifestations with higher occurrence evaluations are considered as having higher propensity. Accordingly, relative rankings were calculated in the descending sequence of mean scores of the twenty measurement statements. Item 8 "I search for information that confirms their assessments" and item 10 "I endorse information that supports their assessments" were ranked within top 3 in all the professional groups. To further explore the underlying construct of the measurement statements, principal component factor analysis (PCFA) was conducted using IBM SPSS version 23.0.

Varimax rotation was conducted to achieve a simpler factor structure [103]. Kaiser–Meyer–Olkin (KMO) and Bartlett's test of sphericity were performed to test the sampling adequacy and suitability of the data [31]. The results are satisfactory and summarized in Table 5 [104]. Hair et al. [83] recommended that only factors with eigenvalue greater than one are to be kept. Factor loadings stand for the correlation between the items and the factors, the larger the factor loading, the more representative the item is representing the factor [83]. According to Hair et al. [83], a factor loading of 0.5 is the baseline of significance both statistically (alpha = 0.05) and practically for a sample of 134 and this was applied accordingly. As a result, a four-factor structure without cross loading was extracted and presented in Table 6.

Upon examining the measurement statements under each factor, it was found that manifestations under the same factor sharing a common enabler. In this regard, four factors were extracted as four constructs of biased behaviors in CDN. With reference to the meaning, nature and underpinning theories of manifestations under each construct, four constructs represent four types of biased behaviour in CDN. The four types of bias were identified as: preconception, self-affirmation, optimism and interest-oriented respectively.

The reliability of measurement statements and correlation coefficients of the four types of bias are shown in Table 7. From Table 7, measurement statements under each factor have satisfactory internal consistency, with Cronbach's alpha values larger than baseline of 0.70 as suggested by Nunnally and Bernstein [151]. Besides, the significant correlation in Table 7 indicates interrelatedness of the four biases. The bias framework in CDN is shown in Fig. 2.

Factor score of each type of bias was then calculated by the average of respective item scores. For example, the factor score of preconception bias is the average of the respective score of item1, item2, item3, item4 and item5. Analysis of variance (ANOVA) multiple comparison was conducted for different background groups to

Table 4 Relative importance ranking of the items

Item	Client (N = 33)			Consultant (N = 61)			Contractor (N = 40)		
	Mean score	S. D	Ranking	Mean score	S.D	Ranking	Mean score	S. D	Ranking
Item 1	3.30	0.728	18	3.10	0.768	16	3.28	0.816	17
Item 2	3.30	0.728	17	2.64	0.895	18	3.20	0.758	19
Item 3	3.94	0.496	6	2.64	0.984	19	3.15	0.622	20
Item 4	3.30	0.684	16	3.08	0.802	17	3.25	0.707	18
Item 5	3.18	0.983	20	2.64	0.753	20	3.38	0.868	14
Item 6	3.94	0.966	5	3.90	1.012	3	4.08	0.997	2
Item 7	4.03	1.185	3	3.56	1.009	5	3.65	0.893	7
Item 8	4.42	1.032	1	4.02	1.057	2	4.30	1.091	1
Item 9	3.70	1.015	10	3.36	0.913	9	3.70	0.823	4
Item 10	4.30	1.045	2	4.03	0.983	1	3.98	1.025	3
Item 11	3.48	0.755	15	3.21	0.839	13	3.40	0.900	12
Item 12	3.97	0.810	4	3.49	0.977	7	3.53	1.062	9
Item 13	3.70	0.984	9	3.18	0.847	15	3.40	0.928	13
Item 14	3.76	1.119	7	3.52	1.010	6	3.65	0.975	6
Item 15	3.55	0.833	13	3.26	0.854	10	3.48	0.847	10
Item 16	3.55	1.252	14	3.25	0.767	11	3.58	0.844	8
Item 17	3.73	1.039	8	3.23	0.783	12	3.38	0.838	16
Item 18	3.64	1.025	12	3.75	1.059	4	3.68	1.023	5
Item 19	3.24	0.936	19	3.18	0.847	14	3.38	0.952	15
Item 20	3.64	0.859	11	3.41	0.990	8	3.40	0.928	11

Table 5 Measures of sampling adequacy and suitability

KMO	Bartlett's test of sphericity				Factor			
	Chi-square	DF	Sig		1	2	3	4
0.842	852.973	190	0.000	Eigenvalue	5.922	1.720	1.675	1.456
				% of Variance	29.609	8.602	8.376	7.280

Note KMO = Kaiser–Meyer–Olkin measure of sampling adequacy; DF = degree of freedom; Sig. = significance

Table 6 Factor matrix of bias in CDN

Manifestations in CDN		Factor			
		1	2	3	4
Preconception bias	1. My final assessment has been influenced by the first offer of the counterpart				0.607
	2. My assessments have been influenced by unsubstantiated figures raised by the counterpart during the resolution process				0.626
	3. My decisions have been adjusted because of the ambitious arguments of the counterpart				0.685
	4. I cannot get away with the assessments made at prior round of resolution of the dispute				0.578
	5. I become immune to alternative reasonable assessments after forming my first assessment about the dispute				0.646
Self-affirmation bias	6. I have paid more attention to the information which is consistent with my prior knowledge of the dispute		0.674		
	7. I incline to interpret further information as evidence to justify my assessments		0.567		
	8. I search for information that confirms my assessments		0.743		

(continued)

Table 6 (continued)

Manifestations in CDN		Factor 1	2	3	4
	9. I consider that my party has contributed more to the positive outcomes of the resolution		0.520		
	10. I endorse information that supports my assessments		0.793		
Optimism bias	11. I am very optimistic about the likelihood of winning irrespective of the arguments of the counterpart	0.570			
	12. I totally believe that the outcome of the resolution will be good for my party	0.687			
	13. I am very confident that my ambitious requests will succeed	0.775			
	14. I believe that my party is able to avoid bias	0.779			
	15. At the conclusion of the dispute, I feel "I know the outcome all along"	0.540			
Interest-oriented bias	16. I think that the counterpart is having bias			0.609	
	17. I think the counterpart should take greater responsibility to the negative outcomes of the resolution			0.717	
	18. I stick to the arguments that are beneficial to my party			0.607	
	19. After knowing the negative outcome of the resolution, I consider the demands of the counterpart during the dispute as unreasonable			0.672	
	20. At the conclusion of the dispute, I consider the failure to settle as inevitable because of the negative attitude of counterpart			0.721	

test whether there was significant difference among factor scores of different types of bias. The results of ANOVA multiple comparisons in Table 8 show that in all the professional groups, the mean factor score of self-affirmation bias is the highest among the four types of bias. That suggests that professionals from all background groups had higher frequency of practicing self-affirmation bias in their construction

Table 7 Reliabilities and correlation coefficients of the four types of bias

Factor	Cronbach's alpha	Correlation coefficient			
		Preconception bias	Self-affirmation bias	Optimism bias	Interest-oriented bias
Preconception bias	0.718	1.000	0.484*	0.419*	0.382*
Self-affirmation bias	0.789	0.484*	1.000	0.445*	0.445*
Optimism bias	0.770	0.419*	0.445*	1.000	0.389*
Interest-oriented bias	0.750	0.382*	0.445*	0.389*	1.000

Note *Correlation is significant at the 0.01 level (2-tailed)

dispute negotiation process. Specifically, for the consultant group, the mean factor score of preconception bias is significantly lower than other types of bias, suggesting that consultants are relatively less prone to have preconception bias among the four types of biases.

Validation

To validate the proposed CDN bias framework, the PCFA factor structure was tested by the confirmatory factor analysis (CFA) (Fig. 3). Error terms were included in the CFA model to represent the proportion of the variance in the variable that is not explained by the factors [74]. These include measurement errors in observed variables and residuals in latent variables [169]. The statistical significance of the CFA model was assessed by goodness-of-fit (GOF) measures, including comparative Chi-square/df (χ^2/df), goodness-of-fit index (GFI), adjusted goodness-of-fit index (AGFI), root mean square error of approximation (RMSEA). comparative fit index (CFI), Tucker–Lewis Index (TLI), parsimony normed-fit index (PNFI) and parsimony comparative fit index (PCFI). The results of the GOF indices are shown in Table 9.

Bootstrapping was conducted to augment the reliability of CFA results in this study [110, 153, 154, 167]. Bootstrapping allows the testing of the significance of parameter estimates by comparing the results from original data set with the bootstrapped estimates [45, 98]. It can be seen from Table 10 that the regression weights generated from the original data set were within the upper and lower bounds generated from bootstrapped data set at 95% confidence level. Besides, all estimates have critical ratio values >1.96, indicating their statistical significance at 95% confidence level [93]. These results collectively indicate that the parameter estimates obtained from the CFA analysis are statistically significant. Confirmatory factor analysis (CFA) was conducted using IBM SPSS Amos version 23.0.

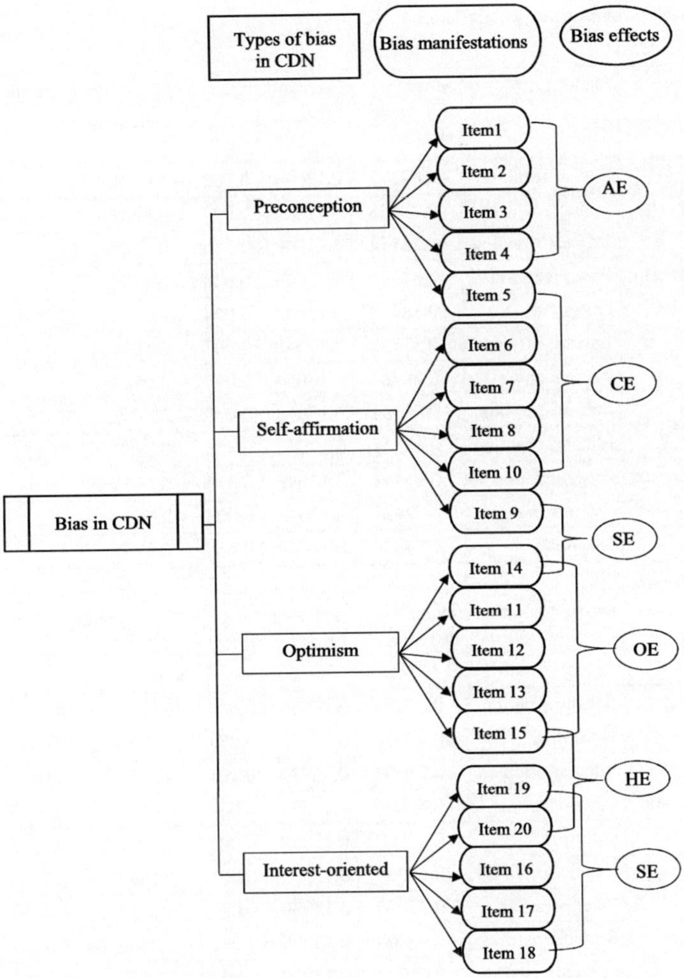

Fig. 2 A framework of bias in CDN

Findings

A CDN bias framework is developed with four types of bias: preconception, self-affirmation, optimism and interest-oriented. The bias framework has also been validated by a statistical-significant CFA.

Under preconception bias, item 3 (0.685), item 5 (0.646), item 2 (0.626), item 1 (0.607), item 4 (0.578) are included and presented in descending order of factor loadings. In CDN, aggressive arguments of a counterpart precipitate as anchors from where a disputing party may develop preconception of the dispute. The preconception renders the disputing party to make compromise to the first offer received.

Table 8 ANOVA multiple comparisons

Client (N = 33)

Factor (I)	Factor (J)	Mean difference (I-J)	Standard error	Sig.	95% confidence interval	
					Lower bound	Upper bound
Preconception	Self-affirmation	−0.67273*	0.17048	0.000	−1.0101	−0.3354
	Optimism	−0.28485	0.17048	0.097	−0.6222	0.0525
	Interest-oriented	−0.15152	0.17048	0.376	−0.4888	0.1858
Self-affirmation	Preconception	0.67273*	0.17048	0.000	0.3354	1.0101
	Optimism	0.38788*	0.17048	0.025	0.0506	0.7252
	Interest-oriented	0.52121*	0.17048	0.003	0.1839	0.8585
Optimism	Preconception	0.28485	0.17048	0.097	−0.0525	0.6222
	Self-affirmation	−0.38788*	0.17048	0.025	−0.7252	−0.0506
	Interest-oriented	0.13333	0.17048	0.436	−0.2040	0.4707
Interest-oriented	Preconception	0.15152	0.17048	0.376	−0.1858	0.4888
	Self-affirmation	−0.52121*	0.17048	0.003	−0.8585	−0.1839
	Optimism	−0.13333	0.17048	0.436	−0.4707	0.2040

Contractor (N = 40)

Factor (I)	Factor (J)	Mean difference (I-J)	Standard Error	Sig.	95% confidence interval	
					Lower bound	Upper bound
Preconception	Self-affirmation	−0.69000*	0.13939	0.000	−0.9653	−0.4147
	Optimism	−0.24000	0.13939	0.087	−0.5153	0.0353
	Interest-oriented	−0.23000	0.13939	0.101	−0.5053	0.0453
Self-affirmation	Preconception	0.69000*	0.13939	0.000	0.4147	0.9653
	Optimism	0.45000*	0.13939	0.002	0.1747	0.7253
	Interest-oriented	0.46000*	0.13939	0.001	0.1847	0.7353
Optimism	Preconception	0.24000	0.13939	0.087	−0.0353	0.5153
	Self-affirmation	−0.45000*	0.13939	0.002	−0.7253	−0.1747
	Interest-oriented	0.01000	0.13939	0.943	−0.2653	0.2853
Interest-oriented	Preconception	0.23000	0.13939	0.101	−0.0453	0.5053
	Self-affirmation	−0.46000*	0.13939	0.001	−0.7353	−0.1847
	Optimism	−0.01000	0.13939	0.943	−0.2853	0.2653

Consultant (N = 61)

Factor (I)	Factor (J)	Mean difference (I-J)	Standard error	Sig.	95% confidence interval	
					Lower bound	Upper bound
Preconception	Self-affirmation	−0.95410*	0.11681	0.000	−1.1842	−0.7240
	Optimism	−0.51475*	0.11681	0.000	−0.7449	−0.2847

(continued)

Table 8 (continued)

	Interest-oriented	−0.54426*	0.11681	0.000	−0.7744	−0.3142
Self-affirmation	Preconception	0.95410*	0.11681	0.000	0.7240	1.1842
	Optimism	0.43934*	0.11681	0.000	0.2092	0.6694
	Interest-oriented	0.40984*	0.11681	0.001	0.1797	0.6399
Optimism	Preconception	0.51475*	0.11681	0.000	0.2847	0.7449
	Self-affirmation	−0.43934*	0.11681	0.000	−0.6694	−0.2092
	Interest-oriented	−0.02951	0.11681	0.801	−0.2596	0.2006
Interest-oriented	Preconception	0.54426*	0.11681	0.000	0.3142	0.7744
	Self-affirmation	−0.40984*	0.11681	0.001	−0.6399	−0.1797
	Optimism	0.02951	0.11681	0.801	−0.2006	0.2596

Note *The mean difference is significant at the 0.05 level

This preconception can thus bring strategic advantage should this perception works favourably to one disputing party and intimidates the counterpart. Thus, offering a high demanding first offer serves as an anchor that may give preconception on the counterpart that there are good reasons to support the offer. The preconception would steer subsequent resolution process as well. In the experiment of Galinsky and Mussweiler [78], it was found that first offer had strong correlation with the final agreed price. Thus, the party making the first offer in general derives more benefits. In CDN, the amount the contractor claims due to culpable acts of the client may influence the final quantum they get in the end. However, the situation may reverse if the client chooses to make a settlement offer first. Very often, disputants choose to ignore rational analysis of evidence and legal opinions in making a first offer with the aim of building room for negotiation. Chapman and Bornstein [32] described this phenomenon as first offer advantage: the more you ask for, the more you get. Besides, early decisions made at prior stage could also give dispute negotiators a stable preconception about the situation, therefore they have the tendency to retain and defend the early assessment.

Self-affirmation bias is represented by item 10 (0.793), item 8 (0.743), item 6 (0.674), item 7 (0.567) and item 9 (0.520) in descending order of factor loadings. Dispute negotiators like to affirm themselves through seeking a positive self-image. Thus, it is quite natural for them to endorse information that supports their assessments. In fact, strong self-affirming disputants would even search for and interpret information that reinforce prior assessments. As a result, greater attention has been paid to information that is consistent with prior knowledge or assessment. In addition, they would amplify their contribution to the successful outcomes of the settlement to affirm self-worth.

Item 14 (0.779), item 13 (0.775), item 12 (0.687), item 11 (0.570) and item 15 (0.540) represent the optimism bias factor. Construction dispute negotiators who are having optimism bias have the following behaviour patterns: they overestimate their ability in assessing the dispute; they raise ambitious requests and are unwarrantedly confident that the same would be met by the counterparts. During dispute negotiation,

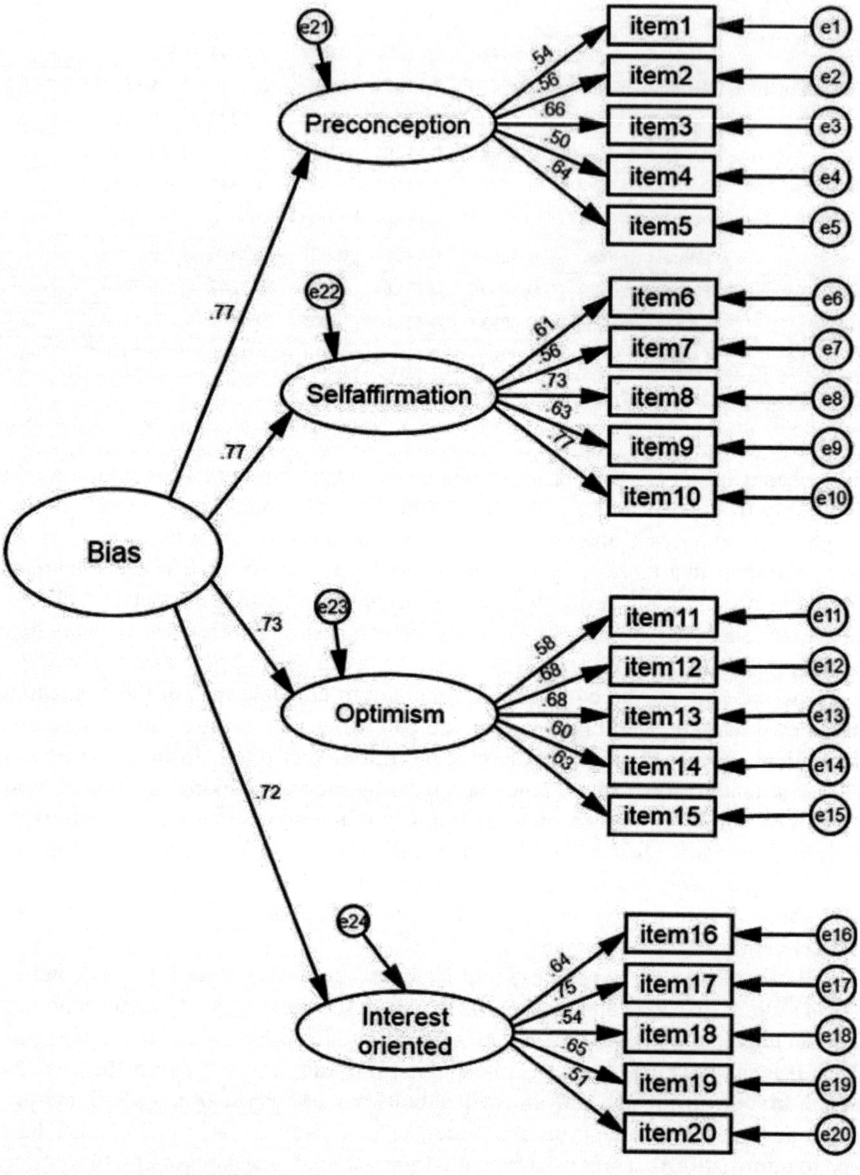

Fig. 3 CFA of bias in CPDR

Table 9 GOF indices results

Fit index	Desired levels	Model results
Absolute fit indices		
χ^2/df	2 or below[a]	1.314 (sufficiently good fit)[b]
GFI	0.8 or above[c]	0.867 (good fit)[d]
AGFI	0.8 or above[e]	0.831 (recommended fit)[e]
RMSEA	0.06 or below[f]	0.049 (excellent fit)[g]
Incremental fit indices		
CFI	0.8 or above[c]	0.927 (good fit)[c]
TLI	0.8 or above[c]	0.917 (good fit)[a]
Parsimonious fit		
PNFI	0.5 or above[h]	0.663 (good fit)[h]
PCFI	0.5 or above[h]	0.810 (good fit)[h]

Note: GOF = goodness-of-fit indexes; χ^2/df = chi square/degree of freedom; GFI = goodness-of-fit index; AGFI = adjusted goodness-of-fit index; RMSEA = root mean square error of approximation; CFI = comparative fit index; TLI = Tucker-Lewis index; PNFI = parsimony normed-fit index; PCFI = parsimony comparative fit index
[a]Hair et al. [83, 84], [b]Xiong et al. [199], [c]Maskarinec et al. [131], [d]Wong et al. [195], [e]Gefen [79], [f]Hu and Bentler [96], [g]Marsh and Hau [129], [h]Chen and Fong [39]

no matter how the counterparts defend, it cannot alleviate the disputing parties' optimistic attitudes about winning and the overconfidence about their requests. Upon completion of the dispute negotiation, under the influence of hindsight effect, they feel they know the outcome all along, which further reinforces their optimism.

Item 20 (0.721), item 17 (0.717), item 19 (0.672), item 16 (0.609) and item 18 (0.607) represent interest-oriented bias. In CDN, when disputants are under the influence of interest-oriented bias, maximizing self-interest characterizes their actions and behaviors. When the negotiation failed to reach an amicable settlement, interest-oriented biased disputants would claim that this outcome is inevitable. With no rethink of their insistence in pursuing their interests that had led to the impasse, they would attribute the failure to settle as the responsibility of the counterparts. They believe the counterpart is having bias and their demands during negotiation are unreasonable. They would flee away from their responsibility of settlement failure by attributing all negative outcomes to the counterpart. In addition, during the negotiation, they would only take note of the arguments that favour them.

The propensity of the biases can also be assessed by the path coefficients of CFA model. The path coefficients represent the relative strength of the four biases. The path coefficients of preconception bias, self-affirmation bias, optimism bias and

Table 10 Standardized regression weights and 1000 sample bootstrapped estimates

Parameter	Estimate*	Mean (bootstrapped)	S.E. (bootstrapped)	Lower	Upper	P
Preconception ← F1	0.771**	0.763	0.091	0.576	0.946	0.001
Self-affirmation ← F1	0.772 (3.891)	0.783	0.085	0.575	0.923	0.005
Optimism ← F1	0.725 (3.673)	0.71	0.111	0.498	0.928	0.001
Interest-oriented ← F1	0.717 (3.805)	0.713	0.103	0.483	0.894	0.002
Item 1 ← Preconception	0.541**	0.543	0.081	0.354	0.692	0.003
Item 2 ← Preconception	0.56 (4.562)	0.549	0.089	0.364	0.719	0.001
Item 3 ← Preconception	0.664 (4.948)	0.663	0.063	0.542	0.783	0.002
Item 4 ← Preconception	0.502 (4.176)	0.503	0.09	0.307	0.655	0.003
Item 5 ← Preconception	0.639 (4.945)	0.635	0.069	0.48	0.753	0.002
Item 6 ← Self-affirmation	0.609**	0.606	0.075	0.451	0.733	0.002
Item 7 ← Self-affirmation	0.555 (5.191)	0.55	0.077	0.39	0.692	0.002
Item 8 ← Self-affirmation	0.726 (6.321)	0.723	0.061	0.578	0.824	0.003
Item 9 ← Self-affirmation	0.63 (5.609)	0.631	0.07	0.474	0.749	0.003
Item 10 ← Self-affirmation	0.768 (6.443)	0.77	0.06	0.618	0.864	0.004
Item 11 ← Optimism	0.583**	0.588	0.076	0.409	0.706	0.004
Item 12 ← Optimism	0.683 (5.616)	0.678	0.065	0.548	0.806	0.001
Item 13 ← Optimism	0.681 (5.558)	0.675	0.072	0.514	0.802	0.002
Item 14 ← Optimism	0.601 (5.168)	0.603	0.094	0.378	0.754	0.004
Item 15 ← Optimism	0.626 (5.204)	0.624	0.073	0.447	0.739	0.003
Item 16 ← Interest-oriented	0.637**	0.639	0.069	0.487	0.766	0.003
Item 17 ← Interest-oriented	0.751 (6.583)	0.741	0.076	0.579	0.873	0.002
Item 18 ← Interest-oriented	0.541 (4.857)	0.547	0.086	0.349	0.698	0.004

(continued)

Table 10 (continued)

Parameter	Estimate*	Mean (bootstrapped)	S.E. (bootstrapped)	Lower	Upper	P
Item 19 ← Interest-oriented	0.649 (5.643)	0.65	0.069	0.492	0.769	0.003
Item 20 ← Interest-oriented	0.507 (4.697)	0.498	0.086	0.316	0.657	0.002

Note S.E.: standard error
*Figures in parentheses are critical ratios from the unstandardized solutions
**The critical ratio is not available, because the regression weight is fixed at 1

interest-oriented bias are 0.77, 0.77, 0.73 and 0.72 respectively (Fig. 3 refers). Self-affirmation bias has the highest path coefficients. In this regard, self-affirmation bias has greater propensity to creep in construction dispute negotiation. It echoes the results of ANOVA multiple comparisons in Table 8. Self-affirmation bias has the highest mean factor score for all the professional groups and thus suggesting higher propensity. Preconception bias has the same path coefficient as self-affirmation bias in the CFA model. The early assessments made by the participants may serve as the preconception influencing their subsequent decisions. The small differences of four biases' path coefficients in CFA together with significant correlation coefficients (see in Table 7) suggest the interrelatedness of these four biases.

Implications on Construction Project Dispute Management

With extensive urbanization and infrastructural developments, globalization and international collaboration become the commonly accepted norm to deliver mega projects. It can be very challenging to work with project participants from different disciplines and cultural background [133, 180]. The existence of bias stifles rational analyses resulting in suboptimal decisions. If biased decisions are minimized, negotiation efficiency would be greatly improved thus saving vast resources.

Furthermore, use of multi-tiered dispute resolution process (MTDR) incorporating alternative dispute resolution (ADR) before arbitration has been the predominant dispute resolution approach [40, 44, 119]. The characterizing feature of MTDR approach is that a dispute will be evaluated repeatedly from negotiation, mediation to adjudication and arbitration [121]. Repeated dispute evaluations might allow the creeping in of all four types of bias. This study posits to raise the awareness of bias and further suggests practical measures to address these biases.

Minimizing Bias in CDN: The Important Role of Project Manager

Project manager (PM) plays a vital role in ensuring that dispute negotiators think and behave in a rational manner [146, 178, 198]. PM should be mindful of the existence of biases when settling construction disputes. In this regard, this bias framework can be used as a checklist of biased behaviors. Self-affirmation bias has been identified with the highest chance to creep in construction dispute negotiation irrespective of the professional background of the negotiators. Therefore, PM should note whether the project team members are keen to confirm themselves and seek positive self-images during construction dispute negotiation, which are the potential traps of self-affirmation bias. In this regard, PM should guide the team members to (i) search complete information about the dispute, not only the supporting evidence to their own arguments, (ii) be open to alternatives irrespective of the assessment already made; (iii) carefully consider the rationality of counterpart's arguments and evidences.

In order to minimize preconception bias, before commencing construction dispute negotiation, PM should remind the team members to forget about their previous preconception about the counterpart and review their assessment about the dispute. When aggressive offers are received from the counterpart, PM should lead the team to carefully consider the counterpart's reservation price based on the conditions of the project. In response to the ambitious arguments and unsubstantiated figures presented by the counterpart, PM should encourage the team to carefully re-estimate the project matter and check the objectivity of the arguments from counterpart. PM should always keep the team alert and re-assess the dispute matter when more and more information is collected and analysed.

To alleviate optimism bias, construction professionals should be reminded of settlement failure. There is no substitute of prudent action in carefully analysing the evidence raised by the counterparts. It is also quite normal for the disputants to focus on their self-interest during dispute negotiations when huge money is at stake. Hence, it is suggested that possibility of future collaboration and a long-term relationship between the project parties should be taken into consideration. Disputants should aim at achieving win–win result through seamless communication. PM should also remind the negotiation team to respectfully listen to both side's grievances, control their emotions and express their opinions in a decent manner. When settlement is not achieved, the team should review what had gone wrong. Besides, in the consultant group it has been found that preconception bias has the lowest mean factor score than other biases (see in Table 8), suggesting that consultants are less prone to or more prepared to deal with preconception bias. It may be the result of the professional training that consultants have received. Professionally, as the neutral 3rd parties between client and contractor, consultants should not have pre-disposition to their own impression or preconception of the situation. Therefore, receiving de-biasing training before commencing dispute resolution process would be helpful to minimize preconception bias.

Optimizing the CDN Mechanism

The study also contributes to the design of construction dispute negotiation procedure. Major industry reviews have called for innovative dispute resolution (CIRC 2001). Use of multi-tiered dispute resolution process (MTDR) incorporating alternative dispute resolution (ADR) is now the mainstream approach [40, 44, 119]. The characterizing feature of MTDR is that disputes will be evaluated repeatedly at each of the tiers [121]. Would this arrangement improve the chance of settlement? This is perhaps the good wish of a MTDR design. However, the issue of bias would aggregate if the same individual or group is doing the repeated evaluations. Under the influence of preconception bias, information collected, or decisions made in the prior tier may become preconception that impedes further rational analysis of the dispute. The way the disputants collect and interpret information could also be biased towards justifying themselves—a form of self-affirmation bias. In entering a new tier of resolution, disputants could be optimistic about the chance of winning thereby refuse to compromise under the influence of optimism bias. Under the influence of interest-oriented bias, should settlement be not achieved, disputants could attribute the undesirable resolution outcome and expensive cost to the counterparts' unwillingness to settle in prior tiers. The study contributes to CDN study by challenging the use of MTDR design because of the happening of biases. Dispute resolution procedure with extended tiers may not materialize the intended outcome due to the creeping in of bias. Thus, this study suggests directing more resources, energy and inputs to resolve disputes in the negotiation stage.

Moreover, it is prudent to be aware of the limitations of this study. The most notable is the social desirability bias, which means respondents tend to reply survey questions in a way to make them look more favourable [75, 148]. People may be loath to admit their practice of bias. Therefore, they may lower their ratings on the frequency of biased behaviours in the questionnaire. Measures to alleviate social desirability bias have been employed in this study.

Summary

Negotiation studies have largely been developed based on the assumption of rational analyses and free-will bargain. However, negotiators are human, and bias appears inevitable [4, 51]. This study conceptualizes bias in CDN by proposing a bias framework. Manifestations of bias in CDN were operationalized after summarising effects of bias from literature. Construction professionals were invited to provide data on their negotiation behaviours. A PCFA further suggested a four-factor bias framework. The four types of bias are preconception, self-affirmation, optimism and interest-oriented. The framework was validated by a CFA. The findings inform construction professionals that the practice of biased behaviours in CDN is real. The irrationality of human decisions as a result of bias is thus highlighted in this study. Practical measures to minimize biases in CDN are proposed. In terms of construction dispute resolution process design, this study timely reminds the caveats in employing MTDR. Repeated dispute evaluations allow biases to creep in. More resources and energy should therefore be deployed to enhance the settlement of dispute through negotiation before embarking on convoluted multi-tiered procedures. Repeated evaluations of a dispute may bring unintended outcomes of hardening of positions and uncompromising attitude.

Acknowledgements The empirical works of this chapter has been reported in a paper entitled "Biases in construction project dispute resolution" of the Journal of Engineering, Construction and Architectural Management. The support of CityU SRG projects (number 7004624, 7004870) and HKSAR GRF (11209118) is duly acknowledged.

References

1. Adam BJ (2016) Training in the mitigation of anchoring bias: a test of the consider-the-opposite strategy. Learn Motiv 53:36–48
2. Adebayo OO, Osmond IC (2011) A search for an acceptable margin of valuation error: a case study of valuers and their clients in Nigeria. Sri Lanka J Real Estate 1(4):54–73
3. Alvarez JFA, Pustina A, Hällgren M (2011) Escalating commitment in the death zone: new insights from the 1996 Mount Everest disaster. Int J Project Manage 29(8):971–985
4. Ariely D (2008) Predictably irrational: the hidden forces that shape our decisions, 1st edn. Harper, New York, NY
5. Arkes HR (1991) Costs and benefits of judgment errors: Implications for debiasing. Psychol Bull 110(3):486
6. Arkes HR, Faust D, Guilmette TJ, Hart K (1988) Eliminating the hindsight bias. J Appl Psychol 73(2):305
7. Babcock L, Loewenstein G (1997) Explaining bargaining impasse: the role of self-serving biases. J Econ Perspect 11(1):109–126
8. Babcock L, Loewenstein G, Issacharoff S, Camerer C (1995) Biased judgments of fairness in bargaining. Am Econ Rev 85(5):1337–1343
9. Barlett JE, Kotrlik JW, Higgins CC (2001) Organizational research: determining appropriate sample size in survey research. Inf Technol Learn Perform J 19(1):43
10. Barnes JH (1984) Cognitive biases and their impact on strategic planning. Strateg Manag J 5(2):129–137
11. Baumgardner AH, Arkin RM (1988) Affective state mediates causal attributions for success and failure. Motiv Emot 12(2):99–111
12. Bazerman MH, Curhan JR, Moore DA, Valley KL (2000) Negotiation. Annu Rev Psychol 51(1):279–314
13. Bazerman MH, Moore DA (2008) Judgment in managerial decision making. Wiley, New York
14. Bazerman MH, Neale MA (1982) Improving negotiation effectiveness under final offer arbitration: the role of selection and training. J Appl Psychol 67(5):543
15. Bendle N, Cotte J (2016) Assumptions of rationality in political marketing: the case of the Republican Autopsy. J Nonprofit Public Sect Mark 28(1):66–83
16. Bergman O, Ellingsen T, Johannesson M, Svensson C (2010) Anchoring and cognitive ability. Econ Lett 107(1):66–68
17. Bernstein DM, Erdfelder E, Meltzoff AN, Peria W, Loftus GR (2011) Hindsight bias from 3 to 95 years of age. J Exp Psychol Learn Mem Cogn 37(2):378

18. Bhattacharya C, Jasper JD (2018) Degree of handedness: a unique individual differences factor for predicting and understanding hindsight bias. Personality Individ Differ 125:97–101
19. Binmore K (2009) Rational decisions. Princeton University Press, Princeton, NJ
20. Bjork RA, Dunlosky J, Kornell N (2013) Self-regulated learning: Beliefs, techniques, and illusions. Annu Rev Psychol 64:417–444
21. Blank H, Nestler S (2007) Cognitive process models of hindsight bias. Soc Cogn 25(1):132–146
22. Blank H, Nestler S, von Collani G, Fischer V (2008) How many hindsight biases are there? Cognition 106(3):1408–1440
23. Bordley R, Licalzi M, Tibiletti L (2017) A target-based foundation for the "hard-easy effect" bias. In: Country experiences in economic development, management and entrepreneurship. Springer, Cham, pp 659–671
24. Bradley GW (1978) Self-serving biases in the attribution process: a re-examination of the fact or fiction question. J Pers Soc Psychol 36(1):56
25. Breaugh JA (2013) Employee recruitment. Annu Rev Psychol 64:389–416
26. Bromiley P, Papenhausen C (2003) Assumptions of rationality and equilibrium in strategy research: the limits of traditional economic analysis. Strateg Organ 1(4):413–437
27. Brown JD, Rogers RJ (1991) Self-serving attributions: the role of physiological arousal. Pers Soc Psychol Bull 17(5):501–506
28. Bukszar E, Connolly T (1988) Hindsight bias and strategic choice: some problems in learning from experience. Acad Manag J 31(3):628–641
29. Campbell WK, Sedikides C (1999) Self-threat magnifies the self-serving bias: a meta-analytic integration. Rev Gen Psychol 3(1):23
30. Carlson KA, Russo JE (2001) Biased interpretation of evidence by mock jurors. J Exp Psychol Appl 7(2):91
31. Cerny BA, Kaiser HF (1977) A study of a measure of sampling adequacy for factor-analytic correlation matrices. Multivar Behav Res 12(1):43–47
32. Chapman GB, Bornstein BH (1996) The more you ask for, the more you get: anchoring in personal injury verdicts. Appl Cogn Psychol 10(6):519–540
33. Chapman GB, Johnson EJ (1999) Anchoring, activation, and the construction of values. Organ Behav Hum Decis Process 79(2):115–153
34. Chapman GB, Johnson EJ (2002) Incorporating the irrelevant: anchors in judgments of belief and value. In: Gilovich T, Griffin D, Kahneman D (eds) Heuristics and biases: the psychology of intuitive judgment. Cambridge university press, pp 120–138
35. Charness G, Dave C (2017) Confirmation bias with motivated beliefs. Games Econom Behav 104:1–23
36. Chase VM, Hertwig R, Gigerenzer G (1998) Visions of rationality. Trends Cogn Sci 2(6):206–214
37. Chaxel AS (2014) The impact of procedural priming of selective accessibility on self-generated and experimenter-provided anchors. J Exp Soc Psychol 50:45–51
38. Chelley-Steeley PL, Kluger BD, Steeley JM (2015) Earnings and hindsight bias: an experimental study. Econ Lett 134:130–132
39. Chen L, Fong PS (2012) Revealing performance heterogeneity through knowledge management maturity evaluation: a capability-based approach. Expert Syst Appl 39(18):13523–13539
40. Chen YQ, Zhang YB, Zhang SJ (2014) Impacts of different types of owner-contractor conflict on cost performance in construction projects. J Construct Eng Manage 140(6). https://doi.org/10.1061/(ASCE)CO.1943-7862.0000852
41. Chertkoff JM, Conley M (1967) Opening offer and frequency of concession as bargaining strategies. J Pers Soc Psychol 7(2):181–185
42. Cheung SO, Yiu TW (2006) Are construction disputes inevitable? IEEE Trans Eng Manage 53(3):456–470
43. Chinyio E, Taiwo A (2016) Psychology in construction (psycon), In: Inted2016 proceedings, p 529

44. Chong HY, Mohamad Zin R (2012) Selection of dispute resolution methods: factor analysis approach. Eng Constr Archit Manag 19(4):428–443
45. Chow PT, Cheung SO, Chan KY (2012) Trust-building in construction contracting: mechanism and expectation. Int J Project Manage 30(8):927–937
46. Christensen-Szalanski JJ, Willham CF (1991) The hindsight bias: a meta-analysis. Organ Behav Hum Decis Process 48(1):147–168
47. Clegg BA, McKernan B, Martey RM, Taylor SM, Stromer-Galley J, Kenski K et al (2015) Effective mitigation of anchoring bias, projection bias, and representativeness bias from serious game-based training. Proc Manuf 3:1558–1565
48. Coleman MD (2011) Emotion and the self-serving bias. Curr Psychol 30(4):345–354
49. Conlisk J (1996) Why bounded rationality? J Econ Lit 34(2):669–700
50. Connolly T, Bukszar EW (1990) Hindsight bias: self-flattery or cognitive error? J Behav Decis Mak 3(3):205–211
51. Cusick W (2009) All customers are irrational. AMACOM, New York, NY
52. Davis T, Pharro R (2003) The relationship manager: the next generation of project management. Gower Publishing, Ltd.
53. Deffains B, Espinosa R, Thöni C (2016) Political self-serving bias and redistribution. J Public Econ 134:67–74
54. Diener E, Oishi S, Lucas RE (2003) Personality, culture, and subjective well-being: emotional and cognitive evaluations of life. Annu Rev Psychol 54(1):403–425
55. Doherty ME, Mynatt CR, Tweney RD, Schiavo MD (1979) Pseudodiagnosticity. Acta Physiol (Oxf) 43(2):111–121
56. Dunbar NE, Miller CH, Adame BJ, Elizondo J, Wilson SN, Lane BL et al (2014) Implicit and explicit training in the mitigation of cognitive bias through the use of a serious game. Comput Hum Behav 37:307–318
57. Dunning D (2011) The Dunning-Kruger effect: on being ignorant of one's own ignorance. Advances in experimental social psychology, Academic Press 44:247–296
58. Englich B, Mussweiler T (2001) Sentencing under uncertainty: anchoring effects in the court room. J Appl Soc Psychol 31(7):1535–1551
59. Englich B, Mussweiler T, Strack F (2006) Playing dice with criminal sentences: the influence of irrelevant anchors on experts' judicial decision making. Pers Soc Psychol Bull 32(2):188–200
60. Epley N, Gilovich T (2001) Putting adjustment back in the anchoring and adjustment heuristic: differential processing of self-generated and experimenter-provided anchors. Psychol Sci 12(5):391–396
61. Epley N, Gilovich T (2006) The anchoring-and-adjustment heuristic why the adjustments are insufficient. Psychol Sci 17(4):311–318
62. Erev I, Wallsten TS, Budescu DV (1994) Simultaneous over- and underconfidence: the role of error in judgment processes. Psychol Rev 101(3):519–527
63. Eriksson T, Kadefors A (2017) Organisational design and development in a large rail tunnel project—influence of heuristics and mantras. Int J Project Manage 35(3):492–503
64. Evans JSB (2008) Dual-processing accounts of reasoning, judgment, and social cognition. Annu Rev Psychol 59:255–278
65. Farnsworth W (2003) Legal regulation of self-serving bias. UC Davis Law Rev 37:567
66. Feld J, Sauermann J, De Grip A (2017) Estimating the relationship between skill and overconfidence. J Behav Exp Econ 68:18–24
67. Fellner G, Krügel S (2012) Judgmental overconfidence: three measures, one bias? J Econ Psychol 33(1):142–154
68. Fellner-Röhling G, Krügel S (2014) Judgmental overconfidence and trading activity. J Econ Behav Organ 107:827–842
69. Ferrell WR (1994) Discrete subjective probabilities and decision analysis: elicitation, calibration and combination. In: Wright G, Ayton P (eds) Subjective probability. Wiley, Chichester, pp 411–451

70. Fischhoff B (1975) Hindsight ≠ foresight: the effect of outcome knowledge on judgment under uncertainty. J Exp Psychol Hum Percept Perform 1(3):288–299
71. Fischhoff B, Beyth-Marom R (1983) Hypothesis evaluation from a Bayesian perspective. Psychol Rev 90(3):239–260
72. Flyvbjerg B (2006) From Nobel Prize to project management: getting risks right. Proj Manag J 37(3):5–15
73. Flyvbjerg B (2017) The iron law of worldwide mega projects. In: The Oxford handbook of megaproject management, Flyvbjerg (Edr), Oxford University, pp1–20
74. Fornell C, Larcker DF (1981) Evaluating structural equation models with unobservable variables and measurement error. J Market Res 39–50
75. Furnham A (1986) Response bias, social desirability and dissimulation. Personality Individ Differ 7(3):385–400
76. Furnham A, Boo HC (2011) A literature review of the anchoring effect. J Socio-Econ 40(1):35–42
77. Furnham A, Boo HC, McClelland A (2012) Individual differences and the susceptibility to the influence of anchoring cues. J Individ Differ 33(2):89–93
78. Galinsky AD, Mussweiler T (2001) First offers as anchors: the role of perspective-taking and negotiator focus. J Pers Soc Psychol 81(4):657
79. Gefen D (2000) E-commerce: the role of familiarity and trust. Omega 28(6):725–737
80. Gelfand MJ, Christakopoulou S (1999) Culture and negotiator cognition: judgment accuracy and negotiation processes in individualistic and collectivistic cultures. Organ Behav Hum Decis Process 79(3):248–269
81. Gigerenzer G, Hoffrage U, Kleinbölting H (1991) Probabilistic mental models: a Brunswikian theory of confidence. Psychol Rev 98(4):506
82. Greenberg J, Pyszczynski T, Solomon S (1982) The self-serving attributional bias: beyond self-presentation. J Exp Soc Psychol 18(1):56–67
83. Hair JF, Anderson RE, Tatham RL, Black WC (1998) Multivariate data analysis, 5th edn. Prentice Hall International, NY
84. Hair JF, Black WC, Babin BJ, Anderson RE, Tatham RL (2006) Multivariate data analysis. Prentice Hall Upper Saddle River
85. Haji-Kazemi S, Andersen B, Klakegg OJ (2015) Barriers against effective responses to early warning signs in projects. Int J Project Manage 33(5):1068–1083
86. Hardt O, Einarsson EÖ, Nader K (2010) A bridge over troubled water: reconsolidation as a link between cognitive and neuroscientific memory research traditions. Annu Rev Psychol 61:141–167
87. Harley EM (2007) Hindsight bias in legal decision making. Soc Cogn 25(1):48
88. Harvey N (1997) Confidence in judgment. Trends Cogn Sci 1(2):78–82
89. Hawkins SA, Hastie R (1990) Hindsight: biased judgments of past events after the outcomes are known. Psychol Bull 107(3):311
90. Hendry SH, Shaffer DR (1989) On testifying in one's own behalf: interactive effects of evidential strength and defendant's testimonial demeanor on jurors' decisions. J Appl Psychol 74(4):539–545
91. Hernandez I, Preston JL (2013) Disfluency disrupts the confirmation bias. J Exp Soc Psychol 49(1):178–182
92. Hilbert M (2012) Toward a synthesis of cognitive biases: how noisy information processing can bias human decision making. Psychol Bull 138(2):211
93. Ho R (2013) Handbook of univariate and multivariate data analysis with IBM SPSS. CRC Press
94. Hoffrage U, Pohl R (2003) Research on hindsight bias: a rich past, a productive present, and a challenging future. Memory 11(4–5):329–335
95. Holstein JA (1985) Jurors' interpretations and jury decision making. Law Hum Behav 9(1):83–99
96. Hu LT, Bentler PM (1999) Cutoff criteria for fit indexes in covariance structure analysis: conventional criteria versus new alternatives. Struct Equ Model 6(1):1–55

97. Huber VL, Neale MA (1986) Effects of cognitive heuristics and goals on negotiator performance and subsequent goal setting. Organ Behav Hum Decis Process 38(3):342–365
98. Janz BD, Prasarnphanich P (2009) Freedom to cooperate: Gaining clarity into knowledge integration in information systems development teams. IEEE Trans Eng Manage 56(4):621–635
99. Jonas E, Schulz-Hardt S, Frey D, Thelen N (2001) Confirmation bias in sequential information search after preliminary decisions: an expansion of dissonance theoretical research on selective exposure to information. J Pers Soc Psychol 80(4):557
100. Juslin P, Winman A, Olsson H (2000) Naive empiricism and dogmatism in confidence research: a critical examination of the hard–easy effect. Psychol Rev 107(2):384–396
101. Kahneman D (1992) Reference points, anchors, norms, and mixed feelings. Organ Behav Hum Decis Process 51(2):296–312
102. Kahneman D (2003) Maps of bounded rationality: psychology for behavioral economics. American Economic Review 93(5):1449–1475
103. Kaiser HF (1958) The varimax criterion for analytic rotation in factor analysis. Psychometrika 23(3):187–200
104. Kaiser HF (1974) An index of factorial simplicity. Psychometrika 39(1):31–36
105. Kassin SM, Dror IE, Kukucka J (2013) The forensic confirmation bias: problems, perspectives, and proposed solutions. J Appl Res Mem Cogn 2(1):42–52
106. Keren G, Teigen KH (2004) Yet another look at the heuristics and biases approach. In: Koehler DJ, Harvey N (eds), Blackwell handbook of judgment and decision making. Wiley, pp 89–109
107. Klayman J (1995) Varieties of confirmation bias. Psychol Learn Motiv 32:385–418
108. Klayman J, Ha YW (1987) Confirmation, disconfirmation, and information in hypothesis testing. Psychol Rev 94(2):211
109. Klayman J, Soll JB, Gonza´ lez-Vallejo C, Barlas S (1999) Overconfidence: it depends on how, what, and whom you ask. Org Behav Hum Decis Process 79(3):216–247
110. Kline RB (1998) Structural equation modeling. The Guilford Press, New York
111. Koriat A, Goldsmith M, Pansky A (2000) Toward a psychology of memory accuracy. Annu Rev Psychol 51(1):481–537
112. Koriat A, Lichtenstein S, Fischhoff B (1980) Reasons for confidence. J Exp Psychol Hum Learn Memory 6(2):107–118
113. Kramer RM, Newton E, Pommerenke PL (1993) Self-enhancement biases and negotiator judgment: effects of self-esteem and mood. Organ Behav Hum Decis Process 56(1):110–133
114. Kriss PH, Loewenstein G, Wang X, Weber RA (2011) Behind the veil of ignorance: self-serving bias in climate change negotiations. Judgm Decis Mak 6(7):602
115. Kruglanski AW (1989) Lay epistemics and human knowledge: cognitive and motivational bases. Plenum Press, New York, NY
116. Kumaraswamy MM (1997) Conflicts, claims and disputes in construction. Eng Constr Archit Manag 4(2):95–111
117. Kutsch E, Hall M (2005) Intervening conditions on the management of project risk: dealing with uncertainty in information technology projects. Int J Project Manage 23(8):591–599
118. Kutsch E, Hall M (2010) Deliberate ignorance in project risk management. Int J Project Manage 28(3):245–255
119. Lee CK, Yiu TW, Cheung SO (2016) Selection and use of alternative dispute resolution (ADR) in construction projects—past and future research. Int J Project Manage 34(3):494–507
120. Lerner JS, Keltner D (2000) Beyond valence: toward a model of emotion-specific influences on judgment and choice. Cogn Emot 14(4):473–493
121. Li K, Cheung SO (2016) The potential of bias in multi-tier construction dispute resolution processes. In: Chan PW, Neilson CJ (eds), Proceedings of the 32nd annual ARCOM conference, vol 1, pp 197–205
122. Libby R, Rennekamp K (2012) Self-serving attribution bias, overconfidence, and the issuance of management forecasts. J Account Res 50(1):197–231
123. Lichtenstein S, Fischhoff B, Phillips LD (1977) Calibration of probabilities: the state of the art. Decision making and change in human affairs. Springer, Netherlands, pp 275–324

124. Lobel IB (1994) Realities of interest based (win-win) bargaining. Labor Law J 45(12):771
125. Lombrozo T, Carey S (2006) Functional explanation and the function of explanation. Cognition 99(2):167–204
126. Lu W, Li Z, Wang S (2017) The role of justice for cooperation and contract's moderating effect in construction dispute negotiation. Eng Constr Archit Manag 24(1):133–153
127. MacCoun RJ (1998) Biases in the interpretation and use of research results. Annu Rev Psychol 49(1):259–287
128. Malouff J, Schutte NS (1989) Shaping juror attitudes: effects of requesting different damage amounts in personal injury trials. J Soc Psychol 129(4):491–497
129. Marsh HW, Hau KT (1996) Assessing goodness of fit: is parsimony always desirable? J Exp Educ 64(4):364–390
130. Marti MW, Wissler RL (2000) Be careful what you ask for: the effect of anchors on personal-injury damages awards. J Exp Psychol Appl 6(2):91
131. Maskarinec G, Novotny R, Tasaki K (2000) Dietary patterns are associated with body mass index in multiethnic women. J Nutr 130(12):3068–3072
132. McCray GE, Purvis RL, McCray CG (2002) Project management under uncertainty: the impact of heuristics and biases. Proj Manag J 33(1):49–57
133. Meng X, Boyd P (2017) The role of the project manager in relationship management. Int J Project Manage 35(5):717–728
134. Merkle C (2017) Financial overconfidence over time: foresight, hindsight, and insight of investors. J Bank Finance 84:68–87
135. Meub L, Proeger TE (2015) Anchoring in social context. J Behav Exp Econ 55:29–39
136. Mezulis AH, Abramson LY, Hyde JS, Hankin BL (2004) Is there a universal positivity bias in attributions? A meta-analytic review of individual, developmental, and cultural differences in the self-serving attributional bias. Psychol Bull 130(5):711
137. Miller DT, Ross M (1975) Self-serving biases in the attribution of causality: fact or fiction? Psychol Bull 82(2):213–225
138. Millward RB, Spoehr KT (1973) The direct measurement of hypothesis-testing strategies. Cogn Psychol 4(1):1–38
139. Mochon D, Frederick S (2013) Anchoring in sequential judgments. Organ Behav Hum Decis Process 122(1):69–79
140. Moore DA, Healy PJ (2008) The trouble with overconfidence. Psychol Rev 115(2):502
141. Morewedge CK, Kahneman D (2010) Associative processes in intuitive judgment. Trends Cogn Sci 14(10):435–440
142. Mullinix G, Gray O, Colado J, Veinott E, Leonard J, Papautsky EL et al. (2013). Heuristica: designing a serious game for improving decision making. In: 2013 IEEE international on games innovation conference (IGIC), pp 250–255
143. Mussweiler T, Strack F (1999) Hypothesis-consistent testing and semantic priming in the anchoring paradigm: a selective accessibility model. J Exp Soc Psychol 35(2):136–164
144. Mussweiler T (1997) A selective accessibility model of anchoring: Linking the anchoring heuristic to hypothesis-consistent testing and semantic priming. Pabst, Lengerich, Germany
145. Mussweiler T, Englich B, & Strack F (2004) Anchoring effect. In: Pohl R (ed) Cognitive illusions—a handbook on fallacies and biases in thinking, judgment, and memory. Psychology Press, pp 183–200
146. Müller R, Turner R (2010) Leadership competency profiles of successful project managers. Int J Project Manage 28(5):437–448
147. Neale MA, Bazerman MH (1992) Negotiator cognition and rationality: a behavioral decision theory perspective. Organ Behav Hum Decis Process 51(2):157–175
148. Nederhof AJ (1985) Methods of coping with social desirability bias: a review. Eur J Soc Psychol 15(3):263–280
149. Newey CA (2016) Fairness as "appropriate impartiality" and the problem of the self-serving bias. Ethical Theory Moral Pract 19(3):695–709
150. Nickerson RS (1998) Confirmation bias: a ubiquitous phenomenon in many guises. Rev Gen Psychol 2(2):175

151. Nunnally JC, Bernstein IH (1994) Psychometric theory, 3rd edn. McGraw-Hill, New York
152. Otto PE, Bolle F (2015) Exploiting one's power with a guilty conscience: an experimental investigation of self-serving biases. J Econ Psychol 51:79–89
153. Ozorhon B, Arditi D, Dikmen I, Birgonul MT (2008) Implications of culture in the performance of international construction joint ventures. J Construct Eng Manage 134(5). https://doi.org/10.1061/(ASCE)0733-9364(2008)134:5(361)
154. Paiva EL, Roth A, Fensterseifer JE (2008) Organizational knowledge and the manufacturing strategy process: a resource-based view analysis. J Oper Manag 26(1):115–132
155. Peer E, Gamliel E (2013) Heuristics and biases in judicial decisions. Court Review 49:114–118
156. Pennington N, Hastie R (1993) The story model for juror decision making. In: Hastie R (ed) Inside the juror: the psychology of juror decision making. Cambridge University Press, New York, pp 192–221
157. Pezzo MV (2011) Hindsight bias: a primer for motivational researchers. Soc Pers Psychol Compass 5(9):665–678
158. Pinto JK (2013) Lies, damned lies, and project plans: recurring human errors that can ruin the project planning process. Bus Horiz 56(5):643–653
159. Pohl RF (2007) Ways to assess hindsight bias. Soc Cogn 25(1):14–31
160. Pohl RF, Hell W (1996) No reduction in hindsight bias after complete information and repeated testing. Organ Behav Hum Decis Process 67(1):49–58
161. Pronin E (2007) Perception and misperception of bias in human judgment. Trends Cogn Sci 11(1):37–43
162. Pronin E, Lin DY, Ross L (2002) The bias blind spot: perceptions of bias in self versus others. Pers Soc Psychol Bull 28(3):369–381
163. Pulford BD, Colman AM (1997) Overconfidence: feedback and item difficulty effects. Personality Individ Differ 23(1):125–133
164. Radzevick JR, Moore DA (2009) Competing to be certain (but wrong): Social pressure and overprecision in judgment. Academy of Management Proceedings, Academy of Management. 2009(1):1–6
165. Roese NJ, Vohs KD (2012) Hindsight bias. Perspect Psychol Sci 7(5):411–426
166. Russo JE, Shoemaker PJH (1989) Decision traps. Simon and Schuster, New York
167. Sadler-Smith E, Smith PJ (2004) Strategies for accommodating individuals' styles and preferences in flexible learning programs. Br J Edu Technol 35(4):395–412
168. Samuelson WF, Bazerman MH (1985) Negotiation under the winner's curse. Res Exp Econ 3:105
169. Schumacker RE, Lomax RG (2016) A beginner's guide to structural equation modeling, 4th edn. Routledge, New York, NY, p 71
170. Schwind C, Buder J (2012) Reducing confirmation bias and evaluation bias: when are preference-inconsistent recommendations effective–and when not? Comput Hum Behav 28(6):2280–2290
171. Sedikides C, Campbell WK, Reeder GD, Elliot AJ (1998) The self-serving bias in relational context. J Pers Soc Psychol 74(2):378
172. Sherif M, Hovland CI (1961) Social judgment: assimilation and contrast effects in communication and attitude change
173. Shore B (2008) Systematic biases and culture in project failures. Proj Manag J 39(4):5–16
174. Simon HA (1955) A behavioral model of rational choice. Q J Econ 69:99–118
175. Simon HA (1979) Rational decision making in business organizations. Am Econ Rev 69(4):493–513
176. Smith AC, Greene E (2005) Conduct and its consequences: attempts at debiasing jury judgments. Law Hum Behav 29(5):505
177. Soll JB (1996) Determinants of overconfidence and miscalibration: the roles of random error and ecological structure. Organ Behav Hum Decis Process 65(2):117–137
178. Sommerville J, Craig N, Hendry J (2010) The role of the project manager: all things to all people? Struct Surv 28(2):132–141

179. Stallard MJ, Worthington DL (1998) Reducing the hindsight bias utilizing attorney closing arguments. Law Hum Behav 22(6):671–683
180. Stingl V, Geraldi J (2017) Errors, lies and misunderstandings: systematic review on behavioural decision making in projects. Int J Project Manage 35(2):121–135
181. Strack F, Bahník Š, Mussweiler T (2016) Anchoring: accessibility as a cause of judgmental assimilation. Curr Opin Psychol 12:67–70
182. Strack F, Mussweiler T (1997) Explaining the enigmatic anchoring effect: mechanisms of selective accessibility. J Pers Soc Psychol 73(3):437
183. Suantak L, Bolger F, Ferrell WR (1996) The hard–easy effect in subjective probability calibration. Organ Behav Hum Decis Process 67(2):201–221
184. Tamborski M, Brown RP, Chowning K (2012) Self-serving bias or simply serving the self? Evidence for a dimensional approach to narcissism. Personality Individ Differ 52(8):942–946
185. Taplin JE (1975) Evaluation of hypotheses in concept identification. Mem Cognit 3(1):85–96
186. Thompson L, Loewenstein G (1992) Egocentric interpretations of fairness and interpersonal conflict. Organ Behav Hum Decis Process 51(2):176–197
187. Thompson L (2008) The truth about negotiations. Upper Saddle River, Pearson Education, NJ
188. Thunström L, Nordström J, Shogren JF (2015) Certainty and overconfidence in future preferences for food. J Econ Psychol 51:101–113
189. Troutman CM, Shanteau J (1977) Inferences based on nondiagnostic information. Organ Behav Hum Perform 19(1):43–55
190. Tsai CI, Klayman J, Hastie R (2008) Effects of amount of information on judgment accuracy and confidence. Organ Behav Hum Decis Process 107(2):97–105
191. Tversky A, Kahneman D (1974) Judgment under uncertainty: heuristics and biases. Science 184:1124–1131
192. Villarroel C, Felton M, Garcia-Mila M (2016) Arguing against confirmation bias: the effect of argumentative discourse goals on the use of disconfirming evidence in written argument. Int J Educ Res 79:167–179
193. Wang L, Kunc M, Bai SJ (2017) Realizing value from project implementation under uncertainty: an exploratory study using system dynamics. Int J Project Manage 35(3):341–352
194. Wilson TD, Houston CE, Etling KM, Brekke N (1996) A new look at anchoring effects: basic anchoring and its antecedents. J Exp Psychol Gen 125(4):387
195. Wong WK, Cheung SO, Yiu TW, Pang HY (2008) A framework for trust in construction contracting. Int J Project Manage 26(8):821–829
196. Wood G (1978) The knew-it-all-along effect. J Exp Psychol Hum Percept Perform 4(2):345–353
197. Wood W (2000) Attitude change: persuasion and social influence. Annu Rev Psychol 51(1):539–570
198. Wu G, Liu C, Zhao X, Zuo J (2017) Investigating the relationship between communication-conflict interaction and project success among construction project teams. Int J Project Manage 35(8):1466–1482
199. Xiong B, Skitmore M, Xia B (2015) A critical review of structural equation modeling applications in construction research. Autom Constr 49:59–70
200. Zuckerman M (1979) Attribution of success and failure revisited, or: the motivational bias is alive and well in attribution theory. J Pers 47(2):245–287

Conceptualising Bias in Construction Dispute Negotiation

Keyao Li and Sai On Cheung

Introduction

Capital investments are characterised by massive resource input, long duration and lasting uses [1, 2]. Quality is one of the key indicators of a successful project because the built facilities are expected to last and function for a long period. Furthermore, infrastructural developments are used quite commonly as economic booster at times of recession. Capital investments have the ripple effect in vitalising other industries within the supply chain. With the advancement in living standard and the concern over preserving the natural environment, sustainable construction is now much advocated. Very often, this change in working paradigm is not met with sufficient commitment and enthusiasm. One reason may well be the lack of environmental concern of the stakeholders of the construction industry. Another possible cause is the profit maximising orientation of construction enterprises. This chapter offers an investigation on biases in construction decisions in general and for dispute in particular. The former provides the theoretical bases that support the conceptualisation of the latter.

Human factor in construction project is very much understudied in the construction project management domain. In fact, the complex contractual network and enormous resources that are at stake make rational analysis very difficult in many construction decisions [3]. Disputes are therefore inevitable in different phases of construction projects [4–6]. Dispute management is one of the key functions of construction professionals. Most professionals consider themselves rational and work according to principles. Observations by dispute facilitators suggest otherwise [7]. Cognitive

K. Li (✉)
Future of Work Institute, Faculty of Business & Law, Curtin University, Perth, Australia
e-mail: Keyao.li@curtin.edu.au

S. O. Cheung
Construction Dispute Resolution Research Unit, City University of Hong Kong, Hong Kong, China
e-mail: Saion.cheung@cityu.edu.hk

S. O. Cheung (ed.), *Construction Dispute Research Expanded*, Springer Tracts in Civil Engineering, https://doi.org/10.1007/978-3-030-80256-1_2

bias is a kind of psychological barrier against dispute negotiation [8–10, 3]. Biases obviate rational decisions that derail proper negotiation courses [11–14]. Li and Cheung [15] first explored the potential of bias happening in construction dispute negotiation (CDN). It was found that repeated evaluations invite biases. Studying biases in CDN should aim to mitigate its effect so that the chance of having negotiated settlement is preserved. If successful, the significance is evidently clear. In addition, construction project can be delivered more efficiently without wasting enormous time and resources. Hence, alleviating bias in CDN would increase sustainability parameters of construction projects in the following aspects: (1) economic aspect, minimizing the expenses and costs of settling construction dispute by smoothing and shortening the protracted dispute resolution processes [3]; (2) environmental aspect, saving enormous resources and materials that would be wasted in the prolonged dispute resolution processes [16, 17], and (3) social aspect, improving the intense relationship between the disputing parties and enhancing partnership collaboration and healthy community in construction industry [1, 18].

This study first offers biases conceptualisation for the purpose of establishing theoretical anchor for further studies on biases in CDN. Accordingly, types of bias in CDN are proposed.

To achieve this aim, five stages of work are involved. First, the constructs of bias are developed. Second, the extent of impact of biases is examined. Third, approaches to minimise biases are studied. Fourth, the usefulness of the bias minimising measures is evaluated. Fifth, a summary is provided. The flow of the study is presented in Fig. 1.

Bias Constructs in Construction Dispute Negotiation (CDN)

The empirical evidence of happening of biases in CDN has been reported in Chapter One. The characteristics and theoretical background of cognitive biases had also been outlined. Repeated evaluations may not always improve the quality of the decisions, with biases taking heel, rational decisions may become more remote. Providing a theory-rich bias conceptualisation underpins and paves the path for further studies on biases in CDN. This study therefore aims to develop a robust bias conceptualization in CDN with different sets of data collected from three sources. The first set of data is self-reflection by the disputants, which was collected in [16] with sixteen identification statements operationalised. The second set of data is self-realization of the respondents who participated in a construction project dispute resolution simulation. The simulation includes contextual information, making the environment closer to reality. In this way, the decisions in the simulation were more tangible and reflecting the real-life situation. The third set of data was collected from practicing third party neutrals. Their assessment on the practice of biased behaviours are based on their observations. It is believed that their assessment would be more objective when compared with self-reflection and self-realization. Further information on the three data sets is given here follows.

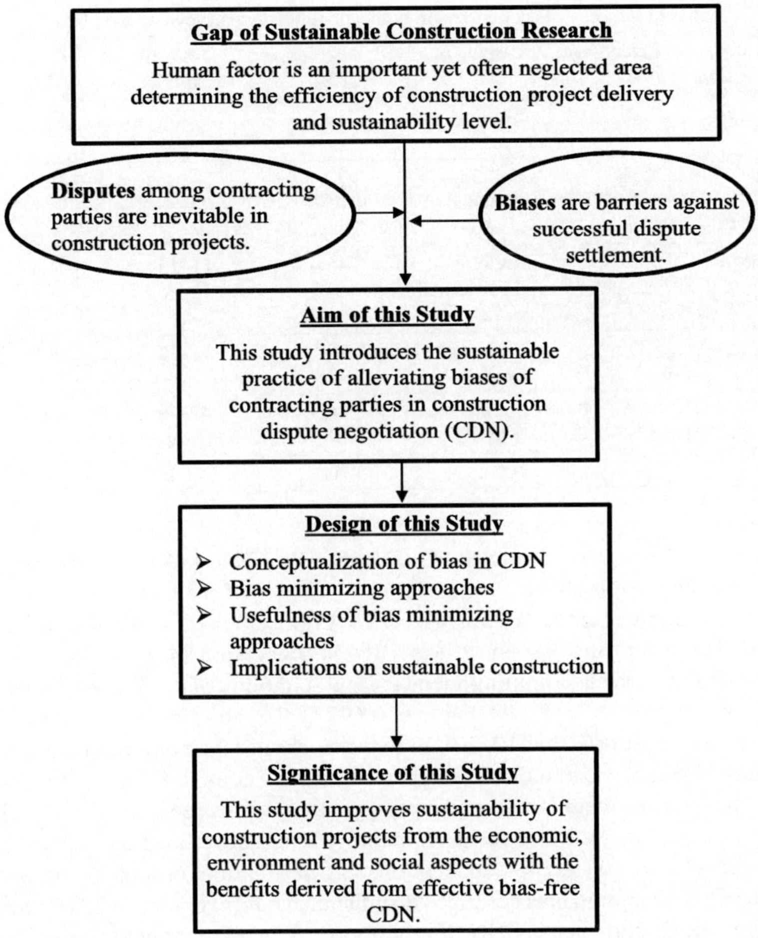

Fig. 1 Flow of the study

Self-reflection of Disputants

First set of data is extracted from self-reflection of disputants collected by [16]. In that study, bias identification statements were developed by operationalizing effects of bias into biased behaviours. Respondents were then asked to rate on the frequency of happening of the bias behaviours according to the reflection of their own CDN practice. A Likert six-point scale was used. For the second set of data, the respondents of the first data set [16] were approached for participation in the simulation (details to follow). Their responses were called self-realization. Only the data provided by those respondents who completed both self-reflection survey and simulation in this study were used for data analysis. Profile of the respondents to both self-reflection

Table 1 Profile of the subjects completed in both self-reflection survey and simulation

Professional organization	Percentage (%)	Dispute involved	Percentage (%)	Years of experience	Percentage (%)
Contractor	26.8	Building services work	12.5	Below 5 years	19.6
Client	37.5	Building (foundation) work	7.1	5–10 years	46.4
Consultant	35.7	Building (superstructure) work	46.4	11–15 years	21.4
Total	100	Civil engineering work	19.6	16–20 years	8.9
		Maintenance work	14.3	Above 20 years	3.6
		Total	100	Total	100

survey and self-realization simulation is shown in the Table 1. A total of 56 responses were obtained for this study.

Principal component factor analysis (PCFA) was applied to the first set of data to unveil the underlying bias constructs. IBM SPSS version 24.0 was used. Varimax rotation was applied and sampling adequacy and suitability of the data were supported by Kaiser–Meyer–Olkin (KMO) value of 0.697 ($\geq$0.6) and significant Bartlett's test of sphericity result (<0.001)[19, 20]. Eigenvalue greater than one was considered as significant for factor extraction as suggested by Hair et al. [21]. Accordingly, only bias manifestations with factor loadings larger than 0.5 were retained [22–24]. The PCFA result points to a four-factor structure without cross loading (Table 2). The four constructs of bias are: preconception, self-affirmation, optimism and interest-oriented. Preconception bias describes that disputants form preconceptions about the dispute before commencing CDN. Furthermore, their subsequent assessments were also heavily influenced by these preconceptions. Once preconceptions were formulated, it is mentally hard to ignore and go back to first principles. Self-affirmation bias occurs when disputants in CDN selectively search information with the aim of supporting their already held positions. This would prevail even other possible options become available. Optimism biased disputants are having unrealistic expectation that their requirements would be satisfied. Very often the expectation has been elevated without reasonable grounds. Interest-oriented bias makes disputants only focus on their own interests even at the expense of neglecting win–win solutions. All four types of biases would render communication ineffective among the disputing parties in CDN.

Table 2 Bias constructs extracted from the three sets of data

	Manifestations in CDN	Data set 1: self-reflection				Data set 2: self-realization				Data set 3: 3rd party neutral			
		Factor				Factor				Factor			
		1	2	3	4	1	2	3	4	1	2	3	4
		1.328*	1.819*	1.540*	5.018*	1.540*	2.857*	1.866*	3.345*	1.313*	2.779*	4.480*	1.778*
Preconception bias	1. My final assessment has been influenced by the first offer of E	0.732				0.580				0.807			
	2. My assessment has been influenced by unsubstantiated figures raised by E during the resolution process	0.774				0.766				0.769			

(continued)

Table 2 (continued)

	Manifestations in CDN	Data set 1: self-reflection				Data set 2: self-realization				Data set 3: 3rd party neutral			
		Factor				Factor				Factor			
		1	2	3	4	1	2	3	4	1	2	3	4
		1.328*	1.819*	1.540*	5.018*	1.540*	2.857*	1.866*	3.345*	1.313*	2.779*	4.480*	1.778*
	3. My decision has been adjusted because of the ambitious arguments of E	0.766				0.586				0.655			
	4. I cannot get away with my claim amount HK$ 1.13 billion made before the mediation stage	0.540				0.611				0.786			

(continued)

Table 2 (continued)

	Manifestations in CDN	Data set 1: self-reflection				Data set 2: self-realization				Data set 3: 3rd party neutral			
		Factor				Factor				Factor			
		1	2	3	4	1	2	3	4	1	2	3	4
		1.328*	1.819*	1.540*	5.018*	1.540*	2.857*	1.866*	3.345*	1.313*	2.779*	4.480*	1.778*
Self-affirmation bias	5. When receiving new information, I have paid more attention to the information that is consistent with my prior knowledge of the dispute		0.585				0.570				0.741		

(continued)

Table 2 (continued)

	Manifestations in CDN	Data set 1: self-reflection				Data set 2: self-realization				Data set 3: 3rd party neutral			
		Factor				Factor				Factor			
		1	2	3	4	1	2	3	4	1	2	3	4
		1.328*	1.819*	1.540*	5.018*	1.540*	2.857*	1.866*	3.345*	1.313*	2.779*	4.480*	1.778*
	6. I have inclined to interpret further information as evidences to justify my assessment of the claim		0.625				0.667				0.732		
	7. I search for information that confirms my assessment		0.840				0.849				0.685		

(continued)

Table 2 (continued)

	Manifestations in CDN	Data set 1: self-reflection				Data set 2: self-realization				Data set 3: 3rd party neutral			
		Factor				Factor				Factor			
		1	2	3	4	1	2	3	4	1	2	3	4
		1.328*	1.819*	1.540*	5.018*	1.540*	2.857*	1.866*	3.345*	1.313*	2.779*	4.480*	1.778*
	8. I would endorse information that supports my assessment		0.688				0.847				0.679		
Optimism bias	9. I have been very optimistic about the likelihood of winning irrespective of the arguments of E			0.556				0.646				0.794	

(continued)

Table 2 (continued)

	Manifestations in CDN	Data set 1: self-reflection				Data set 2: self-realization				Data set 3: 3rd party neutral			
		Factor				Factor				Factor			
		1	2	3	4	1	2	3	4	1	2	3	4
		1.328*	1.819*	1.540*	5.018*	1.540*	2.857*	1.866*	3.345*	1.313*	2.779*	4.480*	1.778*
	10. I totally believe that the outcome of the resolution will be good for my party			0.737				0.679				0.873	
	11. I am very confident that my ambitious arguments will succeed			0.778				0.699				0.770	

(continued)

Table 2 (continued)

	Manifestations in CDN	Data set 1: self-reflection				Data set 2: self-realization				Data set 3: 3rd party neutral			
		Factor				Factor				Factor			
		1	2	3	4	1	2	3	4	1	2	3	4
		1.328*	1.819*	1.540*	5.018*	1.540*	2.857*	1.866*	3.345*	1.313*	2.779*	4.480*	1.778*
	12. If the mediation fails to reach a settlement, I believe I know this outcome all along			0.661				0.605				0.630	
Interest-oriented bias	13. I think that E is having bias				0.712				0.664				0.822

(continued)

Table 2 (continued)

Manifestations in CDN	Data set 1: self-reflection				Data set 2: self-realization				Data set 3: 3rd party neutral			
	Factor				Factor				Factor			
	1	2	3	4	1	2	3	4	1	2	3	4
	1.328*	1.819*	1.540*	5.018*	1.540*	2.857*	1.866*	3.345*	1.313*	2.779*	4.480*	1.778*
14. If the mediation fails to reach a settlement, I believe We should take greater responsibility				0.857				0.862				0.847
15. If the mediation fails to reach a settlement, I believe the arguments of E during the mediation are unreasonable				0.506				0.856				0.631

(continued)

Table 2 (continued)

	Manifestations in CDN	Data set 1: self-reflection				Data set 2: self-realization				Data set 3: 3rd party neutral			
		Factor				Factor				Factor			
		1	2	3	4	1	2	3	4	1	2	3	4
		1.328*	1.819*	1.540*	5.018*	1.540*	2.857*	1.866*	3.345*	1.313*	2.779*	4.480*	1.778*
	16. If the mediation fails to reach a settlement, I believe the failure of mediation is inevitable because of the opportunistic attitude of E				0.822				0.698				0.502

*The initial eigenvalues for each factor

Self-realization of Disputants Through a CDN Simulation

Self-reflection data may be affected by the inherent bias of the respondents. Another method was used to obtain data from the same group of respondents—answering what they would do in a simulated construction project dispute resolution situation. The data collected from the simulation is called self-realization to distinguish from the way data were obtained in the self-reflection survey. Simulation aims to create a decision environment closer to reality by incorporating contextual information. The dispute was related to a simulated land reclamation project. There are four parts in the simulation. Part A introduces particulars of the project, including project scope, contract sum and contract period. Part B explains the dispute and include the issues, arguments presented and the amount in dispute. In Part C, the respondents went through the mediation of the dispute including preparation before mediation, joint caucus and then private caucus. In Part D, respondents were asked to describe their decision-making approaches taken in the simulation by rating the bias identification statements that were developed by Li and Cheung [16] with a seven-point Likert Scale from "1 = Strongly Disagree" to "7 = Strongly Agree". Higher scores would suggest greater chance of happening of the biased behaviours. These bias identification statements have been modified in contexts with due regard for the simulation. For example, "I cannot get away with the assessments made at prior round of resolution of the dispute." was changed to "I cannot get away with my claim amount HK$ 1.13 billion made before the mediation stage."

56 valid responses to the simulation were received (the self-reflection data set has 105 responses). The profile of the subjects participated in the simulation is shown in Table 1. When extracting the factor structure, PCFA suggests the same four bias constructs as shown in Table 2.

Observations of Third-Party Neutrals

To explore the bias constructs from another perspective, the third Data Set was collected from practicing construction dispute third party neutrals, including accredited mediators, arbitrators and adjudicators in CDN. This approach further avoids the influence of bias inherent within the disputants as respondents. Moreover, the observation of third-party neutral can only be useful if the observations are truly reflective of the thinking of the disputants. Input of experienced third-party neutral is thus critical. As an international business and financial centre, Hong Kong offers full range of high-quality professional dispute resolution services. Accredited third-party neutrals listed in globally recognized dispute resolution services providers were approached. The contacts of potential respondents were collected from learned societies, including Society of Construction Law Hong Kong (SCLHK), the Hong Kong International Arbitration Centre (HKIAC), the Hong Kong Mediation Accreditation Association Limited (HKMAAL), the Hong Kong Institute of Arbitrators (HKIAB) and the Hong

Kong Institution of Engineers (HKIE). This group of third-party neutrals are having a good mix of expertise as they come from various professional backgrounds as well as nationality, practice location, jurisdiction of admission and dispute resolution expertise. The validated bias identification statements previously used were distributed to third party neutrals to solicit their opinion on the frequency of disputants having these behaviours with a frequency scale from "1 = Never" to "7 = Always".

The survey was distributed online through email with a cover letter introducing the background information of the study. In total, 66 valid responses were received out of more than 600 questionnaires distributed. Among the respondents, 76% of them have more than 15 years' experience in CDN, nearly 60% of them have worked in CDN for more than 20 years. The profile of the respondents is shown in Table 3. Practice locations of the respondents presented in Fig. 2. This set of data is the third of the study.

PCFA was performed to explore the constructs of bias based on the responses received under Data Set Three. KMO value of 0.68 and significant Bartlett's test of sphericity result supported the sampling adequacy and data suitability [19]. Again, only identifications with factor loadings larger than 0.5 were retained and factor matrix extracted is shown in Table 2. The same four bias constructs were extracted, indicating that third-party neutral group observed the same four types of bias occurring in CDN—preconception, self-affirmation, optimism and interest-oriented. Thus, these four bias constructs were verified by Data Set Three. The robustness of the bias constructs is enhanced by the consistent results obtained from the three data sets.

Table 3 Profile of the third-party neutral respondents

Years of experience	Percentage (%)	Roles in CDN	Percentage (%)
Less than 5 years	6	Mediator	42
5–10 years	11	Arbitrator	53
11–15 years	7	Adjudicator	3
16–20 years	17	Others	2
More than 20 years	59	Total	100
Total	100		
Dispute type	Percentage (%)	Cause of the dispute	Percentage (%)
Building services installations	4.5	Risk uncertainty	7.6
Building (Foundation) Work	7.6	Collaboration among the parties	19.7
Building (Superstructure) Work	36.4	Contract incompleteness	42.4
Civil engineering work	39.4	Opportunistic behaviour	12.1
Maintenance work	9.1	Affective conflict	1.5
Others	3.0	Others	16.7
Total	100	Total	100

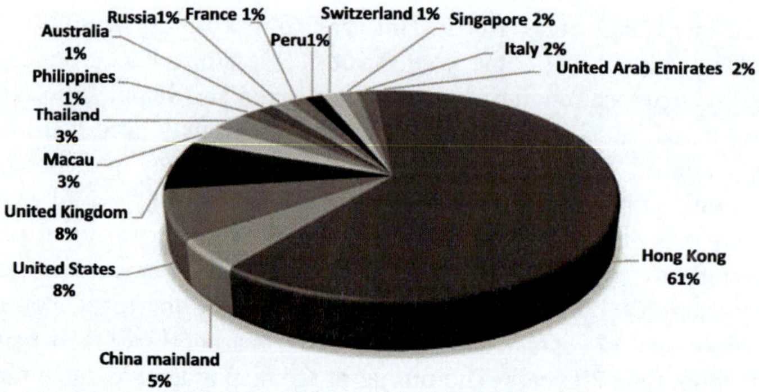

Fig. 2 Practice locations of the respondents

Magnitude of the Biases

Magnitude score (MS) can be used to indicate the potency of the four sources of bias [18]. As the constructs of bias reflect the respective sources of bias, the MS for each source of bias was calculated as the average of the mean scores of the bias identification statements under each bias construct and was calculated according to the Eq. (1):

$$MS_i = \frac{\sum_{j=1}^{n} BS_{ij}}{n} \tag{1}$$

where MS_i is the magnitude score of bias type *i*; BS_{ij} is the mean score of the *j* th bias identification statement of bias type *i*; *n* is the number of bias identification statements in bias type *i*.

The MSs of the sources of bias are listed in Table 4. In Data Set One, the assessment of bias practice was based on a six-point Likert Scale frequency level. In Data Set Two and Three, seven-point Likert Scale was employed. Transformation of the assessments in Data Set One was conducted for easy comparison with the following Eq. (2) as recommended by statistical handbook [25]:

$$\text{R7} = \frac{R6 - 1}{5} \times 6 + 1 \tag{2}$$

where R7 is the rescaled variable, which is 1 to 7 scale in this study; R6 is the original scale, which is 1–6 scale in this study.

After the MSs were transformed into a same metric, it can be noted that the MSs of the biases in Data Set Two (self-realization) are larger than the MSs in Data Set One (self-reflection). The results indicate that with the same group of respondents, use of simulation made biased behaviours more notable. Moreover, the relative rankings of the biases remain unchanged for Data Set Two and Data Set One. Hence, in both Data Set Two and Data Set One, self-affirmation bias was identified as the strongest and happened most frequently. It thus was confirmed by the disputants that they tended to

Table 4 Magnitude of biases in data set one, data set two and data set three

Bias magnitude in CDN									
Data set one				Data set two			Data set three		
Bias type	MS (6-point Scale)	MS (7-point Scale)	Rank	Bias type	MS (7-point Scale)	Rank	Bias type	MS (7-point Scale)	Rank
Pre-conception	2.85	3.22	4	Pre-conception	4.16	4	Pre-conception	4.06	4
Self-affirmation	4.17	4.80	1	Self-affirmation	5.32	1	Self-affirmation	5.05	2
Optimism	3.38	3.86	3	Optimism	4.33	3	Optimism	4.59	3
Interest-oriented	3.40	3.88	2	Interest-oriented	4.38	2	Interest-oriented	5.13	1

defend themselves and did not mind or subconsciously collect and interpret information in pre-disposed manner. Interest-oriented bias was ranked 2nd highest and can be interpreted as confession of the disputants about their interest-maximization strategy. Optimism and preconception were ranked 3rd and 4th, indicating that although the disputants are overly optimistic and affected by previously formed perception, they believe these two types of behaviours happen less frequently than self-affirmation and interest-oriented tendency.

The MSs of the biases based on the Data Set Three were shown in Table 4. The four constructs of bias in Data Set Three are higher than those obtained from Data Set One, suggesting that 3rd party neutrals in CDN observed more frequent happening of biased behaviours of disputants than the self-reported results. Looking into the rankings of MSs obtained from the three data sets, it can be concluded that by the inclusion of contextual information whereby the respondents can more readily relate to their practices. In other words, contextual information of CDN scenario makes biased behaviours more apparent. Third party neutrals' responses were based on their observations of disputing parties' biased practices in real CDN situations and may well be the most objective among the three. Similarly, the third-party neutrals observed more frequent happening of biased behaviours than the self-reflection of the disputants in Data Set One.

It cannot be excluded that the disputants may have the tendency to project positive self-image of being professional and be influenced by biases in their decisions. As such, they were more reluctant to admit that they had made biased decisions [26, 27]. Their self-reflection on their biased behaviours in Data Set One may well have been downplayed. Besides, the bias magnitude ranking in Data Set Three is slightly different from the results in Data Set Two and Data Set One. Third-party neutrals consider that interest-oriented bias rather than self-affirmation is the strongest bias displayed by disputants. As third-party neutral can only deduce the thinking of the disputants through their decisions during the negotiations like proposals and exchange of offers, it is not too surprising to spot self-interest disposition that is more manifest. Interest-oriented bias explains why aggression is used even without justifiable causes. Interest-oriented bias is thus more notable and observable. For example, it is easier for the third party neutral to objectively observe that the disputants are bargaining for their self-interest by insisting on their positions without any will to compromise. Self-affirmation bias focuses on disputants' suboptimal choices in information searching and interpretation, which are more subtle and less detectable from observations. Thus, it is harder to observe disputants' behaviours of biased information analysis as these are mental processes.

To summarize the findings for objective one, with three different data sets, the same four constructs of bias in CDN have been resulted from PCFA. The following section of the chapter deals with the work for the accomplishment of objective two.

Bias Minimizing Approaches

To accomplish objective two, four bias minimizing approaches are identified through a literature review. These are: (1) allow adequate time and effort in making decisions; (2) consider the opposite and question oneself; (3) keep rational and consider long-term benefit; and (4) review design of dispute resolution mechanism. These approaches were further operationalized into twenty bias minimizing measures. The afore-mentioned bias minimizing measures and their respective references are listed in Table 5.

The usefulness of the listed bias minimizing measures was evaluated. First, the measures were incorporated in the CDN simulation as consulting mediators' suggestions. In Part D of the simulation, respondents were asked to consider the usefulness of these bias minimizing measures from "1 = Helpless" to "7 = Absolutely helpful". The practicality of these bias minimizing measures was also considered by the practicing third-party using the afore-mentioned scale. With the ratings by the disputants and third-party neutrals, the relative usefulness of these bias minimizing measures was calculated. The Usefulness Index (UI) of each single bias minimizing measure was calculated by Eq. (3) [68, 69]:

$$\text{Usefulness Index} = \frac{\sum_{i=1}^{7}(a_i * x_i)}{6\sum_{i=1}^{7} x_i} \tag{3}$$

where a_i = constant expressing the weight assigned to the *i*th response; $a_i = 0, 1, 2, 3, 4, 5, 6$ for $I = 1, 2, 3, 4, 5, 6, 7$, respectively; $a_1 = 0$ is assigned to "Helpless"; $a_7 = 6$ is assigned to "Absolutely helpful"; X_i = the percentage of the degree of helpfulness; X_1 = percentage of frequency of "Helpless" responses; X_7 = percentage of frequency of "Absolutely helpful" responses.

The UIs of the bias minimizing measures were calculated and shown in Table 5. Usefulness of each approach was calculated as the average of the UIs of the bias minimizing measures under the approach. The usefulness of these approaches was ranked in Table 5 as well. The usefulness indices were grouped in Table 6 to show the respondents' evaluation.

From Tables 5 and 6, it can be seen that disputants rated the four approaches as "Moderately Useful". Third party neutrals rated Approach 1: Allow adequate time and effort in making decisions, Approach 2: Consider the opposite and question oneself and Approach 3: Be rational and consider long-term benefit as "Reasonably Useful". Approach 4: Dispute resolution mechanism design was rated as "Moderately Useful". Therefore, these bias minimizing approaches were validated by both disputants (Data Set Two) and third-party neutrals (Data Set Three).

Besides, both the disputants and third-party neutrals ranked similarly the usefulness of the four bias minimizing approaches. They believe Approach 3: Be rational and consider long-term benefit as the most useful among the four approaches because uncontrolled emotion invites biases. Staying rational, enhancing mutual understanding and focusing on long-term benefit and reputation were rated as valuable

Table 5 Usefulness of bias minimizing approaches and measures

Bias minimizing approaches and measures	References	UIs data set two (Rank)	UIs data set three (Rank)
• ***Strategy-based (for preconception bias and self-affirmation bias)***			
• ***Approach 1: Allow adequate time and effort in making decisions***		***40.10 (2)***	***47.6 (2)***
1. Disputants should allow adequate time for making decision to avoid premature closure of thinking	[28–31]	40.17	47.17
2. Disputants should review the case and the possible bottom line of the counterpart when a very low offer was forwarded	[29, 32]	42.17	43.50
3. Disputants should check the accuracy of the evidences provided by the counterpart	[29, 32]	43.83	52.00
4. Disputants should delay forming an assessment until all the available information has been considered	[32, 33]	34.33	45.00
5. Disputants should be open to other alternatives even after a first assessment about the dispute has been formed	[34–37]	40.00	50.50
• ***Approach 2: Consider the opposite and question oneself***		***38.38 (3)***	***46.6 (3)***
6. Disputants should consider information that may work against a prior assessment	[32, 38, 39]	37.83	45.83
7. Disputants should readily question the soundness of a prior assessment	[30, 40–42]	37.83	47.00
8. Disputants should review the reasons of the counterpart	[42–45]	42.33	49.50
9. Disputants should ask for feedbacks and assistance from third party neutral	[46, 47]	35.50	44.17
• ***Attitude-based (for Interest-oriented bias and Optimism bias)***			
• ***Approach 3: Be rational and consider long-term benefit***		***40.60 (1)***	***49.3 (1)***

(continued)

Table 5 (continued)

Bias minimizing approaches and measures	References	UIs data set two (Rank)	UIs data set three (Rank)
10. Disputants should consider mutually beneficial trade-offs between the parties	[16, 33, 48–50],	43.33	51.00
11. Disputants should avoid being emotional	[33, 51–54]	45.17	53.33
12. Disputants should try to understand the position of their counterpart by stepping in their roles	[36, 55–57]	40.83	49.83
13. Disputants should respectfully listen to their counterpart's grievances	[36, 55–57]	40.33	49.33
14. Disputants should consider long-term relationship and future collaboration with their counterpart in handling the dispute	[14, 51, 58, 59]	39.83	46.83
15. Disputants should consider the chance of settlement failure	[44, 60–62]	36.83	48.50
16. Disputants should think about their own responsibilities when the dispute fails to settle	[59, 63]	37.83	46.17
• ***Process-based (for Preconception bias and Interest-oriented bias)***			
• ***Approach 4: Review design of dispute resolution mechanism***		***35.38 (4)***	***36.4 (4)***
17. Disputants should receive de-biasing training before participating in resolution processes	[14, 29, 44, 60, 64, 65]	33.50	37.33
18. To start a new round of resolution, the resolution team should include new members	[44, 66]	34.00	30.17
19. Re-assessment and reconstruction of decisions are required to start a new round of resolution	[33, 67]	36.50	38.50
20. A process to ensure needs are reviewed is required at each round of resolution	[33, 67]	37.50	39.67

Table 6 Usefulness groups and indices

Usefulness group	Usefulness Index (UI)
Useless	0–14.3
Slightly useful	14.3–28.6
Moderately useful	28.6–42.9
Reasonably useful	42.9–57.2
Very useful	57.2–71.5
Most useful	71.5–85.8
Absolutely useful	85.8–100

measures because all these underpin rational analysis. Approach 1: Allow adequate time and effort in making decisions was ranked as the 2nd most useful, therefore, adequate time and effort in decision making were confirmed in calming heated disputants and encouraging a considerate and mature decision. Approach 2: Consider the opposite and question oneself and Approach 4: Dispute resolution mechanism design were ranked 3rd and 4th in usefulness respectively.

Grouping of Bias Minimizing Approaches

This part of the chapter analyses bias minimizing approaches based on their nature and with reference to the types of bias identified for objective one. Accordingly, three groups of approach are proposed: strategy-based, attitude-based and process-based. Table 5 gives the tabulated framework together with the UIs.

Strategy-Based

Approach one (allow adequate time and effort in making decisions) and approach two (consider the opposite and question oneself) were grouped into strategy-based group of bias minimizing approach. It is advocated that disputants would obtain a better picture of the current situation and a more holistic view of the dispute through taking enough time to review the case and carefully considering the offer and evidence provided from the counter project team. Assessment should not be hastily taken before available information was considered. This would lower the chance of being affected by preconception of the issue in dispute. Hence, enough time and effort paid in making assessment would avoid a premature formation of opinion and position that will become enduring preconception. Besides, questioning previously held positions before making every major decision would help disputants objectively review their earlier assessments about the issue in dispute. Seeking feedbacks and assistance from third party neutrals (consulting mediators and dispute resolution advisors) would also help disputants to get an outsider point of view whereby avoiding self-affirmation. Therefore, approach one and approach two are strategies helping project contracting parties to obtain a holistic view of the dispute and to keep an open mind to further information. Preconception bias and self-affirmation bias would be minimized correspondingly.

Attitude-Based

Approach three (be rational and consider long-term benefit) minimizes bias by adjusting project contracting parties' attitude and restraining their negative emotions in making decisions. This attitude-based strategy group is effective in alleviating disputants' interest-oriented and optimism biased behaviours. By considering mutual benefits, meaningful trade-offs, long-term relationship and potential future collaboration with the counterpart, disputing parties would restrain from short-term interest-maximizing behaviour. They would love to work for an amicable partnership to seek long-run benefits. Besides, when they try to step in counterpart's shoe and understand their positions and concerns, they may adopt a more collaborative negotiation. In fact, focusing on the possibility of having a win–win solution would be beneficial to the disputing parties. In addition, by reality testing with the negative impact resulting from a negotiation breakdown, the disputants would calm down and be less unrealistically optimistic. All in all, when the disputants can stay away from being too emotional, overly optimistic expectations can be avoided. As a result, they are more ready for rational decisions in construction dispute negotiation (CDN).

Process-Based

Approach four (dispute resolution mechanism design) aims to minimize bias by optimizing the CDN process. This process-based approach points to the minimization of preconception bias and interest-oriented bias. By incorporating pre-negotiation training, disputing parties would be reminded of the happening of biases. They would be trained to detect and skip possible bias minefields. In addition, including new members would also bring fresh new ideas to the CDN team. The input of new member would decrease the obstinate adherence to old positions. Re-framing of the dispute and assessment before the commencement of a new round of CDN would help the disputants to re-organize the strategy. Revisiting the assumptions, expectations etc. would mitigate the influence of preconception bias. A process of reviewing initial needs would help project disputing parties to realize that the current impasse is not conducive in achieving their needs. Disputing parties are encouraged to think about other alternatives that would better serve for their essential interests and at the same time could be accepted by the counterpart.

Implications on Dispute Management

Biases have been identified as one of the major barriers against conducive construction dispute negotiation, thus alleviating biases in CDN should be an integral part of dispute negotiation training. In fact, construction industry is dispute prone, protracted dispute resolution hampers efficiency. In the last few decades, there is clearly a rising use of multi-tiered dispute resolution (MTDR) in construction contracts. Basically, MTDR incorporates alternative dispute resolution (ADR) as pre-condition before arbitration [3, 15]. The design intent of MTDR is to resolve construction disputes in the earlier stages of ADR, without proceeding to more formal proceedings like arbitration and litigation. The advantages of implementing ADR are saving time and cost. However, MTDR may not achieve the intended outcome as repeated evaluations can be breeding ground for biases [16]. In this connection, alleviating bias in CDN as proposed in this study would enhance the efficiency of MTDR processes. Effective dispute negotiation saves substantial resources and materials that would otherwise be wasted in the prolonged dispute resolution processes.

In social aspect, alleviating bias in CDN improves the intense relationship between the construction contracting parties. Minimizing biases enhances the decision-making performance of the disputing parties and keeps them in rational courses [7]. It also reduces their negative view on each other whereby engendering more collaborative effort to seek mutual beneficial win–win positions. When biases are removed, trust relationship, partnership and positive collaboration could be built among the contracting parties [18, 70, 71]. Team efficiency, job satisfaction and employee engagement would also be increased with a positive working environment [72, 73]. Therefore, the practice of alleviating bias in CDN contributes to the building of social sustainability and healthy community in construction industry.

Summary

Biased decisions prohibit effective construction dispute negotiation [16]. Cogent dispute management calls for dispute decisions free from biases. The saving in valuable resources through amicable negotiations can be used in more productive courses. This study contributes to the body of knowledge of dispute management by offering constructs of biases in CDN. This study is robust in going beyond the conventional approach of obtaining self-reflection of biased behaviours by disputants. Instead, data was obtained from three sources: i self-reflection of disputants; ii self-realization of disputants in a dispute negotiation simulation; and iii observations of dispute resolution third party neutrals. Conceptualization of biases in CDN is triangulated by interpreting results of PCFA performed with the three data sets. The use of three sets of data served as triangulation of the empirical findings. The same four bias constructs were extracted as a result. Four major types of biases in CDN were identified as: preconception, self-affirmation, optimism and interest-oriented. This study also suggested bias minimizing measures that address the respective bias sources. Categorically, three groups of bias minimizing measures were proposed: (i) strategy-based approach to deal with preconception bias and self-affirmation bias; (ii) attitude-based approach works to alleviate interest-oriented bias and optimism bias; and (iii) process-based approach is suitable to alleviate the effect of preconception bias and interest-oriented bias minimization. Curbing biases is a prerequisite for effective dispute negotiation and should be conducted by negotiators. Biases hamper rational decisions and derail settlement course. It is also suggested that alleviating bias would improve the relationship between construction contracting parties. Conceptualizing biases in CDN also paves the path for further studies on biases in construction.

Acknowledgements The empirical work of this chapter has been reported in a paper entitled "Alleviating bias to enhance sustainable construction dispute management" of the Journal of Cleaner Production. The texts have been substantially re-written. The support of the HKSAR RGC General Research Fund Project (no. 11209118) is duly acknowledged.

References

1. Carvajal-Arango D, Bahamón-Jaramillo S, Aristizábal-Monsalve P, Vásquez-Hernández A, Botero LFB (2019) Relationships between lean and sustainable construction: positive impacts of lean practices over sustainability during construction phase. J Clean Prod 234:1322–1337
2. Gan X, Zuo J, Ye K, Skitmore M, Xiong B (2015) Why sustainable construction? Why not? An owner's perspective. Habitat Int 47:61–68
3. Cheung SO, Li K (2019) Biases in construction project dispute resolution. Eng Constr Archit Manag 26(2):321–348
4. Cakmak E, Cakmak PI (2014) An analysis of causes of disputes in the construction industry using analytical network process. Procedia Soc Behav Sci 109:183–187
5. Cheung SO, Yiu TW (2006) Are construction disputes inevitable? IEEE Trans Eng Manage 53(3):456–470
6. Love P, Davis P, Ellis J, On Cheung S (2010) Dispute causation: identification of pathogenic influences in construction. Eng Constr Archit Manag 17(4):404–423
7. Li K, Cheung SO (2019) Unveiling cognitive biases in construction project dispute resolution through the lenses of third-party neutrals. J Constr Eng Manag 145(11):04019070
8. Hollander-Blumoff R, Tyler TR (2008) Procedural justice in negotiation: procedural fairness, outcome acceptance, and integrative potential. Law Soc Inq 33(2):473–500
9. Mnookin RH (1992) Why negotiations fail: an exploration of barriers to the resolution of conflict. Ohio State J Dispute Resolut 8(2):235–249
10. Ross L, Ward A (1995) Psychological barriers to dispute resolution. Adv Exp Soc Psychol 27:255–304
11. Brett J, Thompson L (2016) Negotiation. Organ Behav Hum Decis Process 136:68–79
12. Guthrie C, Sally D (2004) The impact of the impact bias on negotiation. Marquette Law Rev 87:817–828
13. Guthrie C, Rachlinski JJ, Wistrich AJ (2002) Judging by heuristic-cognitive illusions in judicial decision making. Judicature 86(1):44–50
14. Thompson L, Nadler J, Lount Jr RB (2000) Judgmental biases in conflict resolution and how to overcome them. The handbook of conflict resolution: theory and practice. pp 213–235
15. Li K, Cheung SO (2016) The potential of bias in multi-tier construction dispute resolution processes. In: Chan PW, Neilson CJ (eds) Proceedings of the 32nd annual ARCOM conference, vol 1, pp 197–205

16. Li K, Cheung SO (2018) Bias measurement scale for repeated dispute evaluations. J Manag Eng 34(4):04018016
17. Vazquez E, Rola S, Martins D, Freitas M, Rosa LP (2011) Sustainability in civil construction applied in the construction site phase. WIT Trans Ecol Environ 144:265–276
18. Wong PSP, Cheung SO (2004) Trust in construction partnering: views from parties of the partnering dance. Int J Project Manage 22(6):437–446
19. Cerny BA, Kaiser HF (1977) A study of a measure of sampling adequacy for factor-analytic correlation matrices. Multivar Behav Res 12(1):43–47
20. Kaiser HF (1958) The varimax criterion for analytic rotation in factor analysis. Psychometrika 23(3):187–200
21. Hair JF, Anderson RE, Tatham RL, Black WC (1998) Multivariate data analysis, 5th. Prentice Hall International, NY
22. Kannan VR, Tan KC (2005) Just in time, total quality management, and supply chain management: understanding their linkages and impact on business performance. Omega 33(2):153–162
23. Matsunaga M (2010) How to factor-analyze your data right: do's, don'ts, and how-to's. Int J Psychol Res 3(1):97–110
24. Reynolds WM (1982) Development of reliable and valid short forms of the marlowe-crowne social desirability scale. J Clin Psychol 38(1):119–125
25. Little TD (2013) Methodology in the social sciences. Longitudinal structural equation modelling. Guilford Press, New York, NY, US
26. Edwards AL (1957) The social desirability variable in personality assessment and research. The Dryden Press, New York
27. Phillips DL, Clancy KJ (1972) Some effects of "social desirability" in survey studies. Am J Sociol 77(5):921–940
28. Chapman GB, Johnson EJ (2002) Incorporating the irrelevant: anchors in judgments of belief and value. In: Gilovich T, Griffin D, Kahneman D (eds) Heuristics and biases: the psychology of intuitive judgment. Cambridge university press, pp 120–138
29. Croskerry P, Singhal G, Mamede S (2013) Cognitive debiasing 2: impediments to and strategies for change, BMJ Qual Saf, bmjqs-2012-001713
30. Epley N, Gilovich T (2006) The anchoring-and-adjustment heuristic why the adjustments are insufficient. Psychol Sci 17(4):311–318
31. Mussweiler T, Englich B, Strack F (2004) "Anchoring effect." In: Pohl R (ed) Cognitive illusions—a handbook on fallacies and biases in thinking, judgment, and memory. Psychology Press, pp 183–200
32. Galinsky AD, Mussweiler T (2001) First offers as anchors: the role of perspective-taking and negotiator focus. J Pers Soc Psychol 81(4):657
33. Fisher R, Ury WL, Patton B (2011) Getting to yes: negotiating agreement without giving in. Penguin
34. Anderson CA (1982) Inoculation and counter explanation: debiasing techniques in the perseverance of social theories. Soc Cogn 1(2):126–139
35. Bentz BG, Williamson DA, Franks SF (2004) Debiasing of pessimistic judgments associated with anxiety. J Psychopathol Behav Assess 26(3):173–180
36. Hammond JS, Keeney RL, Raiffa H (1998) The hidden traps in decision making. Harv Bus Rev 76(5):47–58
37. Heiman VB (1990) Auditors' assessments of the likelihood of error explanations in analytical review. Account Rev 65(4):875–890
38. Kray LJ, Galinsky AD (2003) The debiasing effect of counterfactual mind-sets: increasing the search for disconfirmatory information in group decisions. Organ Behav Hum Decis Process 91(1):69–81
39. Schulz-Hardt S, Jochims M, Frey D (2002) Productive conflict in group decision making: genuine and contrived dissent as strategies to counteract biased information seeking. Organ Behav Hum Decis Process 88(2):563–586

40. Hoch SJ (1985) Counterfactual reasoning and accuracy in predicting personal events. J Exp Psychol Learn Mem Cogn 11(4):719
41. Kennedy J (1995) Debiasing the curse of knowledge in audit judgment. Account Rev 70(2):249–273
42. Koriat A, Lichtenstein S, Fischhoff B (1980) Reasons for confidence. J Exp Psychol Hum Learn Mem 6(2):107–118
43. Bazerman MH, Neale MA (1982) Improving negotiation effectiveness under final offer arbitration: the role of selection and training. J Appl Psychol 67(5):543
44. Larrick RP (2004) "Debiasing". In: Koehler DJ, Harvey N (eds) Blackwell handbook of judgment and decision making. Blackwell Publishing, pp 316–338
45. Mussweiler T, Strack F, Pfeiffer T (2000) Overcoming the inevitable anchoring effect: considering the opposite compensates for selective accessibility. Pers Soc Psychol Bull 26(9):1142–1150
46. Alexander FG (1980) Psychoanalytic therapy: principles and application. University of Nebraska Press
47. Landa Y, Silverstein SM, Schwartz F, Savitz A (2006) Group cognitive behavioral therapy for delusions: helping patients improve reality testing. J Contemp Psychother 36(1):9–17
48. Bazerman MH, Curhan JR, Moore DA, Valley KL (2000) Negotiation. Annu Rev Psychol 51(1):279–314
49. Drolet AL, Morris MW (2000) Rapport in conflict resolution: accounting for how face-to-face contact fosters mutual cooperation in mixed-motive conflicts. J Exp Soc Psychol 36(1):26–50
50. Lyons C (2009) I win, you win: the essential guide to principled negotiation. Bloomsbury Publishing
51. Baron RA (1991) Positive effects of conflict: a cognitive perspective. Emp Responsibilities Rts J 4(1):25–36
52. Forgas JP (1995) Mood and judgment: the affect infusion model (AIM). Psychol Bull 117(1):39
53. Hastie R (2001) Problems for judgment and decision making. Annu Rev Psychol 52(1):653–683
54. Schwenk CR, Cosier RA (1980) Effects of the expert, devil's advocate, and dialectical inquiry methods on prediction performance. Organ Behav Hum Perform 26(3):409–424
55. Babcock L, Loewenstein G (1997) Explaining bargaining impasse: the role of self-serving biases. J Econ Perspect 11(1):109–126
56. Galinsky AD, Ku G (2004) The effects of perspective-taking on prejudice: the moderating role of self-evaluation. Pers Soc Psychol Bull 30(5):594–604
57. Galinsky AD, Moskowitz GB (2000) Perspective-taking: decreasing stereotype expression, stereotype accessibility, and in-group favoritism. J Pers Soc Psychol 78(4):708–724
58. Sedikides C, Campbell WK, Reeder GD, Elliot AJ (1998) The self-serving bias in relational context. J Pers Soc Psychol 74(2):378
59. Soll J, Milkman K, Payne J (2014) A user's guide to debiasing
60. Arkes HR (1991) Costs and benefits of judgment errors: implications for debiasing. Psychol Bull 110(3):486
61. Hirt ER, Markman KD (1995) Multiple explanation: a consider-an-alternative strategy for debiasing judgments. J Pers Soc Psychol 69(6):1069–1086
62. Sanna LJ, Schwarz N (2004) Integrating temporal biases. Psychol Sci 15(7):474–481
63. Bowman Williams J (2017) Accountability as a debiasing strategy: does race matter?
64. Fischhoff B, Beyth-Marom R (1983) Hypothesis evaluation from a Bayesian perspective. Psychol Rev 90(3):239–260
65. Mowen JC, Gaeth GJ (1992) The evaluation stage in marketing decision making. J Acad Mark Sci 20(2):177–187
66. Burke A (2007) Neutralizing cognitive bias: an invitation to prosecutors. NYU J Law Liberty 2:512–530
67. Ashton RH, Kennedy J (2002) Eliminating recency with self-review: the case of auditors' 'going concern' judgments. J Behav Decis Mak 15(3):221–231
68. Ezeh GN, Ogbuehi CN, Eleke N, Diala UH (2013) Severity index analysis of the problems of optical fiber communication in Nigeria: a case study of South Eastern Nigeria. Acad Res Int 4(1):431–438

69. Johnson RA, Bhattacharyya GK (1996) Statistics: principles and methods. Wiley, New York
70. Chow PT, Cheung SO, Chan KY (2012) Trust-building in construction contracting: mechanism and expectation. Int J Project Manage 30(8):927–937
71. Wong PS, Cheung SO, Ho PK (2005) Contractor as trust initiator in construction partnering—prisoner's dilemma perspective. J Constr Eng Manag 131(10):1045–1053
72. Griffin MA, Patterson MG, West MA (2001) Job satisfaction and teamwork: the role of supervisor support. J Organ Behav 22(5):537–550
73. Spence Laschinger HK, Finegan J, Shamian J (2002) The impact of workplace empowerment, organizational trust on staff nurses' work satisfaction and organizational commitment. In: Advances in health care management. Emerald Group Publishing Limited, pp 59–85

A Bias Detection Tool for Construction Dispute Negotiation

Keyao Li and Sai On Cheung

Introduction

The occurrence and conceptualization of biases in construction dispute negotiation (CDN hereafter) have been reported in Chaps. "The Happening of Bias in Construction Dispute Negotiation" and "Conceptualising Bias in Construction Dispute Negotiation", respectively. To further the existing research on biases in CDN, a detection tool is necessary and will prove invaluable. This chapter details the development of a bias detection tool for CDN.

Bias Detection in CDN

While acknowledging the usefulness of dispute negotiation, disputing parties should be mindful that they may unconsciously be influenced by biases. Nevertheless, it is not easy for disputing parties to admit that biases have crept in. Thus, a bias detection tool would be indispensable for bias studies [1]. Raising awareness of biases is the first step to guard against them. Several bias measurement tools have been developed, mostly in the social media [2], cognition and emotion [3], marketing [4], psychology [5, 6] and education [7] domains. Table 1 provides a summary of the aforementioned studies. Notably, bias in construction dispute negotiation has not

K. Li (✉)
Future of Work Institute, Faculty of Business & Law, Curtin University, Perth, Australia
e-mail: Keyao.li@curtin.edu.au

S. O. Cheung
Construction Dispute Resolution Research Unit, City University of Hong Kong, Hong Kong, China
e-mail: Saion.cheung@cityu.edu.hk

S. O. Cheung (ed.), *Construction Dispute Research Expanded*, Springer Tracts in Civil Engineering, https://doi.org/10.1007/978-3-030-80256-1_3

Table 1 Previous study of bias measurement scales

Previous bias measurement studies	Descriptions	References
Assessment of attentional bias towards threats	An index reflecting the magnitude of attentional bias was calculated	[3]
Measurement of media bias	Media bias was measured by estimating ideological scores for several major media outlets	[2]
Estimation of nonresponse bias	The direction and magnitude of nonresponse bias in the context of mail surveys were estimated	[4]
Assessment of optimistic bias	Comparative judgement estimates were employed to identify optimistic bias	[5, 6]
Detection of publication bias	An adjusted rank correlation test was proposed to identify publication bias through a meta-analysis	[8]
Validity scale of recall bias	Proposed that whether and to what extent recall bias is occurring can be assessed by comparing the total validity scale scores of cases with controls	[9]
Measurement of response bias	Eight measures of socially desirable responding (SDR) were compared	[10]
Measurement of bias in search engines	Bias was assessed by measuring deviations from the ideal distributions produced by a particular search engine	[11]
Detection of test-item bias	Statistical techniques for detecting biased items, which worked by identifying items that may measure different things for different groups, were developed	[7]
Measurement of pretrial bias among jurors	A juror bias scale (JBS) was constructed and validated, resulting in an individual-difference measure of juror bias	[12]

been discussed. Thus, this chapter reports the development of a bias detection tool (the tool hereafter) in the context of construction dispute negotiation.

The Development of a Bias Detection Tool for Construction Dispute Negotiation

The most important criterion of any detection tool is its reliability, which is evaluated in terms of consistency over time. A longitudinal approach is therefore appropriate in this context. In essence, the measurement items have to be tested with two rounds of data collected from the same group of respondents. Figure 1 shows the step-by-step procedures applied in this study.

To execute the procedures set out in Fig. 1, several analytical tools were used: a paired t-test, a principal component factor analysis (PCFA), a reliability analysis and a multigroup confirmatory factor analysis (MCFA). A paired t-test is a method used to test whether the means of two sets of data are significantly different [13, 14]. Principal

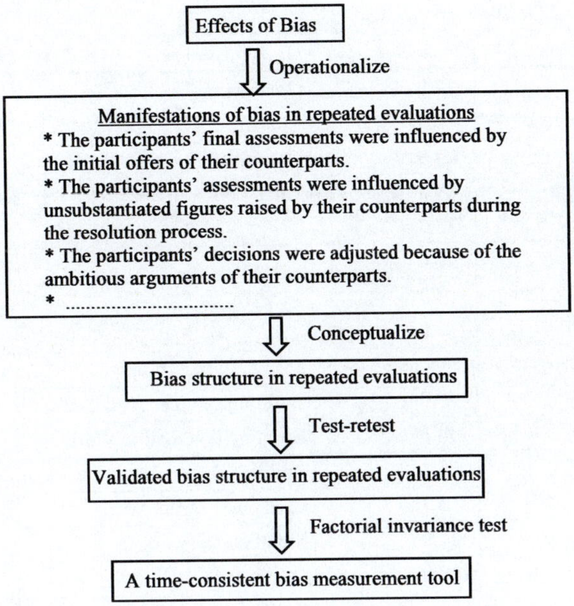

Fig. 1 Procedures used to develop the bias detection tool

component factor analysis (PCFA) is used to reveal the underlying constructs of variables [15, 16]. Reliability analysis examines the level of consistency among detection statements [17, 18]. Multigroup confirmatory factor analysis (MCFA) is used to test the level of factorial invariance across groups [19, 20].

Devising and Conducting the Data Collection Survey

In Chap. "The Happening of Bias in Construction Dispute Negotiation", four types of bias were identified in the context of CDN: preconception, self-affirmation, optimism and interest oriented. To develop a bias detection tool, sixteen detection statements were used to identify behaviours exhibiting bias effects. Therefore, the bias manifestations reported in Chap. "The Happening of Bias in Construction Dispute Negotiation" were the most suitable for this purpose. The test–retest methodology employed involved two rounds of data collection from the same group of respondents. The respondents were first asked to rate biased behaviour statements by indicating the frequency with which they engaged in these behaviours during construction dispute negotiation on a 6-point Likert scale. A high score suggested a high frequency of biased behaviour. It also represented the extent of that bias. The time interval between the two rounds of data collection in such studies is typically three months. This time interval should not be too short, as this could impact the memory effect, but it should

Table 2 Details of the 105 respondents

Profession	Percentage (%)	Organization	Percentage (%)
Architect	2.9	Contractor	22.8
Builder	2.9	Client	30.5
Civil engineer	21.9	Consultant	46.7
Building Surveyor	9.5	Total	100.0
Project manager	13.3		
Quantity Surveyor	46.6		
Others	2.9		
Total	100.0		
Years of experience cons	Percentage (%)	Dispute types	Percentage (%)
Less than 5 years	20.0	Building services Installations	7.6
5–10 years	34.3	Building Foundation Work	6.7
11–15 years	18.1	Building Superstructure work	42.8
16–20 years	5.7	Civil engineering Work	32.4
More than 20 years	21.9	Maintenance work	9.5
Total	100.0	Other	1.0
		Total	100.0

also not be too long, as major contextual changes may take place over time [17, 21]. Applying the test–retest methodology is a challenge, as it requires respondents to be willing to provide data twice. Second, their responses must also be valid for both rounds of data collection. In this study, the contact information of the respondents was collected from open sources such as the websites of government departments and quasi-government organizations as well as the directories of professional institutes and learned societies. Hard copies of the questionnaire were distributed during professional workshops and seminars as and when available. When the first round of questionnaires was distributed, the respondents were informed that they would be approached to participate in the second round after 3 months. In total, 105 valid responses for both rounds of the survey were finally obtained and used in the analysis. The details of these 105 valid responses are summarized in Table 2.

Comparison of the Two Rounds of Survey Responses

A paired t-test was then conducted to investigate any differences between the two rounds of responses. The statistics of the paired differences between the two rounds of responses are shown in Table 3. The paired differences of the mean responses of the two rounds ranged from 0.000 to 0.162. These could be considered nearly zero. All 16 detection statements achieved nonsignificant results ($p > 0.05$). The null hypothesis, namely, that the mean responses in the two rounds are the same,

Table 3 Paired differences of the bias detection items

Item	1st round			2nd round			Paired differences (1st round—2nd round)					
	Mean		Std.	Mean		Std.	Mean	Std.	Std. Error Mean	95% confidence interval of the difference		Sig. (2-tailed)
	Statistic	Std. Error		Statistic	Std. Error					Lower	Upper	
Item 1	3.04	0.088	0.898	2.91	0.100	1.030	0.124	1.124	0.110	−0.094	0.341	0.262
Item 2	2.61	0.104	1.070	2.51	0.111	1.136	0.095	1.213	0.118	−0.140	0.330	0.423
Item 3	2.76	0.103	1.052	2.70	0.097	0.992	0.067	1.031	0.101	−0.133	0.266	0.509
Item 4	3.02	0.099	1.019	2.90	0.095	0.970	0.124	1.016	0.099	−0.073	0.320	0.215
Item 5	4.04	0.098	1.009	4.18	0.090	0.918	−0.143	1.147	0.112	−0.365	0.079	0.205
Item 6	3.85	0.111	1.142	3.69	0.111	1.138	0.162	1.435	0.140	−0.116	0.440	0.250
Item 7	4.36	0.111	1.136	4.39	0.094	0.966	−0.029	1.267	0.124	−0.274	0.217	0.818
Item 8	4.11	0.103	1.059	4.11	0.103	1.059	0.000	1.421	0.139	−0.275	0.275	1.000
Item 9	3.18	0.102	1.045	3.18	0.083	0.852	0.000	1.168	0.114	−0.226	0.226	1.000
Item 10	3.48	0.108	1.110	3.50	0.090	0.921	−0.029	1.274	0.124	−0.275	0.218	0.819
Item 11	3.21	0.098	1.007	3.17	0.089	0.914	0.038	1.046	0.102	−0.164	0.241	0.710
Item 12	3.41	0.105	1.080	3.42	0.084	0.864	−0.010	1.114	0.109	−0.225	0.206	0.930
Item 13	3.40	0.105	1.080	3.50	0.090	0.921	−0.095	1.033	0.101	−0.295	0.105	0.347
Item 14	3.40	0.092	0.947	3.34	0.086	0.886	0.057	1.055	0.103	−0.147	0.261	0.580
Item 15	3.16	0.091	0.932	3.22	0.089	0.909	−0.057	1.150	0.112	−0.280	0.166	0.612
Item 16	3.51	0.106	1.084	3.38	0.088	0.903	0.133	1.152	0.112	−0.090	0.356	0.239

was therefore supported. The data from the respondents were considered stable over the two rounds of data collection. These statistical results supported the use of the sixteen-statement detection tool as a reliable scale for bias studies.

Extension of Factors (1st Round)

A principal component factor analysis (PCFA) was applied to the two data sets separately. IBM SPSS version 21.0 was used. For the first round of 105 responses, the sampling adequacy and suitability of the data was supported by the following tests: Bartlett's Test of Sphericity ($p < 0.001$) and Kaiser–Meyer–Olkin (KMO), which resulted in a value of 0.80; this is greater than threshold of 0.6 [22–24]. To extract factors, an eigenvalue greater than one was used [25]. Varimax rotation was then applied to simplify the factor structure for ease of interpretation [25, 26]. The factor matrix is shown in Table 4, where only factor loadings larger than 0.55 were considered statistically significant (alpha = 0.05) [25]. Applying the effects of the biases that gave rise to the statements, four types of bias (by factor extraction) were obtained. The names of the factors were then given with reference to their respective bias effects. Accordingly, the four types of bias were called preconception bias, self-affirmation bias, optimism bias and interest-oriented bias.

The overall consistency of the sixteen statements of the detection tool was tested with reliability analyses. The results are summarized in Tables 5 and 6. Cronbach's alpha tests the internal consistency of a set of items, and a value higher than 0.70 indicates a high level of intercorrelation among them [27–29]. The overall Cronbach's alpha for the tool was 0.84, and the Cronbach's alpha values for each construct were all above 0.7 (Table 5 refers). Inter-statement correlations pertain to the extent to which one item is related to all other items of the same factor [30]. Inter-item correlation coefficients between 0.2 and 0.7 are considered significant. Accordingly, the internal consistency of the statements was supported [17, 31]. As indicated in Table 5, except for the correlation between statements 1 and 4, the correlations between all the other statements had significant inter-statement coefficients. Item-total correlation was used to determine whether there were statements that were not in line with the overall average of all the detected statements. The baseline value typically used is 0.3 [24, 32]. The item-total correlation coefficients shown in Table 5 range from 0.393 to 0.648, demonstrating satisfactory homogeneity and the high reliability of the detection statements. The median inter-correlations of the statements in the diagonal of this table represent the inter-correlations of the statements under individual factors, and the off-diagonal inter-correlations relate to statements under different factors. The on-diagonal values of Table 6 are greater than the off-diagonal values, indicating that the inter-correlations of the statements under one factor are higher than the inter-correlations of the statements under different factors; this represents the discriminant validity of the items [33].

Table 4 Bias factor matrix of the two rounds of the survey

Detection statements		1st round				2nd round			
		Factor 1	Factor 2	Factor 3	Factor 4	Factor 1	Factor 2	Factor 3	Factor 4
		(1.85)	(1.47)	(1.17)	(4.84)	(1.79)	(3.78)	(2.012)	(1.574)
v1	My final assessment was influenced by the first offer of my counterpart	0.630				0.778			
v2	My assessment was influenced by unsubstantiated figures raised by my counterpart during the resolution process	0.804				0.788			
v3	My decision was adjusted because of ambitious arguments made by my counterpart	0.781				0.767			
v4	I do not agree with the assessments made during prior rounds of dispute resolution	0.576				0.580			
v5	I pay more attention to information that is consistent with my prior knowledge of the dispute		0.712				0.593		
v6	I am inclined to interpret further information as evidence to justify my assessment		0.557				0.632		
v7	I search for information that confirms my assessment		0.856				0.823		
v8	I endorsed information that supported my assessment		0.675				0.798		
v9	I am very optimistic about the likelihood of winning irrespective of the arguments of my counterpart			0.642				0.678	
v10	I completely believe that the outcome of the resolution will be good for my party			0.760				0.796	
v11	I am very confident that my ambitious requests will be granted			0.771				0.819	
v12	When the dispute was concluded, I felt that "I knew the outcome all along"			0.609				0.615	

(continued)

Table 4 (continued)

Detection statements		1st round				2nd round			
		Factor 1	Factor 2	Factor 3	Factor 4	Factor 1	Factor 2	Factor 3	Factor 4
		(1.85)	(1.47)	(1.17)	(4.84)	(1.79)	(3.78)	(2.012)	(1.574)
v13	I think that my counterpart is affected by bias				0.719				0.694
v14	I think that my counterpart should accept greater responsibility for the negative outcomes of the resolution				0.784				0.732
v15	After knowing the negative outcome of the resolution, I viewed the demands that my counterpart made during the dispute as unreasonable				0.584				0.714
v16	When the dispute was concluded, I considered a failure to settle inevitable because of the negative attitude of my counterpart				0.783				0.712

Note Factor 1: Preconception bias; Factor 2: Self-affirmation bias; Factor 3: Optimism bias; Factor 4: Interest-oriented bias; The initial eigenvalues for each factor are reported in parentheses.

Table 5 Reliability test (1st round)

Inter-item correlation									
Preconception bias					Self-affirmation bias				
Items	v1	v2	v3	v4	Items	v5	v6	v7	v8
v1	1.000	0.386[a]	0.437[a]	0.136	v5	1.000	0.364[a]	0.508[a]	0.383[a]
v2	0.386[a]	1.000	0.515[a]	0.360[a]	v6	0.364[a]	1.000	0.436[a]	0.221[b]
v3	0.437[a]	0.515[a]	1.000	0.417[a]	v7	0.508[a]	0.436[a]	1.000	0.485[a]
v4	0.136	0.360[a]	0.417[a]	1.000	v8	0.383[a]	0.221[b]	0.485[a]	1.000
Optimism bias					Interest-oriented bias				
Items	v9	v10	v11	v12	Items	v13	v14	v15	v16
v9	1.000	0.505[a]	0.539[a]	0.308[a]	v13	1.000	0.557[a]	0.384[a]	0.398[a]
v10	0.505[a]	1.000	0.547[a]	0.317[a]	v14	0.557[a]	1.000	0.449[a]	0.491[a]
v11	0.539[a]	0.547[a]	1.000	0.398[a]	v15	0.384[a]	0.449[a]	1.000	0.412[a]
v12	0.308[a]	0.317[a]	0.398[a]	1.000	v16	0.398[a]	0.491[a]	0.412[a]	1.000

[a]Correlation is significant at the 0.01 level (2-tailed)
[b]Correlation is significant at the 0.05 level (2-tailed)

Table 6 Internal consistency, corrected item-total correlations and median inter-correlations of items among the types of bias

Types of bias	Cronbach's alpha	Corrected item-total correlations	Median inter-correlations of items			
			Pre-conception bias	Self-affirmation bias	Optimism bias	Interest-oriented bias
Preconception Bias	0.709	0.393–0.627	0.402	–	–	–
Self-affirmation bias	0.726	0.428–0.642	0.152	0.410	–	–
Optimism bias	0.753	0.410–0.647	0.261	0.275	0.452	–
Interest-oriented bias	0.762	0.512–0.648	0.156	0.237	0.245	0.431

Extraction of Factors (2nd Round)

PCFA was conducted again on the 2nd set of data. The factor extraction method, standards and rotation method used were the same as those used for the 1st data set. The KMO value was 0.72, and Bartlett's test result was significant ($p < 0.001$). The factor matrix of the second-round survey is shown in Table 7. The same four factors that were extracted in the 1st round were extracted again. This result suggests that these four factors are consistent in both sets of data obtained with the 3-month interval. Accordingly, the detection tool was considered consistent over time.

The reliability analyses of the 2nd-round responses are shown in Tables 7 and 8. The inter-statement correlation values ranged from 0.2 to 0.7 and were significant. Cronbach's alpha for the tool overall was 0.77. The Cronbach's alpha values for each type of bias were greater than 0.7. The corrected statement-total correlations were within the range of 0.366–0.698 for the four factors. The on-diagonal median inter-correlations of the items in individual factors were larger than the off-diagonal ones, which represented the inter-correlations of the statements relating to different factors. Collectively, these statistical results support the reliability of the detection statements in terms of representing the different factors [33]. The two sets of data therefore give rise to the same four factors even though the data sets were obtained three months apart from one another.

Longitudinal Factorial Invariance

Since the same four factors were extracted from both data sets, the time consistency of the factor structure could be examined by longitudinal factorial invariance. A multigroup confirmatory factor analysis (MCFA) was used for this purpose [19, 20]. First, a cross-sectional confirmatory factor analysis (CFA) was performed for both sets of data separately. Since statement 5, namely, "I pay more attention to information that is consistent with my prior knowledge of the dispute", and statement 7, namely, "I search for information that confirms my assessment", both address the behavioural tendency of self-affirmation during the process of information collection, the error terms of these two statements were correlated. The model fit was assessed with the relative chi square (χ^2/df), comparative fit index (CFI), the Tucker-Lewis index (TLI) and the root mean square error of approximation (RMSEA). Accordingly, the fit of the cross-sectional CFA models is summarized in Table 9. The results indicated that the extracted bias factors fit the CFA model well. Furthermore, a longitudinal CFA model was then constructed by correlating the same factor over time and fitting the two rounds of data to the two separate cross-sectional models simultaneously. The residuals of each factor in the two models were correlated (Fig. 2 refers) [19, 34]. This longitudinal model with no equality constraints on its parameters was set as the baseline model and named Model A. The satisfactory goodness-of-fit indices (GFI) of Model A are shown in Table 9. If a new model with more equality constraints on its

Table 7 Reliability test (2nd round)

Inter-item correlation									
Preconception bias					Self-affirmation bias				
Items	v1	v2	v3	v4	Items	v5	v6	v7	v8
v1	1.000	0.507[a]	0.473[a]	0.309[a]	v5	1.000	0.202[b]	0.354[a]	0.355[a]
v2	0.507[a]	1.000	0.533[a]	0.285[a]	v6	0.202[b]	1.000	0.428[a]	0.325[a]
v3	0.473[a]	0.533[a]	1.000	0.306[a]	v7	0.354[a]	0.428[a]	1.000	0.595[a]
v4	0.309[a]	0.285[a]	0.306[a]	1.000	v8	0.355[a]	0.325[a]	0.595[a]	1.000
Optimism bias					Interest-oriented bias				
Items	v9	v10	v11	v12	Items	v13	v14	v15	v16
v9	1.000	0.360[a]	0.527[a]	0.288[a]	v13	1.000	0.438[a]	0.306[a]	0.372[a]
v10	0.360[a]	1.000	0.559[a]	0.372[a]	v14	0.438[a]	1.000	0.395[a]	0.364[a]
v11	0.527[a]	0.559[a]	1.000	0.481[a]	v15	0.306[a]	0.395[a]	1.000	0.448[a]
v12	0.288[a]	0.372[a]	0.481[a]	1.000	v16	0.372[a]	0.364[a]	0.448[a]	1.000

[a]Correlation is significant at the 0.01 level (2-tailed)
[b]Correlation is significant at the 0.05 level (2-tailed)

Table 8 Internal consistency, corrected item-total correlations and median inter-correlations of items among the types of bias

Types of bias	Cronbach's alpha	Corrected item-total correlations	Median inter-correlations of items			
			Pre-conception bias	Self-affirmation bias	Optimism bias	Interest-oriented bias
Pre-conception bias	0.731	0.366–0.584	0.391	–	–	–
Self-affirmation bias	0.703	0.378–0.635	0.065	0.355	–	–
Optimism bias	0.754	0.472–0.698	0.141	0.097	0.427	–
Interest-oriented bias	0.716	0.479–0.523	0.118	0.115	0.186	0.384

Table 9 Goodness-of-fit indexes of cross-sectional CFA models and longitudinal CFA models

		Cross-sectional CFA		Longitudinal CFA			Comparison between longitudinal CFA (unrestrained Model A as a baseline model)	
Fit indexes	Desired levels	1st round	2nd round	Model A	Model B	Model C	Model B-Model A	Model C-Model A
χ^2/df	2 or below[a]	1.103	1.054	1.164	1.175	1.172	0.011	0.008
CFI	0.9 or above[a]	0.976	0.985	0.930	0.923	0.924	−0.007	−0.006
zTLI	0.80 or above[b]	0.971	0.982	0.920	0.914	0.916	−0.004	−0.004
RMSEA	0.06 or below[c]	0.032	0.023	0.040	0.041	0.041	0.001	0.001

Note χ^2/df = chi square/degree of freedom; *CFI* = comparative fit index; *TLI* = Tucker-Lewis index
RMSEA = root mean square error of approximation
[a]Hair et al. [25]; [b]Maskarinec et al. [38]; [c]Hu and Bentler [39]

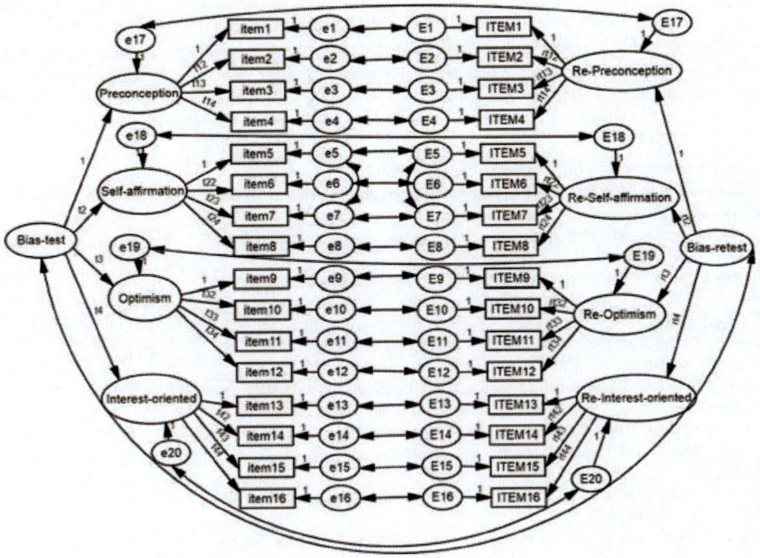

Fig. 2 Longitudinal bias measurement CFA model

parameters achieves a nonsignificant change in terms of its GFI, this new model would be supported, and the equality constraints would hold. In this regard, by adding more constraints to the model parameters, high levels of factorial invariance can be tested [34, 35]. In this study, the invariance of the bias factors was tested by examining the equivalence of all the factor loadings over time [19, 20, 36]. Accordingly, Model B was constructed by equalizing all the first-order factor loadings over time (see Fig. 2, t12 = rt12, t13 = rt13, t14 = rt14, t22 = rt22, t23 = rt23…). Model C was constructed by further requiring the second-order factor loadings to be invariant over time (see Fig. 2, t2 = rt2, t3 = rt3, t4 = rt4). The model fit indices and a comparison of the above longitudinal CFA models are shown in Table 9. The GFIs of the longitudinal CFA models were satisfactory and suggested an excellent model fit for both rounds of responses. In addition, when equality constraints were added to the factor loadings in Models B and C, the GFI did not change significantly. The changes in the CFI values of Models B and C were both less than 0.01. As such, the standard of invariance was met [35, 37]. Hence, it can be inferred from the above results that the structure of the four bias factors is factorial invariant over time. The MCFA was conducted using IBM SPSS Amos version 21.0. A time-consistent bias detection tool with sixteen detection statements was developed, and its time consistency, validity, reliability and factorial invariance were verified. The operationalized bias detection tool is shown in Fig. 3.

A Bias Detection Tool for Construction Dispute Negotiation

Prior studies have suggested that the following behaviours are exhibited by people involved in dispute resolution; please indicate the frequency with which you engage in these dispute negotiation behaviours.

Detection statements	Never (1)	Infrequently (2)	Sometimes (3)	Often (4)	Usually (5)	Always (6)
Preconception (A):						
A1. My final assessment was influenced by the first offer of my counterpart.	1	2	3	4	5	6
A2. My assessment was influenced by unsubstantiated figures raised by my counterpart during the resolution process.	1	2	3	4	5	6
A3. My decision was adjusted because of the ambitious arguments of my counterpart.	1	2	3	4	5	6
A4. I do not agree with the assessments made during prior rounds of dispute resolution.	1	2	3	4	5	6
Self-affirmation (B):						
B1. I pay more attention to information that is consistent with my prior knowledge of the dispute.	1	2	3	4	5	6
B2. I am inclined to interpret further information as evidence to justify my assessment.	1	2	3	4	5	6
B3. I search for information that confirms my assessment.	1	2	3	4	5	6
B4. I endorse information that supports my assessment.	1	2	3	4	5	6
Optimism (C):						
C1. I am very optimistic about the likelihood of winning irrespective of the arguments of my counterpart.	1	2	3	4	5	6
C2. I completely believe that the outcome of the resolution will be good for my party.	1	2	3	4	5	6
C3. I am very confident that my ambitious requests will be granted.	1	2	3	4	5	6
C4. When the dispute was concluded, I felt that "I knew the outcome all along".	1	2	3	4	5	6
Interested-oriented (D):						
D1. I think that my counterpart is affected by bias.	1	2	3	4	5	6
D2. I think the counterpart should accept greater responsibility for the negative outcomes of the resolution.	1	2	3	4	5	6
D3. After knowing the negative outcome of the resolution, I viewed the demands of my counterpart during the dispute as unreasonable.	1	2	3	4	5	6
D4. When the dispute was concluded, I considered a failure to settle inevitable because of the negative attitude my counterpart.	1	2	3	4	5	6

Calculation procedures:

Preconception (A):

Total Score = $\sum_{i=1}^{4} SA_i = \_\_\_\_$, Average Score = $\frac{1}{4}\sum_{i=1}^{4} SA_i = \_\_\_\_$ (SA_i is the score of the i_{th} measurement statement in A)

Self-affirmation (B):

Total Score = $\sum_{i=1}^{4} SB_i = \_\_\_\_$, Average Score = $\frac{1}{4}\sum_{i=1}^{4} SB_i = \_\_\_\_$ (SB_i is the score of the i_{th} measurement statement in B)

Optimism (C):

Total Score = $\sum_{i=1}^{4} SC_i = \_\_\_\_$, Average Score = $\frac{1}{4}\sum_{i=1}^{4} SC_i = \_\_\_\_$ (SC_i is the score of the i_{th} measurement statement in C)

Interested-oriented (D):

Total Score = $\sum_{i=1}^{4} SD_i = \_\_\_\_$, Average Score = $\frac{1}{4}\sum_{i=1}^{4} SD_i = \_\_\_\_$ (SD_i is the score of the i_{th} measurement statement in D)

Fig. 3 The validated time-consistent bias detection tool

Usefulness and Implications of the Tool

The bias detection tool developed in this study can be used as a checklist of biased behaviours. The extent of the impact of the four types of bias can be detected by asking all the participants in a construction dispute to use the tool like an attitude test. A higher score for a particular biased behaviour suggests a greater influence of that behaviour. By calculating the total scores according to Fig. 3, the extent of each bias can be identified. The use of this tool could at least raise an awareness of potential biases. This bias detection tool categorizes the biases that may affect construction dispute negotiations into groups corresponding to preconception, self-affirmation, optimism and interest orientation biases. This categorization enhances the current understanding of bias from the perspective of the incubators of such behaviour. Guarding these incubators would be a de facto bias minimizing measure.

For example, preconceptions work like anchors and thereby limit responses. In construction dispute negotiation, disputing parties often exaggerate their losses and compensation demands. Their inflated opening positions induce an anchoring effect in that the responses of counterparts will not stray far from that position. Furthermore, acting with great confidence also can convince a counterpart and even a third neutral party that a demand is reasonable. When there are several rounds of negotiation, a priori assessments tend to become stronger instead of softening. Previous impressions impact subsequent decisions. To minimize the effect of preconception bias on a priori assessments, disputing parties should be reminded to review the facts and evidence of their case before making further decisions. Managers may consider ways to allow fresh input by introducing new members to a negotiating group. Self-affirmation bias is at work when disputing parties selectively collect information that is consistent with their prior positions. Self-affirming disputing parties interpret information in a way that supports their assertions. Thus, disputing parties with self-affirmation bias are very reluctant to compromise. It is quite common for disputing parties to not move an inch irrespective of what evidence is available. Under the influence of self-affirmation bias, disputing parties only take note of information that supports their assessments, ignoring the evidence and arguments presented by their counterparts. Brain storming sessions involving external consultants are suggested to help break self-affirmation shackles. An optimistic disputing party is likely to be ambitious too, as he or she thinks that all possibilities are within his or her grasp. In fact, overconfident negotiators are likely be ambitious. Likewise, optimism increases the confidence of a disputing party in terms of his or her assessment. The chance of making deals is thus quite remote. To guard against optimism bias, disputing parties should be reminded about the possibilities of settlement failure and losses. Neutral third parties can be instrumental by providing appropriate intervention through the technique of reality testing. When a disputing party is under the influence of interest-oriented bias, they believe that their counterparts are biased. A failed negotiation is the responsibility of both sides. Thus, they have no reason to justify concession. They may even believe that they are the victims. All of these are unfavourable conditions. Again, third-party neutrality may restore rationality to the biased party. By stepping into the shoes of their counterparts, these individuals could obtain a different perspective of the dispute. When disputing parties start to recognize the merits of their counterparts' positions, negotiations can make reasonable progress. In addition, the desires to engage in future collaboration and long-term relationships are useful counterforces to suppress confrontation. To summarize, the bias detection tool is thus helpful, as it facilitates self-evaluations for disputing parties. Mangers can use the tool to understand the status of their negotiating teams in terms of bias propensity. Neutral third parties can use the tool to gauge disputing parties' biased behaviours.

Summary

This chapter details the development of a time-consistent bias detection tool. A longitudinal approach was employed, and two rounds of data collection were conducted through a test–retest approach. A bias detection tool with sixteen measurement statements was obtained. The reliability, time consistency, validity, reliability and factorial invariance of the tool were all statistically significant. The tool is pioneering work intended to detect bias in construction dispute negotiations. The development of this tool enables future research to examine other biases as well.

Acknowledgements The empirical work of this chapter has been reported in a paper entitled "A Bias Measurement Scale for Repeated Dispute Evaluations" of the Journal of Management in Engineering. Where appropriate, editing has been affected to improve the readability and ease for further uses. The support of the HKSAR RGC General Research Fund Project (no. 11209118) is duly acknowledged.

References

1. Bazerman M, Moore DA (2013) Negotiator cognition. In: Judgment in managerial decision making. 8th ed. Wiley, pp 168–178
2. Groseclose T, Milyo J (2005) A measure of media bias. Q J Econ 120(4):1191–1237
3. Mogg K, Mathews A, Eysenck M (1992) Attentional bias to threat in clinical anxiety states. Cogn Emot 6(2):149–159
4. Armstrong JS, Overton TS (1977) Estimating nonresponse bias in mail surveys. J Mark Res 396–402
5. Weinstein ND (1980) Unrealistic optimism about future life events. J Pers Soc Psychol 39(5):806
6. Weinstein ND, Klein WM (1996) Unrealistic optimism: present and future. J Soc Clin Psychol 15(1):1–8
7. Shepard L, Camilli G, Averill M (1981) Comparison of procedures for detecting test-item bias with both internal and external ability criteria. J Educ Stat 6(4):317–375
8. Egger M, Smith GD, Schneider M, Minder C (1997) Bias in meta-analysis detected by a simple, graphical test. BMJ 315(7109):629–634
9. Raphael K (1987) Recall bias: a proposal for assessment and control. Int J Epidemiol 16(2):167–170

10. Paulhus DL (1991) Measurement and control of response bias. In: Robinson JP, Shaver PR, Wrightsman LS (eds) Measures of personality and social psychological attitudes, pp17–59
11. Mowshowitz A, Kawaguchi A (2002) Assessing bias in search engines. Inf Process Manag 38(1):141–156
12. Kassin SM, Wrightsman LS (1983) The construction and validation of a juror bias scale. J Res Pers 17(4):423–442
13. Hölzel BK, Carmody J, Vangel M, Congleton C, Yerramsetti SM, Gard T, Lazar SW (2011) Mindfulness practice leads to increases in regional brain gray matter density. Psychiatry Res Neuroimaging 191(1):36–43
14. Yoshida K, Sanada M, Shiraishi Y, Nowak D, Nagata Y, Yamamoto R, Sato Y, Sato-Otsubo A, Kon A, Nagasaki M, Chalkidis G (2011) Frequent pathway mutations of splicing machinery in myelodysplasia. Nature 478(7367):64–69
15. Larsson J, Eriksson PE, Olofsson T, Simonsson P (2015) Leadership in civil engineering: effects of project managers' leadership styles on project performance. J Manag Eng. https://doi.org/10.1061/(ASCE)ME.1943-5479.0000367
16. Oladinrin OT, Ho CM (2016) Critical enablers for codes of ethics implementation in construction organizations. J Manag Eng. https://doi.org/10.1061/(ASCE)ME.1943-5479.0000385
17. Cheung SO, Wong WK, Yiu TW, Pang HY (2011) Developing a trust inventory for construction contracting. Int J Proj Manag 2(29):184–196
18. Yu AT, Shen GQ (2015) Critical success factors of the briefing process for construction projects. J Manag Eng. https://doi.org/10.1061/(ASCE)ME.1943-5479.0000242
19. Kim H, Ji J (2009) Factor structure and longitudinal invariance of the Maslach Burnout Inventory. Res Soc Work Pract 19(3):325–339
20. Widaman KF, Ferrer E, Conger RD (2010) Factorial invariance within longitudinal structural equation models: Measuring the same construct across time. Child Dev Perspect 4(1):10–18
21. Kline P (2013) Handbook of psychological testing, 2nd edn. Routledge
22. Cerny BA, Kaiser HF (1977) A study of a measure of sampling adequacy for factor-analytic correlation matrices. Multivar Behav Res 12(1):43–47
23. Kaiser HF (1974) An index of factorial simplicity. Psychometrika 39(1):31–36
24. Norusis MJ (1993) SPSS statistical data analysis. SPSS Inc., Chicago, IL
25. Hair JF, Anderson RE, Tatham RL, Black WC (1998) Multivariate data analysis, 5th edn. Prentice Hall International, NY
26. Kaiser HF (1958) The varimax criterion for analytic rotation in factor analysis. Psychometrika 23(3):187–200
27. Cronbach LJ (1951) Coefficient alpha and the internal structure of tests. Psychometrika 16(3):297–334
28. Nunnally JC (1978) Psychometric theory. McGraw-Hill, New York
29. Nunnally JC, Bernstein IH (1994) Psychometric theory, 3rd edn. McGraw-Hill, New York
30. Cohen RJ, Swerdlik ME (2005) Psychological testing and assessment: an introduction to tests and measurement, 6th edn. McGraw-Hill, New York
31. Idvall E, Hamrin E, Unosson M (2002) Development of an instrument to measure strategic and clinical quality indicators in postoperative pain management. J Adv Nurs 37(6):532–540
32. Churchill GA (1979) A paradigm for developing better measures of marketing constructs. J Mark Res 16(1):64–73
33. Rahim MA (1983) Measurement of organizational conflict. J Gen Psychol 109(2):189–199
34. Kim SS, Malhotra NK (2005) A longitudinal model of continued IS use: an integrative view of four mechanisms underlying postadoption phenomena. Manag Sci 51(5):741–755
35. Petersen IT, Hoyniak CP, McQuillan ME, Bates JE, Staples AD (2016) Measuring the development of inhibitory control: the challenge of heterotypic continuity. Dev Rev 40:25–71
36. Horn JL, McArdle JJ, Mason R (1983) When is invariance not invariance: a practical scientist's look at the ethereal concept of factor invariance. South Psychol 1:179–188
37. Cheung GW, Rensvold RB (2002) Evaluating goodness-of-fit indexes for testing measurement invariance. Struct Equ Model 9(2):233–255

38. Maskarinec G, Novotny R, Tasaki K (2000) Dietary patterns are associated with body mass index in multiethnic women. J Nutr 130(12):3068–3072
39. Hu LT, Bentler PM (1999) Cutoff criteria for fit indexes in covariance structure analysis: conventional criteria versus new alternatives. Struct Equ Model 6(1):1–55

Special Forms of Bias: Endowment Effect and Reactive Devaluation

Sai On Cheung and Keyao Li

Introduction

In the preceding three chapters, four forms of bias that would affect dispute decisions have been identified. In this chapter two special forms of bias, endowment effect and reactive devaluation are discussed. It is very common for one to value one's belongings more than what one is willing to pay for the same. This phenomenon is called endowment effect (EE) [88, 103]. Another concept akin to endowment effect is reactive devaluation (RD) that describes the habitual under-valuing of proposal raised by a bargaining counterpart. Both forms of bias are considered as psychological barriers and would impede dispute settlement [74, 76, 96, 119, 121].

Endowment Effect

EE was coined by Professor Richard Thaler, the 2017 Nobel laureate in economics science. EE describes the phenomenon of one usually require more to relinquish their own items than they are willing to pay for the identical items [138]. This is caused by the psychological attachment on one's own item [10]. In more technical terms, people demand more to forgo their belongings as they overvalue the loss in losing them [63, 128]. EE is considered as a kind of cognitive biases that are intuitive in nature [9, 11, 23, 67]. The manifestation of EE in economic exchanges is expressed

S. O. Cheung (✉)
Construction Dispute Resolution Research Unit, City University of Hong Kong, Hong Kong, China
e-mail: Saion.cheung@cityu.edu.hk

K. Li
Future of Work Institute, Faculty of Business & Law, Curtin University, Perth, Australia
e-mail: Keyao.li@curtin.edu.au

S. O. Cheung (ed.), *Construction Dispute Research Expanded*, Springer Tracts in Civil Engineering, https://doi.org/10.1007/978-3-030-80256-1_4

that the amount of money one is willing to pay (WTP) to acquire the good is notably lower than the amount of money one is willing to accept (WTA) to forfeit it [16, 56]. People attribute higher value to their ownings than other people would do [4, 132]. Notwithstanding that happening of EE is often related to physical objects, EE also applies to entitlements, beliefs and ideas [102]. It is proposed that EE also happens in decision making and as a result one would attribute more value and defend their own positions [33]. Liedtka [85] also suggested that EE would lead to attachment to first solution and overlooking alternative options. Extending these propositions to construction dispute negotiations, EE can be a form of psychological barrier against negotiated settlement [88, 102, 138, 142].

Sources of EE

Four sources have been identified from the relevant literature. These are ownership, loss aversion, status quo bias and strategic bargaining habit.

Ownership

Beggan [10] and Morewedge et al. [103] found that ownership would increase the perceived value of a self-owned object that can be physical, values of opinions and positions [33]. It can be explained that one would naturally develop association and attachment with objects that they own. Belongings are considered as part of oneself [4, 132]. Inevitably people would unconsciously ask for more compensation if they were asked to forsake "part of themselves" [144]. Moreover, legal ownership is different from psychological one [132]. Legal ownership starts with the receipt of the object and will be strengthened with time [117, 136]. Psychological ownership on the other hand could develop even without legal possession of the object. One could perceive ownership through imagination or affection [5, 42]. Nevertheless, both legal ownership and psychological ownership fuel unwillingness to forgo. The potential loss of an endowed object is a threat to one's self-image [21, 32]. Besides, ownership would activate and elevate one's recognition of the positive features of the object resulting in attesting higher value [16].

Loss Aversion

Loss aversion also drives endowment effect [62, 132, 138]. Prospect theory explains loss aversion as the human tendency of overvaluing losses and undervaluing gains [63, 125]. In short, there is an asymmetry that losses resonate more than gains of equal magnitude [4, 78]. From psychology point of view, one tends to consider the discomfort provoked by losses more significant than the pleasure brought by gains of equal size [128, 142]. The reference-dependent theory also explains that receiving an item will move an individual from "not-owning" to "owning", whereas giving up an item works the other way [111]. The WTA for giving up an item is perceived to be higher than the WTP to acquire the same [62]. In making decisions, the greater extent people are getting involved, the higher the loss aversion [128]. Furthermore, for a decision that involved much energy and commitment, it is even more difficult to relinquish it [108, 128]. Therefore, loss aversion hampers desire to dispute settlement as negotiating parties are likely averse to concessions because compromises are perceived as losses [25, 62].

Status Quo Bias

Samuelson and Zeckhauser [126] described status quo bias as one's stay put tendency. Humans are change-averse. Status quo bias would lead to EE because the preference to remain status quo would induce them to attribute a higher value to the entitlement they are endowed with, causing a disposition to maintain the ownership [75, 78]. Moreover, status quo bias is triggered by loss aversion [63], because of the inclination to avoid losses. No action is taken as not taking any risk in losing [78]. The disadvantages of abandoning status quo are therefore far more tangible than the prospective gains [59, 62].

Strategic Bargaining Habit

Different strategic bargaining habits of negotiators could also result in variation in the estimated prices [27, 102]. Influenced by strategic bargaining habits, people would unconsciously underestimate their WTP when they are buyers and overestimate their WTA when they are owners [73]. Out of strategic reason, owners usually attribute a higher value to an object with the expectation that the buyers would offer a similar price. Likewise, when buyers value an object with a lower price, they would anticipate the owner would value similarly. Not only will one mis-judge the valuations of others, Van Boven et al. [14] further proposed that people would have unrealistic predictions of what they would offer if they were in the opposite role.

Manifestations of EE in CDN

The possible manifestations of EE in CDN derived from the relevant literature are listed in Table 1.

Questionnaire Survey

Possible manifestations of EE in CDN as listed in Table 1 were included in a questionnaire survey and sent to CDN practitioners. Respondents were asked to rate their agreement of having the listed EE manifestations in a Likert scale of "1 = Strongly Disagree" to "6 = Strongly Agree". Both paper-based and online questionnaire were used. Online questionnaires together with a covering letter introducing the background the study were sent via the survey tool of eSurv. The targeted professionals include architects, building services engineers, building surveyors, project managers, quantity surveyors and structural/civil Engineers. Their contacts were collected from research networks; websites of learned societies, including Hong Kong Institution of Engineers (HKIE), The Hong Kong Institute of Architects (HKIA), Hong Kong Institute of Surveyors (HKIS) and Hong Kong Institute of Construction Managers (HKICM) and HKSAR government department websites, including Buildings Department and Housing Authority and Housing Department. Paper-based questionnaires were distributed at learned society workshops and seminars. A total of 207 questionnaires were distributed, and 112 effective responses were received and used in the data analysis. The response rate is 54%. The demographic information of the respondents, including professional background, nature of their organization, years of experience and the type of dispute are represented in Figs. 1, 2, 3 and 4.

Table 1 Measurement statements of EE in CDN

Sources of EE	Manifestations	References
Ownership	I develop ownership of arguments and positions I use in negotiations	[33]
	I consider my arguments and positions as part of my self-concept	[4, 132, 144]
	I feel good about myself when endorsing my own decision	[16, 132, 144]
	I feel opposition and counter-argumentation from the counterpart as a threat to my self-esteem	[2, 4, 17, 32]
Loss aversion	I consider accepting the offers from the counterpart means losing interests	[62, 102, 111]
	When making trade-offs, I focus on the benefits forgone	[4, 63, 128]
	I don't want to lose my position because I believe it is the best solution	[63, 102, 111]
	Before starting negotiation, I assume my decisions would be accepted by the counterpart	[62, 102, 111]
	In negotiation, I consider making compromise means accepting a less practical solution	[102, 111]

(continued)

Table 1 (continued)

Sources of EE	Manifestations	References
Status Quo Bias	I want to maintain my decisions therefore I am uninterested in negotiating	[59, 62, 78, 75]
	When rethink about my decision, I unconsciously recall reasons why my decisions hold	[78, 75, 126]
	When disagreements occur, I first recall reasons to support my position	[78, 75, 126]
	I believe the counterpart provides premature decisions	[75, 126]
	I believe the counterpart will agree with us if they put more effort in decision making	[59, 62, 78, 75]
	After obtaining much supporting information for my decision, my information searching process terminates	[59, 63]
	I search less information about the counterpart's proposal	[59, 62, 75]
	After forming a reasonable decision, I am immune to other alternatives	[126]
Strategic bargaining habit	When negotiating, I think overvaluing my decision is a strategy to gain more benefits	[14, 73, 120, 122]
	I over-claim the compensation when I have to put up with something negative, such as project overrun due to the fault of the counterpart	[14, 73, 120, 122]
	After forming a decision, I tend to pay more attention to the information that supports my decision	[113]
	I interpret supplementary information as evidence to support my decision	[53]
	I endorse information that supports my decisions	[113]

Findings

Ratings on each of the statements were calculated and summarized according to the categorization of professional background, years of experience and gender of the respondents. Under each category, EE manifestations were ranked according to the level of agreement rated by the respondents. Rating scores and ranks of these EE manifestations are presented in Table 2. It can be seen that most of the statements have average scores more than 3.5, which is higher than the mid-point of the scale. Therefore, the respondents as a whole agreed the happening of EE during construction dispute negotiation. With reference to Table 2, the rankings of the EE manifestations differ across groups of professional background, years of experience

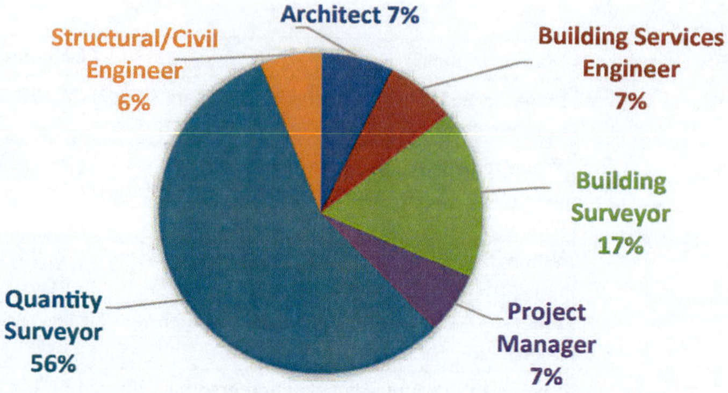

Fig. 1 Professional background of the respondents

Fig. 2 Organisation of the respondents

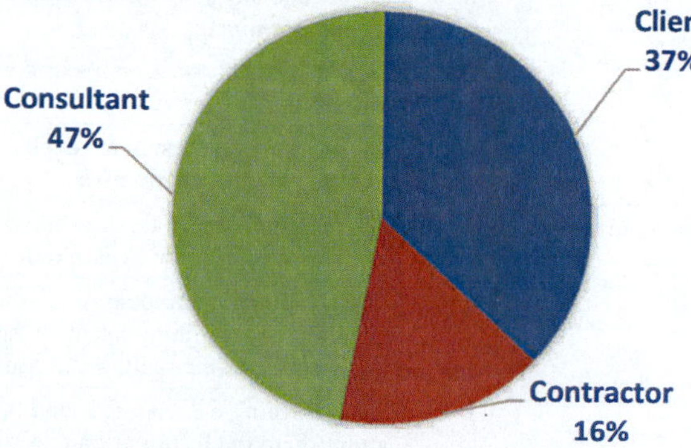

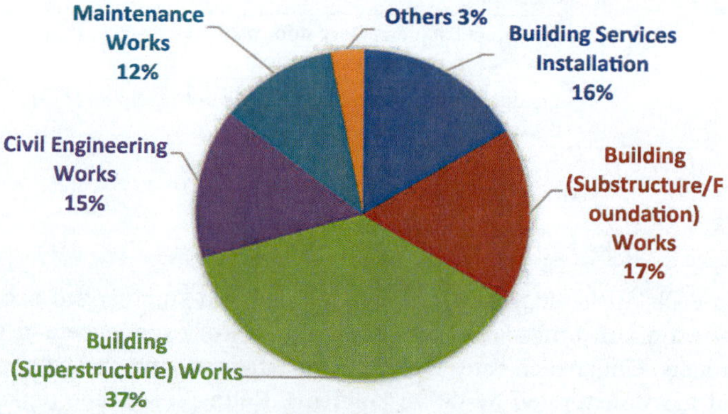

Fig. 3 Dispute type the respondents involved

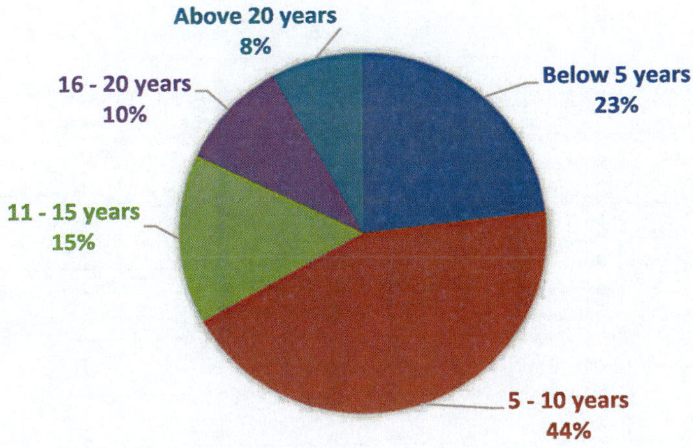

Fig. 4 Years of experience of the respondents

and genders. In order to further explore the influence of age, professional background and gender on the extent of people having EE behaviours, one-way analysis of variance (ANOVA) was conducted. One-way analysis of variance (ANOVA) is used to examine the differences between mean values of different sample groups [8, 57]. The null hypothesis is that there is no significant difference among the mean values of the samples in different groups. A significant *p* value would reject the null hypothesis and indicate significant differences among the groups. In this study, ANOVA was performed to explore the potency of EE with organizational groups (contractor, consultant and client), years of experience group (below 5 years, 5 to 10 years, above 10 years) and gender group (male and female). The ANOVA results are presented in Tables 3, 4, 5 and 6. From Tables 3, 4, 5 and 6, it shows that under each source of EE, the between groups *p* values of organizational groups, years of experience groups and gender groups were all non-significant of values larger than 0.05 (alpha = 0.05). Therefore, no significant differences were found. Accordingly, the potency of EE on dispute negotiators are not affected by organizational perspectives, working experience and genders.

Discussions

The happenings of EE run against rational economic assumptions [138]. The study reported in this chapter examines EE in CDN. Four sources of EE were identified from literature: ownership, loss aversion, status quo bias and strategic bargaining habit. Manifestations of EE in CDN were further operationalized. The opinions from construction practitioners confirmed the happenings of these EE manifestations in practice. The influences of EE in CDN are as follow.

Table 2 Sources of EE

Source of EE	Years of experience (N = 112)						Organization (N = 112)						Gender (N = 112)			
	Below 5 years (N = 26, 23%)		5–10 years (N = 49, 44%)		Above 10 years (N = 37, 33%)		Contractor (N = 18, 16%)		Client (N = 43, 38%)		Consultant (N = 51, 46%)		Male (N = 64, 57%)		Female (N = 48, 43%)	
	Score	Rank	Score	Rank	Score	Rank	Score	Rank	Score	Rank	Score	Rank	Score	Rank	Score	Rank
Ownership	***4.08***	***1***	***4.17***	***2***	***3.99***	***3***	***3.99***	***3***	***4.20***	***1***	***4.04***	***3***	***4.06***	***2***	***4.14***	***2***
I develop ownership of arguments and positions I use in negotiations	3.81	16	4.29	22	4.27	3	3.94	20	4.51	1	3.96	13	4.11	10	4.25	7
I consider my arguments and positions as part of my self-concept	4.31	3	4.24	13	4.16	9	4.28	14	4.21	8	4.24	6	4.28	6	4.17	10
I feel good about myself when endorsing my own decision	4.12	8	4.18	5	3.97	12	4.17	11	4.14	12	4.04	9	4.13	9	4.06	12
I feel opposition and counter-argumentation from the counterpart as a threat to my self-esteem	4.08	9	3.98	6	3.54	21	3.56	2	3.93	15	3.90	17	3.70	19	4.06	13
Loss aversion	***3.95***	***2***	***4.17***	***2***	***4.11***	***1***	***4.09***	***2***	***4.15***	***2***	***4.06***	***2***	***4.06***	***2***	***4.15***	***1***

(continued)

Table 2 (continued)

Source of EE	Years of experience (N = 112)						Organization (N = 112)						Gender (N = 112)			
	Below 5 years (N = 26, 23%)		5–10 years (N = 49, 44%)		Above 10 years (N = 37, 33%)		Contractor (N = 18, 16%)		Client (N = 43, 38%)		Consultant (N = 51, 46%)		Male (N = 64, 57%)		Female (N = 48, 43%)	
	Score	Rank	Score	Rank	Score	Rank	Score	Rank	Score	Rank	Score	Rank	Score	Rank	Score	Rank
I consider accepting the offers from the counterpart means losing interests	3.73	17	3.78	21	3.76	16	3.83	7	3.79	18	3.71	19	3.63	20	3.94	15
When making trade-offs, I focus on the benefits forgone	4.19	7	4.22	12	4.19	8	4.33	15	4.21	9	4.16	7	4.16	8	4.27	6
I believe my proposal is the best solution to the project dispute	3.88	13	4.20	19	4.24	6	4.06	17	4.19	11	4.14	8	4.20	7	4.06	14
Before starting negotiation, I believe my decisions would be accepted by the counterpart	4.35	2	4.35	8	4.11	10	4.22	13	4.30	5	4.25	5	4.31	5	4.21	8

(continued)

Table 2 (continued)

Source of EE	Years of experience (N = 112)						Organization (N = 112)						Gender (N = 112)			
	Below 5 years (N = 26, 23%)		5–10 years (N = 49, 44%)		Above 10 years (N = 37, 33%)		Contractor (N = 18, 16%)		Client (N = 43, 38%)		Consultant (N = 51, 46%)		Male (N = 64, 57%)		Female (N = 48, 43%)	
	Score	Rank	Score	Rank	Score	Rank	Score	Rank	Score	Rank	Score	Rank	Score	Rank	Score	Rank
In negotiation, I am confident that my decisions are more practical than the counterpart's	3.58	20	4.31	4	4.27	4	4.00	19	4.28	7	4.04	10	4.00	12	4.29	4
Status Quo Bias	***3.81***	***4***	3.91	4	***3.81***	***4***	***3.86***	***4***	***3.85***	***4***	***3.87***	***4***	***3.85***	***4***	***3.87***	***4***
I want to maintain my decisions therefore I am uninterested in negotiating	3.19	22	3.71	2	3.68	19	3.67	4	3.53	20	3.59	21	3.38	21	3.85	18
When rethink about my decision, I unconsciously recall reasons why my decisions hold	3.85	15	4.04	9	3.89	14	3.83	9	3.91	16	4.02	12	3.98	13	3.90	16

(continued)

Table 2 (continued)

Source of EE	Years of experience (N = 112)						Organization (N = 112)						Gender (N = 112)			
	Below 5 years (N = 26, 23%)		5–10 years (N = 49, 44%)		Above 10 years (N = 37, 33%)		Contractor (N = 18, 16%)		Client (N = 43, 38%)		Consultant (N = 51, 46%)		Male (N = 64, 57%)		Female (N = 48, 43%)	
	Score	Rank	Score	Rank	Score	Rank	Score	Rank	Score	Rank	Score	Rank	Score	Rank	Score	Rank
When disagreements occur, I first recall reasons to support my position	4.58	1	4.51	7	4.32	2	4.44	21	4.49	2	4.45	2	4.56	1	4.33	2
I believe the counterpart provides premature decisions	3.88	14	3.73	10	3.92	13	3.72	10	4.07	13	3.67	20	3.88	16	3.77	19
I believe the counterpart will agree with us if they put more effort in decision making	3.96	11	4.24	15	4.03	11	4.06	12	4.21	10	4.04	11	3.97	14	4.29	5
After obtaining much supporting information for my decision, my information searching process terminates	4.00	10	4.24	20	3.62	20	3.94	3	4.05	14	3.94	14	3.89	15	4.10	11

(continued)

Table 2 (continued)

Source of EE	Years of experience (N = 112)						Organization (N = 112)						Gender (N = 112)			
	Below 5 years (N = 26, 23%)		5–10 years (N = 49, 44%)		Above 10 years (N = 37, 33%)		Contractor (N = 18, 16%)		Client (N = 43, 38%)		Consultant (N = 51, 46%)		Male (N = 64, 57%)		Female (N = 48, 43%)	
	Score	Rank	Score	Rank	Score	Rank	Score	Rank	Score	Rank	Score	Rank	Score	Rank	Score	Rank
I search less information about the counterpart's proposal	3.38	21	3.29	1	3.27	22	3.39	1	3.30	21	3.27	22	3.34	22	3.25	22
After forming a reasonable decision, I am immune to other alternatives	3.65	19	3.55	17	3.78	15	3.83	8	3.23	22	3.94	15	3.81	18	3.44	21
Strategic bargaining habit	***4.08***	***1***	***4.29***	***1***	***4.08***	***2***	***4.17***	***1***	***4.14***	***3***	***4.20***	***1***	***4.22***	***1***	***4.11***	***3***
When negotiating, I think overvaluing my decision is a strategy to gain more benefits	3.69	18	4.10	14	3.76	17	4.22	6	3.81	17	3.84	18	4.06	11	3.67	20

(continued)

Table 2 (continued)

Source of EE	Years of experience (N = 112)						Organization (N = 112)						Gender (N = 112)			
	Below 5 years (N = 26, 23%)		5–10 years (N = 49, 44%)		Above 10 years (N = 37, 33%)		Contractor (N = 18, 16%)		Client (N = 43, 38%)		Consultant (N = 51, 46%)		Male (N = 64, 57%)		Female (N = 48, 43%)	
	Score	Rank	Score	Rank	Score	Rank	Score	Rank	Score	Rank	Score	Rank	Score	Rank	Score	Rank
I over-claim the compensation when I have to put up with something negative, such as project overrun due to the fault of the counterpart	3.96	12	3.94	3	3.73	18	3.94	5	3.79	19	3.92	16	3.88	17	3.88	17
After forming a decision, I tend to pay more attention to the information that supports my decision	4.23	6	4.37	11	4.27	5	4.17	18	4.30	6	4.35	4	4.38	3	4.21	9
I interpret supplementary information as evidence to support my decision	4.27	4	4.53	16	4.38	1	4.44	22	4.42	3	4.41	3	4.38	4	4.48	1

(continued)

Table 2 (continued)

Source of EE	Years of experience (N = 112)						Organization (N = 112)						Gender (N = 112)			
	Below 5 years (N = 26, 23%)		5–10 years (N = 49, 44%)		Above 10 years (N = 37, 33%)		Contractor (N = 18, 16%)		Client (N = 43, 38%)		Consultant (N = 51, 46%)		Male (N = 64, 57%)		Female (N = 48, 43%)	
	Score	Rank	Score	Rank	Score	Rank	Score	Rank	Score	Rank	Score	Rank	Score	Rank	Score	Rank
I endorse information that supports my decisions	4.27	5	4.51	18	4.24	7	4.06	16	4.37	4	4.47	1	4.41	2	4.31	3

Table 3 Ownership ANOVA results

	Source	Sum of squares	df	Mean square	F	Sig.
Organizational groups	Between groups	0.851	2	0.425	1.438	0.242
	Within groups	32.256	109	0.296		
	Total	33.107	111			
Years of experience groups	Source	Sum of squares	df	Mean square	F	Sig.
	Between groups	0.742	2	0.371	1.250	0.291
	Within groups	32.365	109	0.297		
	Total	33.107	111			
Gender groups	Source	Sum of squares	df	Mean square	F	Sig.
	Between groups	0.179	1	0.179	0.597	0.441
	Within groups	32.928	110	0.299		
	Total	33.107	111			

df degree of freedom; *Sig.* significance value

Table 4 Loss aversion ANOVA results

	Source	Sum of squares	df	Mean square	F	Sig.
Organizational groups	Between groups	0.212	2	0.106	0.526	0.593
	Within groups	21.948	109	0.201		
	Total	22.160	111			
Years of experience groups	Source	Sum of squares	df	Mean square	F	Sig.
	Between groups	0.872	2	0.436	2.233	0.112
	Within groups	21.288	109	0.195		
	Total	22.160	111			
Gender groups	Source	Sum of squares	df	Mean square	F	Sig.
	Between groups	0.246	1	0.246	1.237	0.268
	Within groups	21.914	110	0.199		
	Total	22.160	111			

df degree of freedom; *Sig.* significance value

Under the influence of ownership, dispute negotiators would take their positions as personal belongings. As such, they are reluctant to make compromises because these are threats to their self-identity. Their positions are therefore endowed. This would further hinder the negotiator from exploring alternative options and making attempt to understand the standpoints of the counterpart [146]. Loss aversion makes dispute negotiators more sensible to the downsides of a compromise than their prospective gains from a prospective settlement. Construction disputes mostly involve monetary disagreement between parties, if negotiators are buried with potential losses, any breakthrough decisions like accepting the offer by the counterpart would require

Table 5 Status quo bias ANOVA results

Organizational groups	Source	Sum of squares	df	Mean square	F	Sig.
	Between groups	0.007	2	0.003	0.016	0.984
	Within groups	23.022	109	0.211		
	Total	23.029	111			
Years of experience groups	Source	Sum of squares	df	Mean square	F	Sig.
	Between groups	0.289	2	0.144	0.691	0.503
	Within groups	22.758	109	0.209		
	Total	23.047	111			
Gender groups	Source	Sum of squares	df	Mean square	F	Sig.
	Between groups	0.007	1	0.007	0.032	0.858
	Within groups	23.040	110	0.209		
	Total	23.047	111			

df degree of freedom; *Sig.* significance value

Table 6 Strategic bargaining habit ANOVA results

Organizational groups	Source	Sum of squares	df	Mean square	F	Sig.
	Between groups	0.086	2	0.043	0.164	0.849
	Within groups	28.463	109	0.261		
	Total	28.549	111			
Years of experience groups	Source	Sum of squares	df	Mean square	F	Sig.
	Between groups	1.222	2	0.611	2.437	0.092
	Within groups	27.327	109	0.251		
	Total	28.549	111			
Gender groups	Source	Sum of squares	df	Mean square	F	Sig.
	Between groups	0.334	1	0.334	1.304	0.256
	Within groups	28.214	110	0.256		
	Total	28.549	111			

df degree of freedom; *Sig.* significance value

bold efforts. This seems very unlikely for EE influenced negotiators. The occurrence of status quo bias would make the situation worse. Due to the complex nature of construction project and the current mainstream multi-tiered dispute resolution approach makes CDN fairly convoluted. Protracted negotiations may make the situation worse as the positions of the negotiators might have become holdfast. EE affected negotiators would stick with their assessments and even attach greater weights, resulting in negotiation impasse. Besides, strategically misrepresenting one's positions about the dispute matter, either over-valuing it or under-valuing it, adds difficulty in reaching an agreement. Disputing parties would strategically overclaim their bottom line of bargaining and under-estimate their willingness to accept the offers provided by the counterpart. This inflexible bargaining habit would destroy mutual trust and channel of communication. These effects of EE would no doubt make settlement remote.

From the ANOVA results summarized in Tables 3, 4, 5 and 6, it can be said that the tendency to practise EE is not related to one's organizational background, working experience and gender. Both male and female respondents, no matter they work for consultants, contractors or clients and their years of experience, they would likely all influenced by EE. This is consistent with previous reported studies that, repetitions of trade-offs and opportunities to learn would not eliminate the impact of EE [51, 72].

Managing EE

Dispute negotiators should be made aware of the psychological barriers they may have in handling disputes. Endowment effect is anchored on ownership and self-enhancement tendency of the negotiator. Enhanced objectivity would help disputing parties to refrain from psychological effect like endowment. With rational analysis, the positions taken by the negotiators can be reviewed and adjusted if necessary. There is no substitute to recognise that maximizing mutual benefits make the most commercial sense. When negotiators are focusing on mutual benefits, they would be more prepared to move away from the status quo. It is anticipated that better attention would then be placed on evidence and facts rather than emotional feelings. Negotiators would need external help at times to evaluate the relevancy of the available information, including those put forward by the counterpart. Moreover, there are caveats like opportunism and exploitation of one's good faith. To this end, appropriate intervention from neutral advisors can be useful. It has been suggested that advice from a third party neutral may be sensible than purely technical experts [35, 120, 122]. Practising meditators, facilitators and dispute resolution advisors are recognized groups of third-party neutrals who can facilitate the negotiation process by enhancing communication among the disputing parties [1]. Instead of the typical use of third-party neutral advisor after a dispute has arisen, his services can be standing throughout the project period. The standing arrangement would allow them to keep abreast with the knowledge and progress of the project so that they could provide impartial suggestions. Third party neutral advisors are in the suitable positions to give timely reminders. It is thus important that third party neutral advisors could participate throughout the construction project to remind the disputing parties about their EE tendencies. Reality checks could also be conducted with the assistance of third-party neutral advisor to help the disputing parties objectively review their assessments, their expectation and their attitudes about the dispute. One example in Hong Kong is the use of dispute resolution and avoidance advisor. Apart from disputing parties' irrational evaluation of their positions, it is also suggested that disputing parties' self-perception and emotions before entering the dispute resolution would moderate the extent of EE during CDN [80, 87]. If disputing parties are threatened or having a relatively negative emotion, they would attribute a lower value to their later assessments and present a lower level of EE in the coming resolution process. With regard to construction dispute resolution training, psychological barriers are not yet standard component. In the light of the actual happening of biased behaviours, this omission should be addressed without delay.

Reactive Devaluation

Professor Lee Ross and his colleagues pioneered the study of reactive devaluation (RD) [119] and offered three major RD manifestations: (i) a proposal raised by the counterpart is rated less positively than the one raised by themselves; (ii) a proposal is rated less positively than those yet to be offered, the prospect of further proposals leads to dissatisfaction with the current offer; and (iii) a proposal raised unilaterally is rated less positively than one that was brainstormed by the parties together [101, 119, 120, 122].

Theoretical Explanation of Reactive Devaluation

Loss aversion, attitude polarization and naïve realism are the three theoretical explanation of reactive devaluation. These are discussed seriatim.

Loss Aversion

Loss aversion is a source to not only endowment effect (as discussed in the prior section of this chapter) but also to reactive devaluation. During negotiation, once an assessment of the dispute was made, changing one's position to accept a proposal by the counterpart is often treated as making a loss. A loss-averse offeree will delay or even avoid deciding on proposals on offer. Loss aversion effect would motivate a negotiator to devaluate and reject suggestions made by the counterpart, resulting in an impasse [25, 60, 62, 128].

Attitude Polarization

People have the tendency to selectively pay attention to information that supports their views and dismiss those otherwise [70, 89, 110]. This inclination will get more and more entrenched and even polarized when more supporting information/evidence become available [15, 95, 137]. Cheung and Li [23] described this as self-affirmation bias that makes one firmly believe that his proposal is the most appropriate [37, 118, 123]. Hence, biased information processing strengthens one's already held opinions and polarizes their positions. It is therefore not difficult to find negotiators overvaluing their own assessment about the dispute and at the same time devaluating those of their counterpart [61, 150]. Attitude polarization is a major barrier against effective communication [89, 102].

Naïve Realism

Naïve Realism (NR) is the third source of RD [84, 107, 123, 124] and make negotiators defying objectivity and fairness [114, 120, 122]. NR makes one believe that he knows the real issues underpinning a dispute and his belief is the most sensible [134, 135]. Negotiators having NR would rate their proposals as the most practical [112, 114]. As a result, they would discredit any other views [13, 119, 120, 122]. Besides, negotiators having NR expect his counterpart should share the same conclusion [104, 123]. Thus, his counterpart should agree with him [97, 118]. If this is not the case, it must be that his counterpart is not working hard enough or lacking the ability to do so.

Manifestations of RD in CDN

To operationalize the concept of RD in CDN, 25 manifestations of RD are developed with due reference to the literature. Table 7 summarizes these manifestations and the corresponding references.

Taxonomies of RD Behaviors in CDN

RD behaviors in CDN are conceptualized by developing taxonomies. Furthermore, examining taxonomies of RD in CDN would enlist ways to mitigate RD effects. A data collection questionnaire was designed to collect the opinion of construction dispute resolution practitioners about their practice of RD behaviours. The respondents were asked to indicate their extent of practice of the RD behaviours as listed in Table 7 on a 6-point Likert scale from "1 = Strongly Disagree" to "6 = Strongly Agree".

The questionnaire was distributed to potential respondents who members the Hong Kong Institution of Engineers (HKIE), the Hong Kong Institution of Architects (HKIA), the Hong Kong Institute of Surveyors (HKIS). In total, 115 valid responses were received out of 350 questionnaires distributed, representing a response rate of 32.9%. The profile of the respondents is shown in Figs. 5, 6, 7 and 8.

To identify the underlying dimensions of RD behaviours in CDN, principal component factor analysis (PCFA) was used. PCFA was performed using IBM SPSS version 25.0. Varimax rotation was used to obtain a simplified factor structure [47]. KMO value of 0.80 and significant result of Bartlett's test of sphericity supported

Table 7 Manifestations of RD in CDN

Manifestations of RD in CDN	Sources			References
	LA	AP	NR	
(1) I want to maintain my decisions therefore I am uninterested in negotiation	✓	✓		[61, 63, 76]
(2) Before negotiation, I believe my proposal will be accepted			✓	[112, 114, 120, 122]
(3) After forming a reasonable opinion, I tend to devaluate other possibilities	✓	✓	✓	[13, 119, 120, 122]
(4) When rethink about my decision, I unconsciously recall reasons why my decision should be upheld	✓	✓		[70, 89, 110]
(5) I think the resolution proposal raised by the counterpart is based on incomplete information			✓	[13, 119, 120, 122]
(6) I think the proposal from the counterpart cannot resolve the project dispute	✓		✓	[13, 119, 120, 122]
(7) I think the counterpart is biased			✓	[13, 119, 120, 122]
(8) I doubt the capability of the counterpart			✓	[114, 119, 122]
(9) I doubt the effectiveness of the suggestions raised by the counterpart			✓	[107, 114, 122, 120]
(10) I think the counterpart provides premature decisions			✓	[107, 114, 122, 120]
(11) I think my proposal is more practical		✓	✓	[134, 135]
(12) I think my proposal is the best solution to the project dispute			✓	[118, 119, 131, 132]
(13) I believe the counterpart will agree with us if they put more effort in decision making		✓	✓	[97, 118]
(14) I always think there is still room for bargaining with the proposal from the counterpart			✓	[104, 123]
(15) I think only my proposal can tackle the project dispute			✓	[114, 120, 122]
(16) I endorse information that supports my decisions		✓		[70, 89]
(17) I search for information that confirms my assessment		✓		[70, 110]
(18) After forming a decision, I tend to pay more attention to the information that supports my decision	✓	✓		[70, 89, 110]

(continued)

Table 7 (continued)

Manifestations of RD in CDN	Sources			References
	LA	AP	NR	
(19) My information searching process terminates after I consider I have found enough supporting information for my decision		✓	✓	[70, 110, 120, 122]
(20) I tend to think the choices provided by the counterpart are disadvantageous to my side	✓		✓	[63, 114, 120, 122]
(21) I tend to believe the choices not provided by the counterpart are more advantageous to my side	✓			[63, 76]
(22) I tend to think the counterpart wants to gain interests from me with his offer	✓			[60, 128]
(23) I think the compromise of the counterpart is just their negotiation strategy	✓			[60, 76, 128]
(24) I tend to think the counterpart is behaving opportunistically	✓		✓	[25, 60, 62, 128]
(25) I tend to devaluate the proposal if I know that it was based on information out of my knowledge	✓			[60, 76, 128]

LA loss aversion; *AP* attitude polarization; *NR* Naïve realism

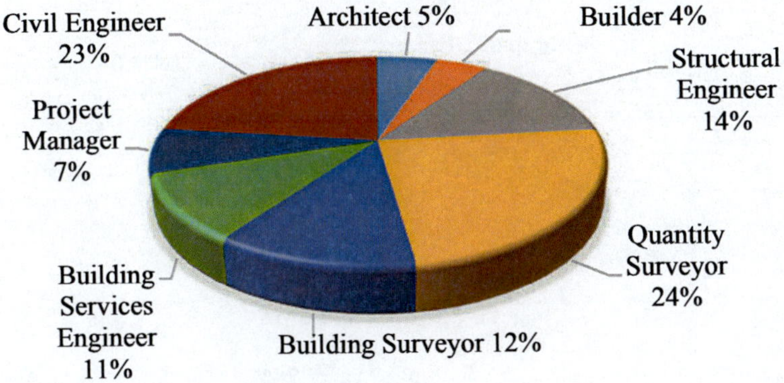

Fig. 5 Professional background of the respondents

the adequacy and suitability of the data [19, 64]. Eigenvalue larger than one was used as the factor extraction baseline [47]. Factor loadings larger than 0.5 were kept [65, 98]. As a result, a structure with five factors without cross loading was obtained and presented in Table 8. The five factors extracted in this study can explain 60% of

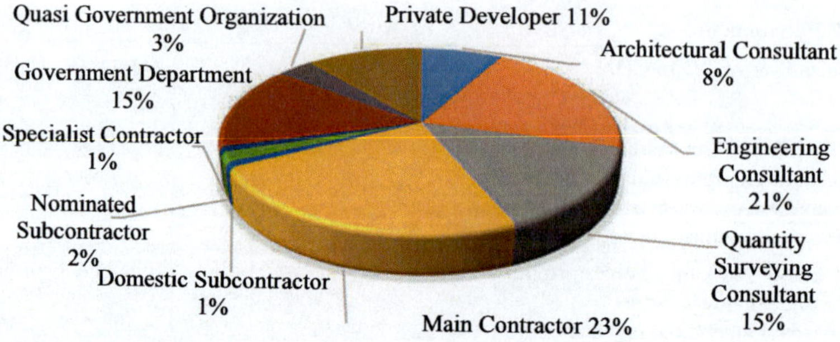

Fig. 6 Organization background of the respondents

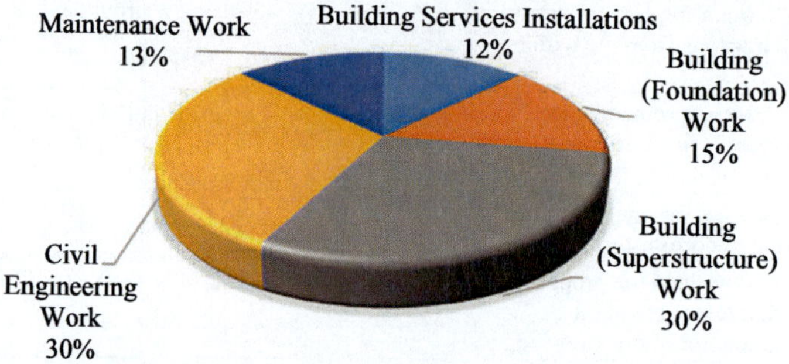

Fig. 7 Major dispute type involved by the respondents

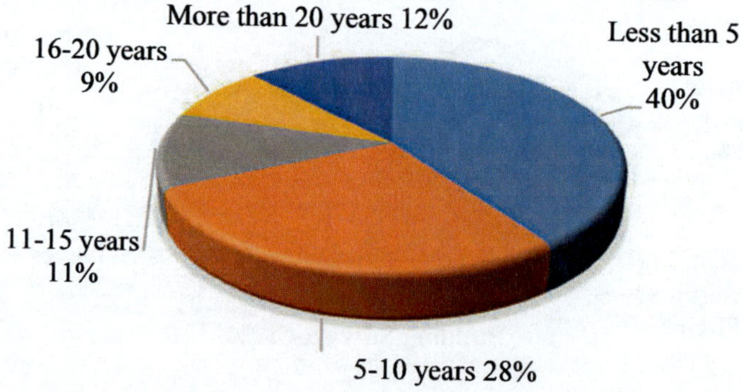

Fig. 8 Years of working experience of the respondents

the total variance. Based on the PCFA results, five taxonomies of RD behaviours in CDN are proposed.

Table 8 Taxonomies of RD manifestations in CDN

Manifestations of RD in CDN	Factor 1	Factor 2	Factor 3	Factor 4	Factor 5
	(1.360)	(2.414)	(1.401)	(1.977)	(7.835)
Reluctance to change					
I want to maintain my decisions therefore I am uninterested in negotiation	0.641				
Before negotiation, I believe my proposal will be accepted	0.718				
After forming a reasonable opinion, I tend to devaluate other possibilities	0.696				
When rethink about my decision, I unconsciously recall reasons why my decision should be upheld	0.631				
Doubts about counterpart's ability					
I think the resolution proposal raised by the counterpart is based on incomplete information		0.747			
I think the proposal from the counterpart cannot resolve the project dispute		0.758			
I think the counterpart is biased		0.585			
I doubt the capability of the counterpart		0.793			
I doubt the effectiveness of the suggestions raised by the counterpart		0.708			
I think the counterpart provides premature decisions		0.580			
Overconfidence					
I think my proposal is more practical			0.524		
I think my proposal is the best solution to the project dispute			0.660		
I believe the counterpart will agree with us if they put more effort in decision making			0.708		
I always think there is still room for bargaining with the proposal from the counterpart			0.583		
I think only my proposal can tackle the project dispute			0.643		
Biased information processing					
I endorse information that supports my decisions				0.631	
I search for information that confirms my assessment				0.726	

(continued)

Table 8 (continued)

Manifestations of RD in CDN	Factor 1 (1.360)	Factor 2 (2.414)	Factor 3 (1.401)	Factor 4 (1.977)	Factor 5 (7.835)
After forming a decision, I tend to pay more attention to the information that supports my decision				0.737	
My information searching process terminates after I consider I have found enough supporting information for my decision				0.663	
Mistrust towards the counterpart					
I tend to think the choices provided by the counterpart are disadvantageous to my side					0.776
I tend to believe the choices not provided by the counterpart are more advantageous to my side					0.691
I tend to think the counterpart wants to gain interests from me with his offer					0.722
I think the compromise of the counterpart is just their negotiation strategy					0.605
I tend to think the counterpart is behaving opportunistically					0.559
I tend to devaluate the proposal if I know that it was based on information out of my knowledge					0.697

Note Eigenvalues of the factors were presented in the parenthesis

Validation of the Taxonomies

Confirmatory factor analysis (CFA) was then conducted to validate the five RD taxonomies. Goodness-of-fit (GOF) measures were used to evaluate the fitness and parsimony of the proposed CFA model [81, 99, 149]. The baseline and results of GOF indices are summarised in Table 9. The regression weights of the CFA model are presented (Table 10), it can be seen that all of the regression weights have *p*

Table 9 GOF indices results

Fit index	Threshold	Model result
χ^2/df	≤2.00	1.54
GFI	≥0.80	0.80
RMSEA	≤0.08	0.07
CFI	≥0.80	0.87
TLI	≥0.80	0.85
PNFI	≥0.50	0.63
PCFI	≥0.50	0.78

Table 10 Regression Weights of CFA model

Parameter	Regression eights	Regression weights (Standardized)	S.E	C.R	P
Reluctance ← Reactive devaluation	1.000	0.700			
Doubts ← Reactive devaluation	0.975	0.642	0.273	3.576	***
Overconfidence ← Reactive devaluation	1.498	0.826	0.366	4.097	***
Bias ← Reactive devaluation	0.990	0.777	0.267	3.704	***
Mistrust ← Reactive devaluation	1.107	0.768	0.289	3.833	***
Item 1 ← Reluctance	1.000	0.535			
Item 2 ← Reluctance	0.881	0.661	0.193	4.558	***
Item 3 ← Reluctance	1.168	0.731	0.247	4.732	***
Item 4 ← Reluctance	0.826	0.527	0.206	4.008	***
Item 5 ← Doubts	1.000	0.686			
Item 6 ← Doubts	0.999	0.750	0.141	7.060	***
Item 7 ← Doubts	1.121	0.647	0.181	6.194	***
Item 8 ← Doubts	1.031	0.749	0.146	7.057	***
Item 9 ← Doubts	1.256	0.813	0.167	7.538	***
Item 10 ← Doubts	0.676	0.570	0.123	5.513	***
Item 11 ← Overconfidence	1.000	0.863			
Item 12 ← Overconfidence	0.785	0.671	0.108	7.233	***
Item 13 ← Overconfidence	0.604	0.570	0.100	6.025	***
Item 14 ← Overconfidence	0.550	0.541	0.097	5.675	***
Item 15 ← Overconfidence	0.502	0.473	0.102	4.898	***
Item 16 ← Bias	1.000	0.660			
Item 17 ← Bias	1.160	0.716	0.189	6.145	***
Item 18 ← Bias	1.237	0.779	0.191	6.471	***
Item 19 ← Bias	0.929	0.566	0.182	5.099	***
Item 20 ← Mistrust	1.000	0.716			
Item 21 ← Mistrust	0.984	0.728	0.139	7.106	***

(continued)

Table 10 (continued)

Parameter	Regression eights	Regression weights (Standardized)	S.E	C.R	P
Item 22 ← Mistrust	1.160	0.734	0.162	7.160	***
Item 23 ← Mistrust	0.963	0.704	0.140	6.888	***
Item 24 ← Mistrust	0.820	0.612	0.136	6.024	***
Item 25 ← Mistrust	1.040	0.670	0.158	6.567	***

S.E. Approximate standard error; *C.R.* Critical ratio; *** mean that the *p*-value is less than 0.001

values less than 0.001, indicating their statistical significance [55]. CFA model of RD in CDN is presented in Fig. 9. Construct validity in the CFA model was assessed by average of variance extracted (AVE) and composite reliability (CR). From Table 11, it can be seen that all the constructs have AVE values larger than acceptable level of 0.4, given that in this study CR values are larger than satisfactory standard of 0.7 [40, 41, 148]. Besides, square root of the AVE (numbers at diagonal in Table 11) for each construct is greater than the correlations with other constructs. These results support good convergent validity and discriminant validity of CFA model.

Discussion

Five taxonomies of RD in CDN are proposed: reluctance to change; doubts about counterpart's ability; overconfidence; biased information processing and mistrust towards the counterpart. As seen in Fig. 9 and Table 10, overconfidence has the highest standardized regression weight of 0.826 and is therefore the most important motivator of RD behaviour in CDN. Unrealistic expectations are germinated by overconfidence [38, 71, 140]. Overconfident negotiators are also over-optimistic about the outcome. When they only see their version of the solution is the best, compromise is no longer an acceptable option [118, 119, 131, 132]. Optimistic negotiators believe that the proposal of the counterpart can be much improved with better effort [104, 123]. Hence, overconfident negotiators always respond negatively towards offers proposed by the counterpart. Biased information processing has the second highest standardized regression weight of 0.777. Negotiators having self-affirming tendency will only attend to evidences that support their own positions [70, 83, 110]. Unjustified weightings are attached to those information that supports their perspective [70, 89]. Mistrust towards the counterpart is also an indicator of the presence of RD in CDN with a standardized regression weight of 0.768. Lack of mutual trust is a common problem in construction dispute resolution industry. Mistrust destroys collaboration and partnership between the contracting parties [145, 147]. Under a mistrust condition, negotiators would view even good faith behavior from the counterpart as opportunistic. This skepticism makes settlement almost impossible [25, 62, 128]. With the standardized regression weight of 0.700, reluctance to change

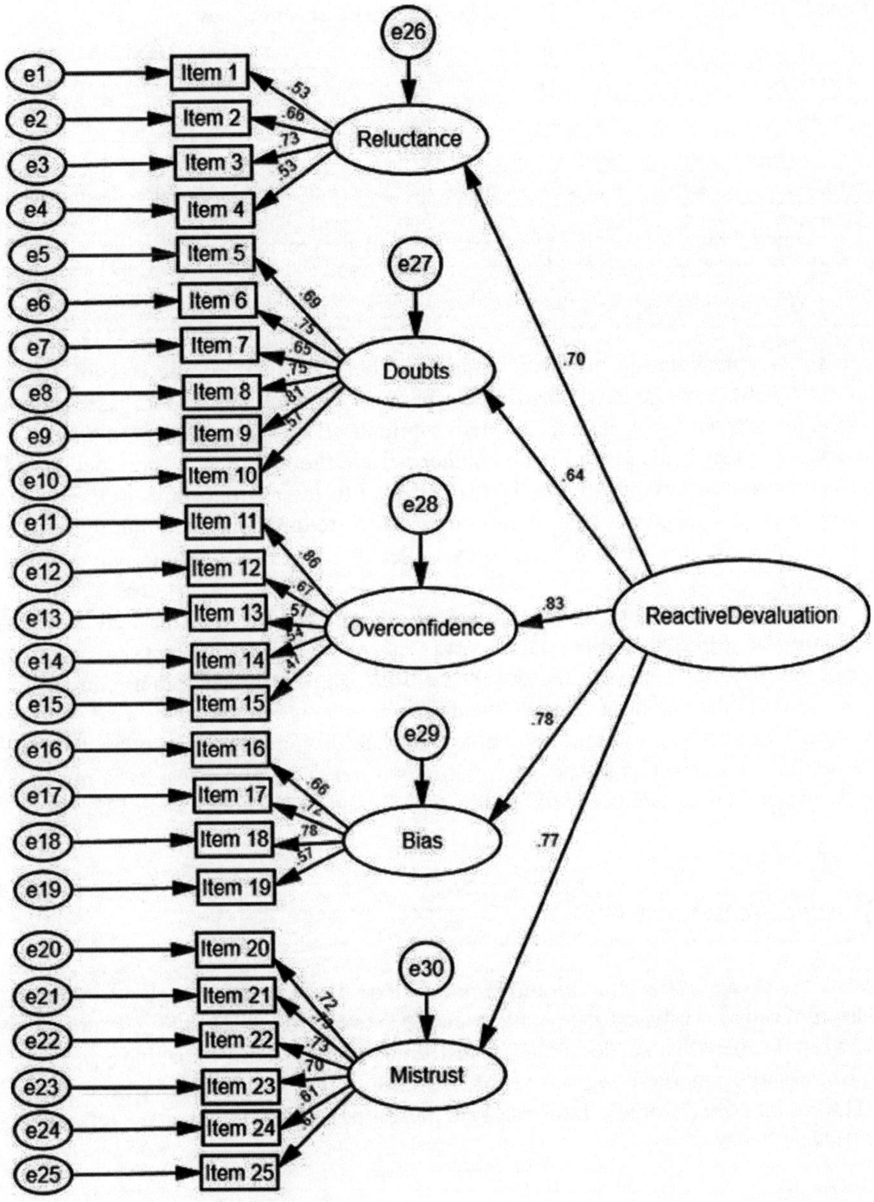

Fig. 9 CFA model of RD in CDN

Table 11 Construct validity of CFA model and inter-construct correlations

	AVE	CR	Reluctance	Doubts	Overconfidence	Bias	Mistrust
Reluctance	0.4	0.7	**0.6**				
Doubts	0.5	0.9	0.4	**0.7**			
Overconfidence	0.4	0.8	0.5	0.4	**0.6**		
Bias	0.5	0.8	0.3	0.3	0.5	**0.7**	
Mistrust	0.5	0.8	0.3	0.5	0.4	0.5	**0.7**

AVE average of variance extracted; *CR* composite reliability; The numbers at diagonal are square roots of AVE

is a good representation of RD. This is the other side of the same coin of status quo [59, 126]. Strengthened by attitude polarization effect, negotiators' assessment would be hardened [89, 120, 122]. Being influenced by naïve realism, negotiators consider the proposals made by the counterpart are premature and developed based on inadequate information [114, 120, 122]. Due to their strong faith in their own assessment, the proposals from their counterparts are downplayed. Framing effects also explain disputing parties' tendency to devalue and reject counterpart's offers [46, 69, 141, 143].

In CDN, sticking with one's assessment is considered a risk-free option while accepting the offers from the counterpart is risky. Having RD inclination, adopting a risk averse attitude is very likely [58, 82, 109]. Hence, negotiators are unwilling to make tradeoffs and turning down the offers from their counterpart is a safe option to him. Worse still, doubting the counterpart's ability and mistrust so created will frame the counterpart as a 'bad' negotiating partner. This frame results in negative evaluations on the practicality of all things proposed by the counterpart [22, 82].

Managing RD in CDN

To alleviate reactive devaluation and improve dispute management, the psychological barrier of mistrust towards the counterpart needs to be cleared. Building relationship between the disputing parties would be instrumental in fostering bona fide exchanges during dispute negotiations [24, 133, 147]. Measures that would mitigate the effect of RD were identified from literature and are presented with their respective references in Table 12.

Summary

Endowment effect (EE) describes the phenomenon that people would demand more to relinquish items that they own than they would be willing to pay for the same. This

Table 12 RD mitigation measures

Taxonomies of RD in CDN	Mitigation measures	References
Reluctance to change	Resolving misunderstandings	[90, 93, 116]
	Negotiation	[34, 77, 116, 129]
	Participation	[30, 77, 91, 129]
Doubts about counterpart's ability	Opening to the counterpart's narrative	[49, 48, 68, 107]
	Awareness and identification of bias	[86, 107, 115]
	Perspective taking	[29, 43, 44, 68, 131]
Overconfidence	Playing devil's advocate	[26, 92, 94, 130]
	Considering the opposite	[45, 79, 89, 105]
	Warnings and reminders	[7, 66, 100]
	Considering possible failures	[6, 54, 79, 127]
Biased information processing	Considering the alternatives	[3, 12, 50, 52]
	Allowing adequate time and effort in making decisions	[20, 28, 36, 106]
	Group discussion	[18, 79, 139]
	Training and Education	[66, 79, 139]
Mistrust towards the counterpart	Encouraging communication	[133, 139, 147]
	Relationship building	[24, 39, 139, 145]
	Encouraging goodwill and benevolence	[31, 39, 77, 91]

can be explained by the tendency of people over-valuing their belongings, properties, opinions and decisions. Thus, the happening of EE undermine rational choices. Four sources of EE were identified from literature: ownership, loss aversion, status quo bias and strategic bargaining habit. With data collected from construction practitioners, the occurrence of these manifestations was found to be real. It was further unveiled that construction disputing parties from different construction sectors displayed a similar extent of EE behaviours in CDN. It is suggested that by utilising the skills of third-party neutrals as standing advisors. The effects of EE can be minimized through appropriate intervention of the standing neutrals as deemed appropriate.

Reactive devaluation (RD) is another well-recognized psychological barrier against dispute settlement. Twenty-five RD manifestations in CDN were developed from literature. Through a principal component factor analysis, five taxonomies of RD behaviours in CDN were extracted: reluctance to change; doubts about counterpart's ability; overconfidence; biased information processing and mistrust towards the counterpart. The potency of these taxonomies was validated with confirmatory factor analysis (CFA). Overconfidence was identified as the most important indicator of RD with the highest standardized regression weight. The findings timely remind dispute negotiators that RD would stifle proposal exchanges and lead to rejection of constructive proposals. Correspondingly, measures to curb RD behaviours were recommended.

Acknowledgements The empirical work of this chapter has been reported in "Study of Endowment Effect in Construction Dispute Resolution" of the ASCE Journal of Legal Affairs and Dispute Resolution in Engineering and Construction and "Reactive Devaluation as a Psychological Impediment to Construction Dispute Negotiation" in the ASCE Journal of Management in Engineering. The texts have been substantially re-written. The support of the HKSAR RGC General Research Fund Project (no. 11209118) is duly acknowledged. Special thanks go to Yui Hang NG and On Yi CHOW for helping in the study especially in data collection.

References

1. Agapiou A (2018) Mediating planning disputes: opportunities, experiences, and challenges. J Leg Aff Dispute Res Eng Constr. https://doi.org/10.1061/(ASCE)LA.1943-4170.0000256
2. Alexopoulos T, Šimleša M, Francis M (2015) Good self, bad self: initial success and failure moderate the endowment effect. J Econ Psychol 50:32–40
3. Anderson CA (1982) Inoculation and counter explanation: debiasing techniques in the perseverance of social theories. Soc Cogn 1(2):126–139
4. Ariely D, Huber J, Wertenbroch K (2005) When do losses loom larger than gains? J Mark Res 42(2):134–138
5. Ariely D, Simonson I (2003) Buying, bidding, playing, or competing? Value assessment and decision dynamics in online auctions. J Consumer Psychol 13(1):113–123
6. Arkes HR (1991) Costs and benefits of judgment errors: implications for debiasing. Psycho Bull 110(3):486–498
7. Arkes HR, Christensen C, Lai C, Blumer C (1987) Two methods of reducing overconfidence. Organ Behav Hum Decis Process 39(1):133–144
8. Armstrong RA, Hilton AC (2010) One-way analysis of variance (Anova). Statist Anal Microbiol: Statnotes 33–37
9. Barnes JH (1984) Cognitive biases and their impact on strategic planning. Strategic Manage J 5(2):129–137
10. Beggan JK (1992) On the social nature of nonsocial perception: the mere ownership effect. J Pers Soc Psychol 62(2):229
11. Bendle N, Cotte J (2016) Assumptions of rationality in political marketing: the case of the republican autopsy. JNPSM 28(1):66–83
12. Bentz BG, Williamson DA, Franks SF (2004) Debiasing of pessimistic judgments associated with anxiety. J Psychopathol Behav Assess 26(3):173–180
13. Bland B, Powell B, Ross L (2012) Barriers to dispute resolution: reflections on peacemaking and relationships between adversaries. In: Goodman R, Jinks D, Woods AK (eds) Understanding social action, promoting human rights. Oxford University Press, pp 265–297
14. Van Boven L, Dunning D, Loewenstein G (2000) Egocentric empathy gaps between owners and buyers: Misperceptions of the endowment effect. J Pers Soc Psychol 79:66–76
15. Brauer M, Judd CM, Gliner MD (1995) The effects of repeated expressions on attitude polarization during group discussions. J Pers Soc Psychol 68(6):1014–1029
16. Carmon Z, Ariely D (2000) Focusing on the forgone: how value can appear so different to buyers and sellers. J Consum Res 27(3):360–370
17. Carmon Z, Wertenbroch K, Zeelenberg M (2003) Option attachment: when deliberating makes choosing feel like losing. J Consum Res 30(1):15–29
18. Carnevale PJ (2008) Positive affect and decision frame in negotiation. Group Decis Negot 17(1):51–63

19. Cerny BA, Kaiser HF (1977) A study of a measure of sampling adequacy for factor-analytic correlation matrices. Multivar Behav Res 12(1):43–47
20. Chapman GB, Johnson EJ (2002) Incorporating the irrelevant: anchors in judgments of belief and value. In: Gilovich T, Griffin D, Kahneman D (eds) Heuristics and biases: the psychology of intuitive judgment. Cambridge University Press, Cambridge, pp 120–138
21. Chatterjee P, Irmak C, Rose RL (2013) The endowment effect as self-enhancement in response to threat. J Consumer Res 40(3):460–476
22. Cheng FF, Wu CS (2010) Debiasing the framing effect: the effect of warning and involvement. Decis Support Syst 49(3):328–334
23. Cheung SO, Li K (2019) Biases in construction project dispute resolution. Eng Constr Archit Manage 26(2):321–348
24. Chow PT, Cheung SO, Chan KY (2012) Trust-building in construction contracting: mechanism and expectation. Int J Proj Manage 30(8):927–937
25. Cohen D, Knetsch JL (1992) Judicial choice and disparities between measures of economic values. Osgoode Hall LJ 30(3):737–770
26. Corey G (2011) Group techniques. Brooks/Cole, Belmont, CA
27. Coursey DL, Hovis JJ, Schulze WD (1987) The disparity between willingness to accept and willingness to pay measures of value. Q J Econ 102(3):679–690
28. Croskerry P, Singhal G, Mamede S (2013) Cognitive debiasing 2: impediments to and strategies for change. BMJ Qual Saf bmjqs-2012-001713
29. Davis MH, Conklin L, Smith A, Luce C (1996) Effect of perspective taking on the cognitive representation of persons: a merging of self and other. J Pers Soc Psychol 70(4):713–726
30. Dent EB, Goldberg SG (1999) Challenging 'resistance to change.' J Appl Behav Sci 35(1):25–41
31. Dervitsiotis KN (2006) Building trust for excellence in performance and adaptation to change. Total Qual Manage Bus Excel 17(7):795–810
32. Dommer SL, Swaminathan V (2013) Explaining the endowment effect through ownership: the role of identity, gender, and self-threat. J Consum Res 39(5):1034–1050
33. De Dreu CK, van Knippenberg D (2005) The possessive self as a barrier to conflict resolution: effects of mere ownership, process accountability, and self-concept clarity on competitive cognitions and behavior. J Pers Soc Psychol 89(3):345
34. Dubrin AJ, Ireland RD (1993) Management and organization, 2nd edn. South- Western Publishing, Cincinnati, OH
35. Elliott MLP (1999) The role of facilitators, mediators, and other consensus building practitioners. The consensus building handbook: A comprehensive guide to reaching agreement, pp 199–240
36. Epley N, Gilovich T (2006) The anchoring-and-adjustment heuristic Why the adjustments are insufficient. Psychol Sci 17(4):311–318
37. Ericson KMM, Fuster A (2011) Expectations as endowments: evidence on reference-dependent preferences from exchange and valuation experiments. Q J Econ 126(4):1879–1907
38. Feld J, Sauermann J, De Grip A (2017) Estimating the relationship between skill and overconfidence. J Behav Exp Econ 68:18–24
39. Ford JD, Ford LW, D'Amelio A (2008) Resistance to change: the rest of the story. Acad Manag Rev 33(2):362–377
40. Fornell C, Larker D (1981) Evaluating structural equation models with unobserved variables and measurement errors. J Res Mark 27(3):445–466
41. Fraering M, Minor MS (2006) Sense of community: an exploratory study of US consumers of financial services. Int J Bank Mark 24(5):284–306
42. Fuchs C, Prandelli E, Schreier M (2010) The psychological effects of empowerment strategies on consumers' product demand. J Mark 74(1):65–79
43. Galinsky AD, Ku G (2004) The effects of perspective-taking on prejudice: the moderating role of self-evaluation. Pers Soc Psychol Bull 30(5):594–604

44. Galinsky AD, Moskowitz GB (2000) Perspective-taking: decreasing stereotype expression, stereotype accessibility, and in-group favoritism. J Pers Soc Psychol 78(4):708
45. Galinsky AD, Mussweiler T (2001) First offers as anchors: the role of perspective-taking and negotiator focus. J Pers Soc Psychol 81(4):657–669
46. Gonzalez C, Dana J, Koshino H, Just M (2005) The framing effect and risky decisions: examining cognitive functions with fMRI. J Econ Psychol 26(1):1–20
47. Hair JF, Anderson RE, Tatham RL, Black WC (1998) Multivariate data analysis, 5th edn. Prentice Hall International, NY
48. Hameiri B, Bar-Tal D, Halperin E (2014) Challenges for peacemakers: how to overcome socio-psychological barriers. Policy Insights Behav Brain Sci 1(1):164–171
49. Hameiri B, Halperin E (2015) Sociopsychological barriers to peacemaking and overcoming them: a review of new psychological interventions. The Social Psychology of Intractable Conflicts. Springer, Cham, pp 173–187
50. Hammond JS, Keeney RL, Raiffa H (1998) The hidden traps in decision making. HBR 76(5):47–58
51. Harbaugh WT, Krause K, Vesterlund L (2001) Are adults better behaved than children? Age, experience, and the endowment effect. Econ Lett 70(2):175–181
52. Heiman VB (1990) Auditors' assessments of the likelihood of error explanations in analytical review. Account Rev 65(4):875–890
53. Hendry SH, Shaffer DR (1989) On testifying in one's own behalf: interactive effects of evidential strength and defendant's testimonial demeanour on jurors' decisions. J Appl Psychol 74(4):539–545
54. Hirt ER, Markman KD (1995) Multiple explanation: a consider-an-alternative strategy for debiasing judgments. J Pers Soc Psychol 69(6):1069–1086
55. Ho R (2013) Handbook of univariate and multivariate data analysis with IBM SPSS. CRC Press, New York, NY
56. Horowitz JK, McConnell KE (2002) A review of WTA/WTP studies. J Environ Econ Manag 44(3):426–447
57. Howell DC (2009) Statistical methods for psychology. Cengage Learning
58. Huang HH, Su HJ, Chang CJ (2015) The moderating effect of a no-choice option on risky framing effect. Asia Pac Manage Rev 20(1):18–23
59. Jacques S (1992) The endowment effect and the Coase theorem. Am J Agric Econ 74(5):1316–1323
60. Johnson EJ, Keinan A, Haubl G (2007) Aspects of endowment: a query theory of value construction. J Exp Psychol Learn Mem Cogn 33(3):461–474
61. Johnson-Cartee K, Copeland G (1997) Inside political campaigns: theory and practice. Praeger, Westport, CT
62. Kahneman D, Knetsch JL, Thaler RH (1991) Anomalies: the endowment effect, loss aversion, and status quo bias. J Econ Perspect 5(1):193–206
63. Kahneman D, Tversky A (1984) Choices, values, and frames. Am Psychol 39(4):341–350
64. Kaiser HF (1958) The varimax criterion for analytic rotation in factor analysis. Psychometrika 23(3):187–200
65. Kannan VR, Tan KC (2005) Just in time, total quality management, and supply chain management: understanding their linkages and impact on business performance. Omega 33(2):153–162
66. Kaustia M, Perttula M (2012) Overconfidence and debiasing in the financial industry. Rev Behav Finan 4(1):46–62
67. Keren G, Teigen KH (2004) Yet another look at the heuristics and biases approach. Blackwell handbook of judgment and decision making, pp 89–109
68. Keysar B, Barr DJ, Balin JA, Brauner JS (2000) Taking perspective in conversation: the role of mutual knowledge in comprehension. Psychol Sci 11(1):32–38
69. Kim J, Kim JE, Marshall R (2014) Search for the underlying mechanism of framing effects in multi-alternative and multi-attribute decision situations. J Bus Res 67(3):378–385
70. Klayman J (1995) Varieties of confirmation bias. Psychol Learn Motiv 32:385–418

71. Klayman J, Soll JB, Gonzalez-Vallejo C, Barlas S (1999) Overconfidence: it depends on how, what, and whom you ask. Organ Behav Hum Decis Process 79(3):216–247
72. Knetsch JL, Sinden JA (1987) The persistence of evaluation disparities. Q J Econ 102(3):691–696
73. Knez P, Smith VL, Williams AW (1985) Individual rationality, market rationality, and value estimation. Am Econ Rev 75(2):397–402
74. Korkobkin R (2005) Psychological impediments to mediation success: THEORY and practice. Ohio St J Disp Resol 21:281–307
75. Korkobkin R (2003) Bounded rationality, standard form contracts, and unconscionability. U Chi L Rev 70(1203):1223–32
76. Korobkin R, Guthrie C (1994) Psychological barriers to litigation settlement: an experimental approach. MLR 93:107–192
77. Kotter JP, Schlesinger LA (2008) Choosing strategies for change. HBR 86(7):130–139
78. Kőszegi B, Rabin M (2006) A model of reference-dependent preferences. Q J Econ 121(4):1133–1165
79. Larrick RP (2004) Debiasing. In: Koehler DJ, Harvey N (eds) Blackwell handbook of judgment and decision making. Blackwell Publishing Ltd., Malden, MA, pp 316–338
80. Lerner JS, Small DA, Loewenstein G (2004) Heart strings and purse strings–Carryover effects of emotions on economic decisions. Psychol Sci 15:337–341
81. Leung MY, Ng ST, Skitmore M, Cheung SO (2005) Critical stressors influencing construction estimators in Hong Kong. Constr Manage Ezon 23(1):33–44
82. Levin IP, Schneider SL, Gaeth GJ (1998) All frames are not created equal: a typology and critical analysis of framing effects. Organ Behave Hum Decis Process 76(2):149–188
83. Li K, Cheung SO (2018) Bias measurement scale for repeated dispute evaluations. J Manage Eng 34(4):04018016
84. Liberman V, Minson JA, Bryan CJ, Ross L (2012) Naïve realism and capturing the 'wisdom of dyads.' J Exp Soc Psychol 48(2):507–512
85. Liedtka J (2015) Perspective: Linking design thinking with innovation outcomes through cognitive bias reduction. J Prod Innov Manag 32(6):925–938
86. Lilienfeld SO, Lohr JM, Olatunji BO (2008) Encouraging students to think critically about psychotherapy: overcoming Naïve realism. In: Dana SD, Jane SH, Randolph AS (eds) Teaching critical thinking in psychology: A handbook of best practices. Blackwell Publishing Ltd, Malden, MA, pp 267–271
87. Lin CH, Chuang SC, Kao DT, Kung CY (2006) The role of emotions in the endowment effect. J Econ psychol 27(4):589–597
88. Loewenstein G, Issacharoff S (1994) Source dependence in the valuation of objects. J Behav Decis Mak 7:157–168
89. Lord CG, Ross L, Lepper MR (1979) Biased assimilation and attitude polarization: the effects of prior theories on subsequently considered evidence. J Pers Soc Psychol 37(11):2098–2109
90. Lozano R (2006) Incorporation and institutionalization of SD into universities: breaking through barriers to change. J Clean Prod 14(9–11):787–796
91. Lozano R (2013) Are companies planning their organisational changes for corporate sustainability? An analysis of three case studies on resistance to change and their strategies to overcome it. Corp Soc Resp Environ Manage 20(5):275–295
92. Lunenburg F (2012) Devil's advocacy and dialectical inquiry: antidotes to groupthink. Int J Scholar Acad Intellect Diver 14:1–9
93. Luthans F (2002) Organizational behavior. McGraw-Hill, New York
94. MacDougall C, Baum F (1997) The devil's advocate: a strategy to avoid groupthink and stimulate discussion in focus groups. Qual Health Res 7(4):532–541
95. Mackie D, Cooper J (1984) Attitude polarization: effects of group membership. J Pers Soc Psychol 46(3):575–585
96. Maoz I, Ward A, Katz M, Ross L (2002) Reactive devaluation of an 'Israeli' vs 'Palestinian' peace proposal. J Confl Resolut 46(4):515–546

97. Marks G, Miller N (1987) Ten years of research on the false-consensus effect: an empirical and theoretical review. Psychol bull 102(1):72–90
98. Matsunaga M (2010) How to factor-analyze your data right: do's, don'ts, and how-Do's. Int J Psychol Res 3(1):97–110
99. McDonald RP, Ho MHR (2002) Principles and practice in reporting structural equation analyses. Psychol Methods 7(1):64–82
100. McGraw AP, Mellers BA, Ritov I (2004) The affective costs of overconfidence. J Behav Decis Mak 17(4):281–295
101. Mnookin R, Ross L (1995) Strategic, psychological, and institutional barriers: an introduction. Barriers to the negotiated resolution conflict. Norton, New York, pp 3–27
102. Morewedge CK, Giblin CE (2015) Explanations of the endowment effect: an integrative review. Trends Cogn Sci 19(6):339–348
103. Morewedge CK, Shu LL, Gilbert DT, Wilson TD (2009) Bad riddance or good rubbish? Ownership and not loss aversion causes the endowment effect. J Exp Soc Psychol 45(4):947–951
104. Mullen B, Atkins JL, Champion DS, Edwards C, Hardy D, Story JE, Vanderklok M (1985) The false consensus effect: A meta-analysis of 115 hypothesis tests. J Exp Soc Psychol 21(3):262–283
105. Mussweiler T, Strack F, Pfeiffer T (2000) Overcoming the inevitable anchoring effect: considering the opposite compensates for selective accessibility. Pers Soc Psychol Bull 26(9):1142–1150
106. Mussweiler T, Englich B, Strack F (2004) Anchoring effect. In: Pohl R (ed) Cognitive illusions—A handbook on fallacies and biases in thinking, judgment, and memory. Psychology Press, pp 183–200
107. Nasie M, Bar-Tal D, Pliskin R, Nahhas E, Halperin E (2014) Overcoming the barrier of narrative adherence in conflicts through awareness of the psychological bias of naïve realism. Pers Soc Psychol Bull 40(11):1543–1556
108. Nayakankuppam D, Mishra H (2005) The endowment effect: rose-tinted and dark-tinted glasses. J Consum Res 32:390–395
109. Neale MA, Bazerman MH (1985) The effects of framing and negotiator overconfidence on bargaining behaviors and outcomes. Acad Manage J 28(1):34–49
110. Nickerson RS (1998) Confirmation bias: a ubiquitous phenomenon in many guises. Rev Gen Psychol 2(2):175–220
111. Novemsky N, Kahneman D (2005) The boundaries of loss aversion. J Market Res 42:119–128
112. Pallier G, Wilkinson R, Danthir V, Kleitman S, Knezevic G, Stankov L, Roberts RD (2002) The role of individual differences in the accuracy of confidence judgments. J Gen Psychol 129(3):257–299
113. Peer E, Gamliel E (2013) Heuristics and biases in judicial decisions. Ct Rev 49:114
114. Pronin E, Gilovich T, Ross L (2004) Objectivity in the eye of the beholder: divergent perceptions of bias in self versus others. Psychol rev 111(3):781–799
115. Pronin E, Kugler MB (2007) Valuing thoughts, ignoring behavior: the introspection illusion as a source of the bias blind spot. J Exp Soc Psychol 43:565–578
116. Quinn RE, Spreitzer GM, Brown MV (2000) Changing others through changing ourselves: the transformation of human systems. J Manage Inquiry 9(2):147–164
117. Reb J, Connolly T (2007) Possession, feelings of ownership, and the endowment effect. Judgm Decis Mak 2:107–114
118. Ross L, Greene D, House P (1977) The 'false consensus effect': an egocentric bias in social perception and attribution processes. J Exp Soc Psychol 13(3):279–301
119. Ross L, Stillinger C (1991) Barriers to conflict resolution. Negotiation J 7(4):389–404
120. Ross L, Ward A (1995) Psychological barriers to dispute resolution. Adv Exp Soc Psychol 27:255–304
121. Ross L (1995) Reactive devaluation in negotiation and conflict resolution. In: Arrow K, Mnookin RH, Ross L, Tversky A, Wilson R (eds) Barriers to conflict resolution. W.W. Norton, New York, pp 26–42

122. Ross L, Ward A (1995) Psychological barriers to dispute resolution. In: Advances in experimental social psychology. Academic Press, San Diego, CA, pp 255–304
123. Ross L, Ward A (1996) Naive realism in everyday life: Implications for social conflict and misunderstanding. In; Values and knowledge. Erlbaum, Hillsdale, NJ, pp 103–135
124. Ross L, Lepper M, Ward A (2010) History of social psychology: insights, challenges, and contributions to theory and application. In: Fiske ST, Gilbert DT, Lindzey G (eds) Handbook of social psychology, vol 1. Wiley, New Jersey, pp 3–50
125. Rozin P, Royzman EB (2001) Negativity bias, negativity dominance, and contagion. Pers Soc Psychol Rev 5:296–320
126. Samuelson W, Zeckhauser R (1988) Status quo bias in decision making. J Risk Uncertain 1(1):7–59
127. Sanna LJ, Schwarz N (2004) Integrating temporal biases. Psychol Sci 15(7):474–481
128. Saqib NU, Frohlich N, Bruning E (2010) The influence of involvement on the endowment effect: the moveable value function. J Consum Psychol 20(3):355–368
129. Schermerhorn JR Jr (1989) Management for productivity, 3rd edn. Wiley, New York
130. Schweiger DM, Sandberg WR, Ragan JW (1986) Group approaches for improving strategic decision making: A comparative analysis of dialectical inquiry, devil's advocacy, and consensus. Acad Manage J 29(1):51–71
131. Shih MJ, Stotzer R, & Gutiérrez AS (2013) Perspective-taking and empathy: Generalizing the reduction of group bias towards Asian Americans to general outgroups. Asian Am J Psychol 4(2):79–83
132. Shu SB, Peck J (2011) Psychological ownership and affective reaction: emotional attachment process variables and the endowment effect. J Consumer Psychol 21(4):439–452
133. Siau K, Shen Z (2003) Building customer trust in mobile commerce. Commun ACM 46(4):91–94
134. Smallman HS, Cook MB (2011) Naïve realism: folk fallacies in the design and use of visual displays. Top Cogn Sci 3(3):579–608
135. Smallman HS, John MS (2005) Naïve realism: misplaced faith in realistic displays. Ergon Des 13(3):6–13
136. Strahilevitz MA, Loewenstein G (1998) The effect of ownership history on the valuation of objects. J Consum Res 25(3):276–289
137. Tesser A, Conlee MC (1975) Some effects of time and thought on attitude polarization. J Pers Soc Psychol 31(2):262–270
138. Thaler R (1980) Toward a positive theory of consumer choice. J Econ Behav Organ 1(1):39–60
139. Thompson LL, Lucas BJ (2000) Judgmental biases in conflict resolution and how to overcome them. In: Deutsch M, Coleman P T (eds) The handbook of conflict resolution: theory and practice. Jossey-Bass, San Francisco, CA, pp 255–282
140. Tsai CI, Klayman J, Hastie R (2008) Effects of amount of information on judgment accuracy and confidence. Organ Behav Hum Decis Process 107(2):97–105
141. Tversky A, Kahneman D (1989) Rational choice and the framing of decisions. Multiple criteria decision making and risk analysis using microcomputers. Springer, Berlin, Heidelberg, pp 81–126
142. Tversky A, Kahneman D (1991) Loss aversion in riskless choice: a reference dependent model. Q J Econ 106(4):1039–1061
143. Wallin A, Paradis C, Katsikopoulos KV (2016) Evaluative polarity words in risky choice framing. J Pragma 106:20–38
144. Weiss L, Johar GV (2013) Egocentric categorization and product judgment: seeing your traits in what you own (and their opposite in what you don't). J Consum Res 40(1):185–201
145. Wong PSP, Cheung SO (2004) Trust in construction partnering: views from parties of the partnering dance. Int J Proj Manage 22(6):437–446
146. Wong PS, Cheung SO, Ho PK (2005) Contractor as trust initiator in construction partnering—Prisoner's dilemma perspective. J Constr Eng Manag 131(10):1045–1053
147. Wong WK, Cheung SO, Yiu TW, Pang HY (2008) A framework for trust in construction contracting. Int J Proj Manage 26(8):821–829

148. Wong PP, Teoh K (2015) The influence of destination competitiveness on customer-based brand equity. J Destinat Mark Manag 4(4):206–212
149. Xiong B, Skitmore M, Xia B (2015) A critical review of structural equation modeling applications in construction research. Automat Constr 49:59–70
150. Zillmann D, Bryant J (1985) Affect, mood, and emotion as determinants of selective exposure. In: Selective exposure to communication. Erlbaum, Hillsdale, NJ, pp 157–190

Minimising Biases in Construction Dispute Negotiation

Keyao Li and Sai On Cheung

Introduction

The happenings of biases in construction dispute negotiation have been reported to be real in Chapter "The Happening of Bias in Construction Dispute Negotiation". The sources of biases have been unveiled through conceptualizing biased behaviors (Chapter "Conceptualising Bias in Construction Dispute Negotiation"). The study on biases in CDN is further enhanced with the development of a detection tool (Chapter "A Bias Detection Tool for Construction Dispute Negotiation"). Endowment and Reactive Devaluation as special forms of bias are discussed in Chapter "Special Forms of Bias: Endowment Effect and Reactive Devaluation". This chapter looks into approaches and measures to minimize the happening and effect of biases. As it is not easy for dispute negotiators to admit having biases, this study canvassed the expert knowledge of practicing dispute resolution third-party neutrals instead. In Hong Kong, mediation is the most commonly used alternative dispute resolution for construction disputes. A mediator facilitates the dispute negotiation and hence mediation is also called assisted negotiation. This chapter first presents an overview of construction mediation in Hong Kong then followed by a study of the usefulness of bias minimizing approaches and measures.

K. Li (✉)
Future of Work Institute, Faculty of Business & Law, Curtin University, Perth, Australia
e-mail: Keyao.li@curtin.edu.au

S. O. Cheung
Construction Dispute Resolution Research Unit, City University of Hong Kong, Hong Kong, China
e-mail: Saion.cheung@cityu.edu.hk

S. O. Cheung (ed.), *Construction Dispute Research Expanded*, Springer Tracts in Civil Engineering, https://doi.org/10.1007/978-3-030-80256-1_5

Overview of Mediation Development in Hong Kong Construction Industry

Resolving dispute through arbitration and litigation is costly and lengthy. Mediation is the most widely used alternative dispute resolution (ADR) process to resolve construction dispute in Hong Kong. Essentially, voluntary use of mediation has been specified in construction conditions of contracts in Hong Kong since 1999. Voluntary use of mediation was promoted by the Government of the Hong Kong Special Administrative Region by incorporating it in the Government Standard Forms of Contract for in 1999. Voluntary mediation has reached new heights with the Civil Justice Reform (CJR hereafter) came into effect in 2009. Under the CJR, disputants of civil cases, including construction, are encouraged to attempt mediation before trial.

Specifically for construction cases reaching the Hong Kong High Court Arbitration and Construction List, Practice Direction 6.1 (PD 6.1 hereafter) that was released in April 2009 will apply. Under PD 6.1, where a party wishes to adopt mediation, he or she can do so by serving a Mediation Notice upon the other party and file the copy with the Court. Upon receiving the Mediation Notice, the other party should respond in writing within 14 days stating whether s/he is willing to mediate, whether he or she agrees to the rules identified, whether s/he agrees with the proposed timetable and minimum amount of participation. If s/he refuses to mediate, s/he would have to state the reasons. Even if no party requests mediation at the proceeding, the Court may ask the parties whether they have attempted mediation. Cost sanction is used to deter unreasonable refusal to mediate. Thus, the use of mediation has been boosted indirectly by discouraging "unreasonable refusal to mediate" and "failing to attempt to mediate" [17, 47].

After the Civil Justice Reform came into effect, in 2012 the Hong Kong Mediation Accreditation Association Limited (HKMAAL) was established to provide mediation training. Furthermore, Mediation Ordinance (Laws of Hong Kong CAP 620) was enacted and came into effect in 2013. CAP 620 provides the vital legal backing on mediation practice and most significantly the confidentiality of the mediation communications. It can thus be said that the mediation landscape in Hong Kong has undergone some fundamental changes as a result of the CJR.

Research Design of the Study

To examine bias minimizing strategies in construction dispute negotiations, literature on bias minimizing approaches was first reviewed. Bias minimizing measures were then operationalized in the context of construction dispute negotiation. Third party neutrals who are members of the Hong Kong Society of Construction Law was approached for their assessment on the usefulness of the measures. In view of the non-discrete nature of usefulness evaluation, fuzzy sets based linguistic evaluations were

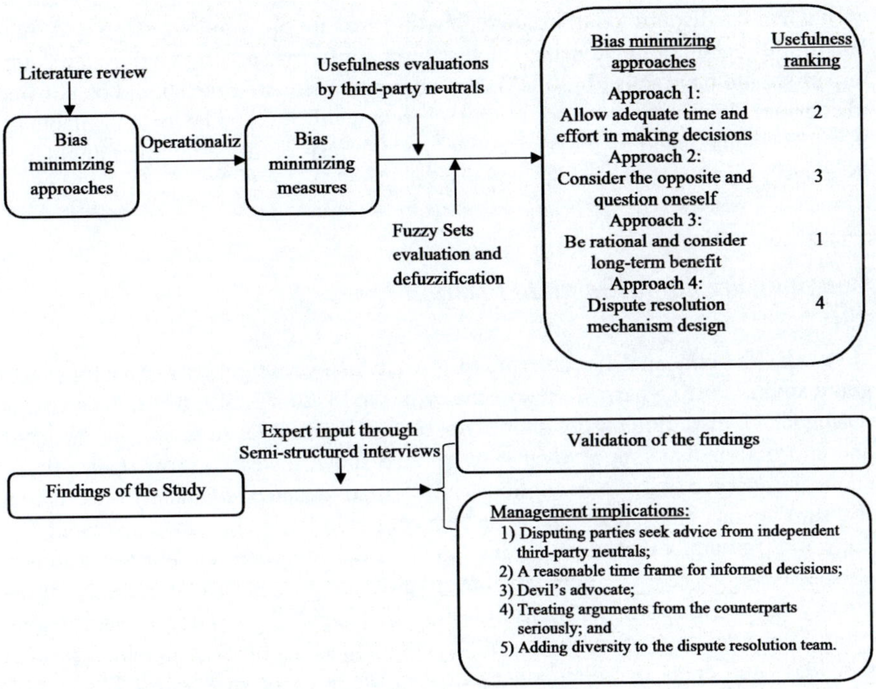

Fig. 1 Research design

used. Rankings of usefulness were calculated from the data collected. Further details in this regard are given in the data collection section. Semi-structured interviews were conducted to validate the findings. Figure 1 gives the research tasks involved and the major findings of the study.

Bias Minimizing Approaches

The following bias minimizing approaches are developed through a review of literature on biases minimizing and avoiding.

Allow Adequate Time and Effort in Making Decisions

Optimal decisions are not commonly made due to time pressure or insufficient analysis [15, 23, 57]. In construction dispute negotiation, similar situation happens when disputing parties opt for a quick and early decision [20]. This not desirable. Instead, even when the disputing parties are facing the pressure of ambitious offers and challenges, duly and comprehensively assessing the information provided by the counterpart should be practiced [20, 27]. For example, adequate time should be allowed to consider alternative scenarios as and when new information become available [2, 9, 31, 33]. Understanding the perspectives and interests of the counterpart might help to identify blind spot one might have missed [25, 27].

Consider the Opposite and Question Oneself

"Consider-the-opposite" has been put forward by many researchers as a useful way to guard against casting narrow view of the evidence [3, 20, 51, 58]. It is suggested that during CDN, disputing parties should not restrict themselves to search for evidence and information that support their view of the dispute. Instead, a more open attitude should be adopted so that other options and alternatives can be evaluated. Information that runs against one's assessment can be useful as more options are considered [27, 43, 63]. Arguments that work against one's initial judgment could offer different perspectives [46, 58]. Undoubtedly, questioning the soundness and currency of the already-formed assessments is a useful approach [23, 35, 41, 42]. By the same token, careful consideration of reasoning of the counterpart's assessments would prompt new ideas and break overconfidence effect of the decision makers [7, 42, 46, 58]. Applying reality testing with the assistance of professional third-party neutrals would also keep the disputing parties' expectations more sensible [1, 44].

Be Rational and Consider Long-Term Benefit

Avoiding negative emotion is useful to minimize the influence of cognitive biases [6, 25, 26, 32, 64]. Trying to understand the standpoints of the counterparts is useful to counter the happening of anchoring and self-serving effects [5, 28, 29, 31]. Extreme behaviors can be suppressed when relationship breakdown is to be avoided [6, 65, 68]. Baron [6] and Thompson and Lucas [70] further suggested that collegial atmosphere is conducive for achieving win–win solutions. To avoid interest-oriented bias, disputing parties should prioritize mutual benefit over one's demand [8, 22, 25, 50]. Being rational and paying attention to the possible failure to settle could alleviate unjustified optimism [3, 34, 62]. To reduce self-serving bias, means should be derived to make everyone's responsibilities apparent especially the accountability for failing to reach a settlement [46, 68, 71]. Lyons [52] and Fisher et al. [25] suggested open communication to cultivate win–win negotiation.

Dispute Resolution Mechanism Design

Multi-tiered dispute resolution (MTDR) process incorporating alternative dispute resolution (ADR) is the prevailing contractual arrangement for construction dispute resolution [48]. Under MTDR arrangement, a dispute may undergo several rounds of resolution unless an early settlement is achieved [50]. Repeated dispute evaluations are conducive for bias like anchoring and self-serving. Thus, bias minimizing measures should be embedded in the MTDR arrangements. Soll et al. [68] echoed this and advocated that de-biasing measures should be an integral part of the resolution process. Nonetheless, this cannot replace addressing the source, de-biasing should be included as part of dispute resolution study and training [3, 20, 24, 46, 56]. Proper training shall equip decision makers with sufficient awareness to avoid contracting any form of bias avoiding [46, 70]. Croskerry et al. [20] further stated that pre-resolution would make decision maker mindful of the detriments that biases can bring about. In this regard, it would be good practice to ask disputing parties to review their interests and prior decisions before commencing a new tier of resolution [4]. Self-reflection would encourage rethinking of alternative options that would preserve their demand as well as satisfying those of the counterpart [25]. At organizational level, introducing new team members or changing the negotiators in charge may also bring new perspectives and ideas [13, 46].

The afore-stated four types of approach to minimize cognitive bias as summarized from relevant literature should be applicable to CDN. For this purpose, the four approaches were operationalized into twenty bias minimizing measures in CDN context. The bias minimizing measures and their respective theoretical bases are listed in Table 1. The effectiveness of these bias-minimizing measures was evaluated by experienced third-party neutrals.

Usefulness of the Bias Mitigating Approaches

The usefulness of the afore-mentioned bias minimizing measures (Table 1) was evaluated by third party neutrals through a data collection form. Each data set includes some basic information such as the role of the respondent, years of experience practicing CDN, the majority of dispute type involved and the most common causes of disputes.

The respondents were practicing third-party neutrals including accredited mediators, arbitrators and adjudicators. They were members of the Society of Construction Law Hong Kong (SCLHK), the Hong Kong International Arbitration Centre (HKIAC), the Hong Kong Mediation Accreditation Association Limited (HKMAAL), the Hong Kong Institute of Arbitrators (HKIAB) and the Hong Kong Institution of Engineers (HKIE). 66 valid responses were finally obtained with 600 distributed. Among the respondents, 76% have more than 15 years in CDN. In fact, nearly 60% of respondents have practiced in construction dispute resolution for more

Table 1 Bias minimizing approaches and measures

	Bias minimizing approaches and measures	References
Approach 1: Allow adequate time and effort in making decisions		
Measures	(1) Disputants should avoid premature closure of thinking by allowing adequate time for decision-making	[15, 20, 23, 57]
	(2) Disputants should carefully re-estimate the case and reservation price of the counterpart when given ambitious offers	[20, 27]
	(3) Disputants should check the accuracy of the evidences provided by the counterpart	[20, 27]
	(4) Disputants should delay forming an assessment until all the available information is considered	[25, 27]
	(5) Disputants should be open to other alternatives even after a first assessment about the dispute has been formed	[2, 9, 31, 33]
Approach 2: Consider the opposite and question oneself		
Measures	(6) Disputants should search for and consider information against the previously formed assessment	[27, 43, 63]
	(7) Disputants should question the soundness of the previously formed assessments routinely	[23, 35, 41, 42]
	(8) Disputants should list reasons why the assessment of their counterpart can hold	[7, 42, 46, 58]
	(9) Disputants should ask for feedbacks and assistance from the third-party neutrals	[1, 44]
Approach 3: Be rational and consider long-term benefit		
Measures	(10) Disputants should consider mutually beneficial trade-offs between the parties	[8, 22, 25, 50, 52],
	(11) Disputants should try not to be emotional	[6, 25, 26, 32, 64]

(continued)

Table 1 (continued)

	Bias minimizing approaches and measures	References
	(12) Disputants should play the role of their counterpart to understand his position	[5, 28, 29, 31]
	(13) Disputants should respectfully listen to their counterpart's grievances	[5, 28, 29, 31]
	(14) Disputants should consider maintaining relationships and future collaboration with their counterpart	[6, 65, 68, 70]
	(15) Disputants should consider the chance of settlement failure	[3, 34, 46, 62]
	(16) Disputants should think about their own responsibilities when settlement fails	[68, 71]
Approach 4: Dispute resolution mechanism design		
Measures	(17) Disputants should receive de-biasing training and education before entering resolution processes	[3, 20, 24, 46, 56, 70]
	(18) New members in the resolution team are required to start a new round of resolution	[13, 46]
	(19) Re-assessment and reconstruction of the decisions are required to start a new round of resolution	[4, 25]
	(20) A process of reviewing initial needs is required in each round of resolution	[4, 25]

than 20 years (Figs. 2 and 3 refer). Dispute types and causes of the dispute that the respondents are involved are presented in Table 2. The disputes are mainly handled in building (superstructure) work and civil engineering work (Table 2) and the mostly happened cause of dispute is incomplete contract.

Data Analyses and Findings

The usefulness evaluation is not suitable for discrete measures because of the subjective nature. To minimize the potential distortion, fuzzy sets based linguistic evaluations were used [67, 73]. Fuzzy linguistic terms "Useless", "Weakly Useful",

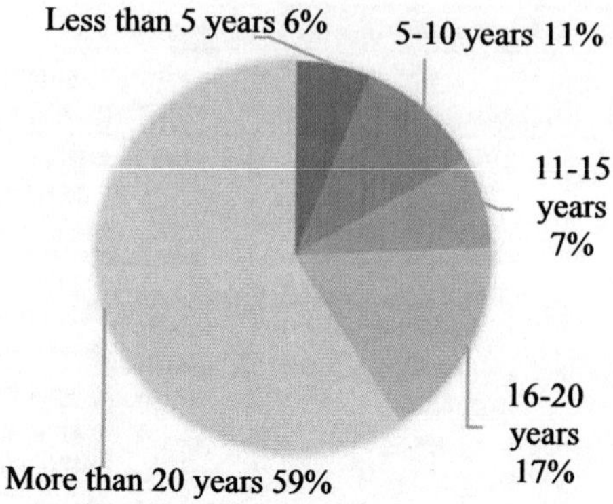

Fig. 2 Years of experience of the respondents

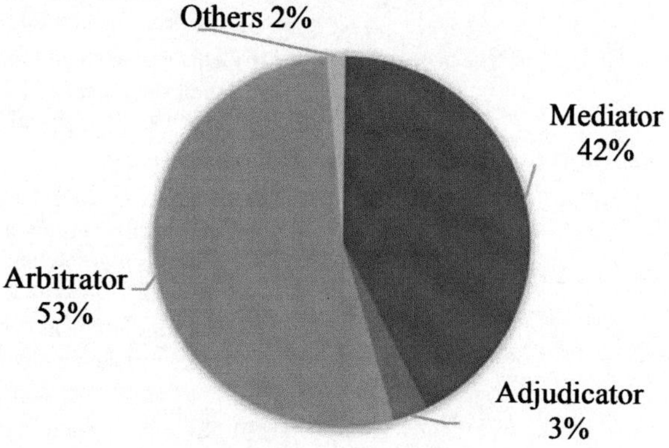

Fig. 3 Roles of the respondents

Table 2 Types and causes of the dispute

Disputes type	Percentage (%)	Cause of the disputes	Percentage (%)
Building services installations	4.5	Risk uncertainty	7.6
Building (foundation) work	7.6	Collaboration among the parties	19.7
Building (superstructure) Work	36.4	Contract incompleteness	42.4
Civil engineering work	39.4	Opportunistic behavior	12.1
Maintenance work	9.1	Affective conflict	1.5
Others	3.0	Others	16.7
Total	100	Total	100

Table 3 Linguistic variables in triangular membership functions

Linguistic variables	Fuzzy number (a, b, c)
Useless	(0.00, 0.10, 0.30)
Weakly useful	(0.00, 0.20, 0.40)
Slightly useful	(0.20, 0.35, 0.50)
Useful	(0.30, 0.50, 0.70)
Moderately useful	(0.50, 0.65, 0.80)
Very useful	(0.60, 0.80, 1.00)
Absolutely useful	(0.70, 0.90, 1.00)

"Slightly Useful", "Useful", "Moderately Useful", "Very Useful" and "Absolutely Useful" were employed to represent the usefulness of the listed bias minimizing measures. Triangular membership functions were chosen for relative ease of mathematical treatment. The pre-defined fuzzy linguistic frequency groups and the triangular membership functions and linguistic variables of usefulness are presented in Table 3 [37, 74]. The graphical representation of the fuzzy numbers is shown in Fig. 4.

For the ith bias minimizing suggestion, aggregation on fuzzy numbers are worked out as per Eqs. 1, 2 and 3 from [19]:

$$\tilde{A}_{\mathrm{i}} = \left(\frac{1}{p}\right) \otimes (\tilde{a}_1 \oplus \tilde{a}_2 \oplus \tilde{a}_3 \oplus \cdots \oplus \tilde{a}_{\mathrm{p}}) \tag{1}$$

$$\tilde{B}_{\mathrm{i}} = \left(\frac{1}{p}\right) \otimes (\tilde{b}_1 \oplus \tilde{b}_2 \oplus \tilde{b}_3 \oplus \cdots \oplus \tilde{b}_{\mathrm{p}}) \tag{2}$$

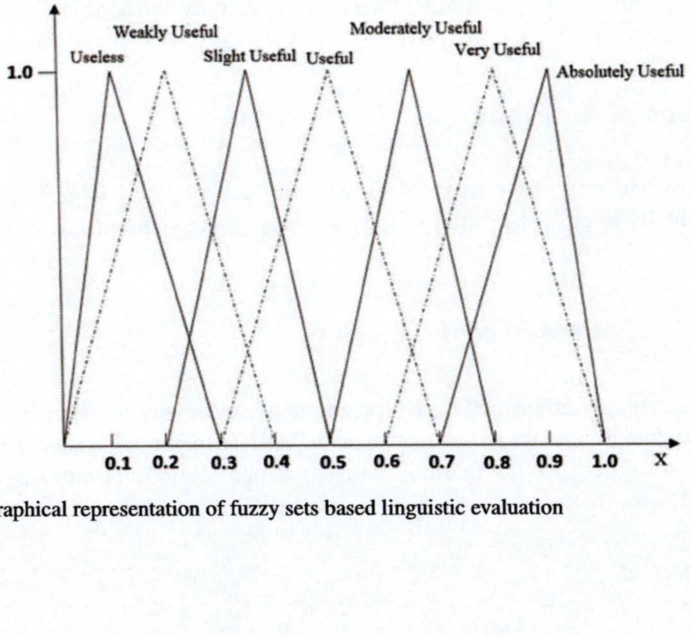

Fig. 4 Graphical representation of fuzzy sets based linguistic evaluation

$$\tilde{C}_i = \left(\frac{1}{p}\right) \otimes (\tilde{c}_1 \oplus \tilde{c}_2 \oplus \tilde{c}_3 \oplus \cdots \oplus \tilde{c}_p) \quad (3)$$

where $\tilde{a}$ is first fuzzy parameter of selected linguistic variable; $\tilde{b}$ is second fuzzy parameter of selected linguistic variable; $\tilde{c}$ is third fuzzy parameter of selected linguistic variable; and p = number of respondents.

Defuzzification was then conducted to interpret the membership into a non-fuzzy value to express the "expected" value of the fuzzy number [66]. The "expected value" makes it easier to rank and compare the fuzzy numbers. Triangular fuzzy number was defuzzified using Eq. 4 [67]:

$$e_i = (\tilde{A}_i + 2\tilde{B}_i + \tilde{C}_i)/4 \quad (4)$$

where $\tilde{A}_i$, $\tilde{B}_i$ and $\tilde{C}_i$ are aggregated fuzzy parameters.

The aggregated fuzzy parameters and defuzzified values of the twenty bias minimizing measures are represented in Table 4. For comparison reason, all the defuzzified values were kept to four decimal places. It can be seen in Table 4 that, except item 18, all the listed measures have defuzzified values larger than 0.5. In fact, measures 1–16 have defuzzified values larger than 0.6. Therefore, the bias minimizing measures collected from literature were evaluated to be useful by the third-party neutrals. Among the four types of bias minimizing approach, being rational and consider long-term benefit ranks the highest with a defuzzified value of 0.7075 (Table 4). Hence, third-party neutral respondents believe that being rational and consider long-term benefit is an effective way for disputing parties to minimize the influence of bias. Allowing adequate time and effort in making decisions was ranked the 2nd most useful with a defuzzified value of 0.6868. Consider the opposite and question oneself was the 3rd useful with a defuzzified value of 0.6756 and dispute resolution mechanism ranks the lowest among the four and the defuzzified value is 0.5474.

Validation of the Findings

Semi-structured interviews with ten expert third-party neutral were conducted to validate the findings. The profiles of the interviewees are summarized in Table 5.

On the Practice of Biased Behaviors

All the interviewees confirmed the happening of biases in the cases they handled and believe that bias minimizing measures listed in Table 4 can be useful. First, controlling emotions would help in curbing irrational assessments. "Emotion controlling is tricky as the burst of emotion can come spontaneously", said by one interviewee. The

Table 4 Usefulness of bias minimizing approaches and measures

Bias minimizing approaches and measures	Aggregated fuzzy parameters	Defuzzified value	Ranking of approach
Approach 1: Allow adequate time and effort in making decisions	(0.50, 0.69, 0.86)	0.6868	2
Disputants should avoid premature closure of thinking by allowing adequate time for decision-making	(0.50, 0.69, 0.86)	0.6830**	
Disputants should carefully re-estimate the case and reservation price of the counterpart when given ambitious offers	(0.45, 0.64, 0.81)	0.6352**	
Disputants should check the accuracy of the evidence provided by the counterpart	(0.55, 0.75, 0.91)	0.7386**	
Disputants should delay forming an assessment until all the available information is considered	(0.47, 0.66, 0.83)	0.6534**	
Disputants should be open to other alternatives even after a first assessment about the dispute has been formed	(0.54, 0.73, 0.90)	0.7239**	
Approach 2: Consider the opposite and question oneself	(0.49, 0.68, 0.85)	0.6756	3
Disputants should search for and consider information against the early formed assessment	(0.48, 0.67, 0.85)	0.6682**	
Disputants should question the soundness of the previously formed assessments routinely	(0.49, 0.68, 0.85)	0.6784**	
Disputants should list reasons why the assessment of their counterpart can hold	(0.53, 0.72, 0.89)	0.7136**	
Disputants should ask for feedbacks and assistance from the third-party neutrals	(0.46, 0.65, 0.82)	0.6420**	

(continued)

Table 4 (continued)

Bias minimizing approaches and measures	Aggregated fuzzy parameters	Defuzzified value	Ranking of approach
Approach 3: Be rational and consider long-term benefit	(0.52, 0.71, 0.88)	0.7075	1
Disputants should consider mutually beneficial trade-offs between the parties	(0.55, 0.74, 0.91)	0.7318**	
Disputants should try not to be emotional	(0.56, 0.76, 0.91)	0.7455**	
Disputants should play the role of their counterpart to understand his position	(0.53, 0.72, 0.89)	0.7170**	
Disputants should respectfully listen to their counterpart's grievances	(0.52, 0.71, 0.88)	0.7091**	
Disputants should consider long-term relationships and future collaboration with their counterpart	(0.50, 0.68, 0.85)	0.6773**	
Disputants should consider the chance of settlement failure	(0.51, 0.70, 0.88)	0.7000**	
Disputants should think about their own responsibilities when settlement fails	(0.49, 0.67, 0.85)	0.6716**	
Approach 4: Dispute resolution mechanism design	(0.37, 0.55, 0.73)	0.5475	4
Disputants should receive debiasing training and education before entering resolution processes	(0.38, 0.56, 0.73)	0.5580*	
New members in the resolution team are required to start a new round of resolution	(0.29, 0.46, 0.64)	0.4648	
Re-assessment and reconstruction of the decisions are required to start a new round of resolution	(0.39, 0.58, 0.76)	0.5761*	

(continued)

Table 4 (continued)

Bias minimizing approaches and measures	Aggregated fuzzy parameters	Defuzzified value	Ranking of approach
A process of reviewing initial needs is required in each round of resolution	(0.41, 0.59, 0.77)	0.5909*	

* Defuzzified value above 0.5. **Defuzzified value above 0.6

Table 5 Profiles of the interviewees

Person	Years of experience	Primary role as third-party neutral	The majority of disputes involved	Countries of practice
No. 1	More than 30 years	Mediator	Civil engineering works	UK, HK
No. 2	More than 30 years	Arbitrator	Building works	HK
No. 3	More than 20 years	Expert witness and representative	Civil engineering works	HK
No. 4	More than 40 years	Mediator and arbitrator	Building works	HK, Mainland China
No. 5	More than 60 years	Arbitrator and mediator and adjudicator and consulting engineer	Building works and civil engineering works	UK
No. 6	More than 20 years	Mediator and arbitrator	Building works	UK, US, HK
No. 7	More than 45 years	Mediator and adjudicator and expert witness and dispute resolution advisor	Building works	UK, HK
No. 8	More than 20 years	Arbitrator and in-house expert	Civil engineering works	HK, Macau
No. 9	More than 20 years	Mediator	Civil engineering works	HK
No. 10	More than 30 years	Arbitrator	Civil engineering works	HK, UK

interviewees also thought that disputing parties having emotion would miss important and critical details and their decisions were therefore suboptimal. Reality checks were suggested by the interviewees as useful tactics that they often use to help disputing parties to stay away from unnecessary emotional reactions. "When the disputing parties were reminded about long-term benefits and company reputation, they could better control their emotions and behave more appropriately", raised by a mediator.

Besides, allowing enough time for disputing parties to review the dispute would improve the chance of coming up with more considerate decisions. One interviewee who has practiced both mediation and arbitration for more than 40 years said: "A cooling-off period is instrumental for heated disputing parties to remain in rational course and consider carefully the information from both sides". It is not standard practice for dispute negotiators to estimate both the accuracy of information provided by their own team and those presented by the counterpart. Objective assessments would only be obtained with comprehensive information and a holistic view of the dispute from both sides' perspective. The interviewees also agreed that dispute negotiators should be critical towards their reasoning to keep themselves distance from preconception bias. From a broader perspective, dispute resolution training and process design can play a part to guard against cognitive bias although this is not often taken on broad. An arbitrator commented that: "Education and training would be helpful if they could keep the disputing parties mindful of the biases". He further suggested that: "Having arrangements to remind disputing parties of their real needs would be helpful".

Suggestions from the Interviewees

Humans are reluctant to admit that their decisions can be influenced by biases or in fact are biased. Construction dispute negotiators are human. Nonetheless, biased behaviors had been observed by third-party neutrals who also concurred that biases were detrimental to rational construction dispute resolution and made settlement more difficult. Minimizing the happening of and effect of biases through appropriate means should be integral part of construction dispute management.

First, the interviewees emphasized they could offer advice or assistance from independent third-party neutrals. The respondents of this study are quite adamant that they can offer professional advice in helping disputing parties to evaluate their cases objectively, review their assessments to determine realistic expectations. Second, the interviewees pointed out that reasonable time frame should be allowed for negotiators to make informed decisions. While, timely but not hastily decisions are welcome, adequate time for reasonable research on both their own and the counterparts' requests should be allowed. Disputing parties should review the options available with an open mind whenever new information becomes available. Moreover, this does not mean allowing protracted decisions. Where appropriate, momentum can be maintained through the expert input of third-party neutrals. Procrastination is sign of failure. Third, introducing a devil's advocate in dispute evaluation can improve objectivity of the assessments. It should also be coupled with mandatory review of different perspectives. Ideally, the devil's advocate should be someone who has not been involved in the dispute. He should be allowed complete freedom to raise different views on the line taken, challenge the validity of the evidence as well as the logic of the arguments. Fourth, arguments given by the counterparts should be treated seriously. It may be worthwhile to require certain time must be devoted to considering arguments presented by the counterparts. Fifth, adding diversity to the dispute resolution team can also be affected through having team members having different professional background and experience. For dispute involving international disputing parties, nationality and cultural background should also be taken into account. Sixth, senior management should be involved as early as possible. As they are not those directly involved in the dispute, the chance of being emotional and perceptive would be less. It is believed that, with the collaboration of third-party neutrals and disputing parties, the potency of cognitive biases would be minimized, and the efficiency of construction dispute management would be enhanced.

Follow Up on the Suggestion of the Third-Party Neutrals

In Hong Kong, mediation is by far the most widely used alternative dispute resolution (ADR) method to resolve construction disputes [18, 12, 30, 60, 69]. Mediators are neutral third parties who facilitate settlement of dispute [47, 55]. Mediators seek to reconcile the concerns of the disputing parties so that common grounds can be identified [25, 39, 59]. With the help of professional mediators, the chance of reaching successful settlement would be improved. As dispute facilitator, a mediator would also be instrumental in helping disputing parties to stay away from biases. This part of the study aims to incorporate de-biasing arrangements in the mediators' toolkit. A review of the development history of mediation in Hong Kong is first outlined. The mediation rules commonly used in Hong Kong are examined to understand whether sufficient provisions have been included or can be used to guard against biases. As reality testing is the most instrumental tactics that mediators use to facilitate dispute negotiation, the potential of using reality testing to minimize biases is examined in detail.

Insights from Mediation Rules

In Hong Kong, most construction related learned societies are offering dispute resolution services with respective rules and regulations. These rules aim to ensure fair procedures that conform to legal requirements. Two mediation rules for construction disputes that are commonly used in Hong Kong are: the Hong Kong Special Administrative Region Construction Mediation Rules (1999 Edition) (the HKSARCM Rules 1999 hereafter) and the Hong Kong Construction Arbitration Centre Construction Mediation Rules 2015 (the HKCACCM Rules 2015 hereafter). The next section outlines the de-biasing arrangements, if any.

The two mediation rules were reviewed article by article. Both mediation rules have detailed the procedures on initiating and terminating mediation, selection of mediators, confidentiality as well as costs of mediations. The rules also stipulate that mediators should not have pecuniary or proprietary interest in relation to the mediation. For the disputing parties, both of the two mediation rules encouraged good faith and co-operation. Both rules also restrict the same mediator to be appointed as arbitrator or representative or counsel in any of the subsequent dispute resolution processes. It is believed that if the same mediator is appointed in the subsequent proceedings, his views of the dispute formed during the mediation may influence his assessment in subsequent proceedings. This provision can serve to guard against preconception bias of the mediator. The details of the bias avoidance arrangements in these two rules are summarized in Table 6. These rules basically aim to address procedural injustice and opportunism. Moreover, during mediation it is possible that disputing parties may be influenced by biases that would lead to irrational decisions and suboptimal assessments. Unfortunately, measures against disputing parties' biases cannot be traced in the mediation rules. Formalized bias avoidance arrangements are therefore insufficient.

Embracing De-biasing in Reality Testing

In the absence of formalized bias avoidance arrangements incorporated in the mediation rules, attention is now turned to the toolkits of the mediators. As suggested by one of the interviewees, the tactic of realty testing can be instrumental in addressing biases. Reality testing is a tactic of mediator, which stimulates the disputing parties to review and re-estimate their decision-making process so as to improve the quality of their decisions [47, 53, 54, 61, 72]. Reality testing involves tactfully asking the disputing parties questions about their opinions to the issues in dispute, their expectations of the resolution, their assessments of the case, their attitude and feelings, etc. [16, 21, 36, 40, 45, 47]. To properly ask these questions, disputing parties need to review their decisions including the assumptions and the process. This accords a renewed opportunity for the disputing parties to unveil any mistakes, misinterpretation and of course effects of biases on their decisions. In view of the versatility of the tactic of reality testing, this part of the study focuses on incorporating de-biasing elements into the tactic of reality testing.

Using Reality Testing to De-bias Disputing Parties' Decision-Making Strategies

It is useful to remind the disputing parties about the potential sources of bias. Disputants have the tendency to rely too much on their first formed assessment. They often found their first assessment satisfying and would not take further effort to review the assumptions and logics again [10, 14, 48]. When a mediator triggers reality testing, the disputing parties are more likely to revisit their assessments. Allowing reasonable time for the disputing parties to carry out the necessary research could avoid mistakes due to rushing decisions and reduce the preconception bias [9, 25, 27]. When challenged by reality testing, a sensible disputing party would seek more

Table 6 Potential bias minimizing provisions in the HKSARCM Rules 1999 and the HKCACCM Rules 2015

Mediation rules	Bias minimizing arrangements			
	Prerequisites of the mediator	Role of the mediator	Role of the parties	Mediator's role in subsequent proceedings
HKSARCM Rules 1999 (Pre-Civil Justice Reform)	• The mediator shall not have any financial or personal interest in the outcome of the mediation • The mediator shall disclose any circumstances likely to create a presumption of bias or prevent a prompt resolution of the dispute	• The mediator may conduct the mediation in such manner as he considers appropriate, taking into account the circumstances of the case, the wishes of the parties and the need for a speedy settlement of the dispute. The mediator may communicate with the parties together or with each party separately	• Each party shall co-operate in good faith with the mediator. Either party may request a private meeting with the mediator at any time	• The parties undertake that the mediator shall not be appointed as arbitrator or representative or counsel of either party in any subsequent arbitration or judicial proceedings whether arising out of the mediation or any other dispute in connection with the same contract

(continued)

Table 6 (continued)

Mediation rules	Bias minimizing arrangements			
	Prerequisites of the mediator	Role of the mediator	Role of the parties	Mediator's role in subsequent proceedings
HKCACCM Rules 2015 (Post- Civil Justice Reform)	• The mediator shall not have any financial or personal interest in the result of the mediation • The mediator shall disclose any circumstances likely to create a presumption of bias or prevent a prompt and properly procured settlement of the dispute	• The mediator may conduct the mediation in such manner as he considers appropriate, taking into account the special and technical circumstances of the case, the wishes of the parties, and the need for a speedy settlement of the dispute without unnecessary cost • The mediator shall conduct the Mediation appropriate to the nature of the dispute and the needs of the parties	• Each party shall co-operate with the Mediator and act in good faith to resolve the dispute • A party may request a private meeting with the Mediator at any time. The parties shall give full assistance to enable the Mediation to proceed and be concluded within the time stipulated	• No party shall appoint the Mediator as an adjudicator, representative, counsel or expert witness of any party in any subsequent adjudication or judicial proceedings arising out of the Mediation

information to study his prior assessment. When being asked about their reactions of receiving ambitious offer from the counterpart, disputing parties would be reminded and suggested to re-estimate the case and the reservation price of the counterpart, which are the measures that could reduce the influence of preconception resulted from the ambitious offer from the counterpart [20, 27, 49]. Moreover, when being asked about the supporting information of the counterpart, disputing parties have to evaluate the case from the perspective of their counterpart. Paying attention to the evidence and assessments from the opposite side becomes a good way to identify one's own blind spots that would in turn alleviate the impact of self-affirmation bias [3, 20, 51, 58].

Using Reality Testing to De-bias Disputing Parties' Attitudes

Emotion control is one of the most effective means to eliminate biases [25, 26, 32, 38]. Reality testing could help the disputing parties to recognize their behavioral shortcomings [11, 16, 47, 61]. Adjusting the demands and expectations through logrolling would cut off the sources of optimism and interested-oriented biases [8, 22, 65, 68]. When being asked to stand in the roles of their counterpart, the disputing parties would develop empathy and understand better the situation of their counterpart. The important thing is to listen carefully the counterpart's deliberation. Deeper mutual understanding curbs selfish interest-maximizing [28, 29, 31]. When being asked about the manners to be adopted if collaboration is desired, one could realize insisting on self-interest is not conducive to resolve the dispute. Furthermore, warm reminder of the consequences of failing to reach a settlement in tangible terms like legal costs and unproductive use of scare resources would help overcoming overconfidence and optimistic biases [62, 65].

Using Reality Testing to De-bias the Mechanism of Mediation

Reality testing could be employed to optimize mediation mechanism. By reality testing, mediator would ask the disputing parties to consider the advantages of including new members with fresh ideas in the mediation process. The mediator would ask whether the disputing parties would opt for a new caucus to consider re-assessment and reconstruction of their positions. By asking the advantages of these processes, mediator is actually suggesting these ways to optimize the mediation mechanism. By including these helpful processes into mediation, disputing parties' biased assessments could receive reasonable questioning, if met with appropriate responses, self-defending would be reduced [4, 13, 25, 46]. Besides, by asking about the initial needs of the disputing parties, disputing parties would have their initial needs reviewed. With the help of mediator, disputing parties would separate their currently held position with their initial needs and develop alternative feasibility [47,

53]. By conducing reality testing about the mediation process, experienced mediator would provide their suggestions on optimizing the mechanism, through which preconception bias and interest-oriented bias would be reduced.

Reality testing has no specific question styles. The proposed reality testing questions are suggestions and can be modified to suit the contexts. Mediators can use these questions to help disputing parties self-realize and correct their biased behaviors. The reality testing questions developed in this study and their respective theoretical bases are listed in Table 7.

Views from the Experts

The validity and practicality of the reality testing questions proposed were verified in this section of the study. The questions were arranged in a questionnaire survey and were distributed to experienced third-party neutrals who had participated in the validation of the bias minimizing measures. These experts have at least twenty years' experience in construction dispute resolution. They were asked to rate on the usefulness of these reality testing questions based on a 7-point Likert scale from "1 = Not at all" to "7 = Absolutely useful". The experts' ratings on the usefulness score and usefulness ranking of these reality testing questions are shown in Table 7. The profile of the experts is shown in Table 8.

It can be seen in Table 7 that all these reality testing questions have usefulness scores larger that mid-point (3.5) of the usefulness scale. Almost all these reality testing questions have usefulness score larger than 4.0, except Question 4 "Are the current set of assumptions exhaustive?" with the usefulness score of 3.9. Therefore, the experienced construction dispute resolution experts validated the usefulness of these questions. Among these questions, Question 6, Question 17 and Question 5 were ranked the top three most useful. Hence, reminding the disputing parties to assess the case from the opposite side and exhaust alternative positions to achieve initial needs have been pinpointed as the most instrumental.

Generally speaking, strategy-related questions were rated highly. The respondents considered that through reminding the disputants to re-think and re-examine their cases from an opposite perspective could address the primary sources of bias. Attitude-related questions were ranked the second most useful. Keeping the disputing parties focus on the mutual beneficial trade-offs and future collaboration were useful to overcome overconfidence effect. Controlling emotion is also important but has to be exercised tactfully and skillfully simply because very few people are willing to accept that they are having emotion. Process-related questions seek to remind disputing parties to search for alternative positions to achieve their initial needs. These are rated as the third most useful. Mediators suggesting a review of the case is considered a good practice whenever appropriate and particularly at each round of negotiation. Having new members should not be neglected in particular when new information becomes available.

Table 7 Reality testing questions

References	Reality testing questions	Purpose of the questions	Usefulness score	Usefulness ranking
Questions about decision making strategies			*5.0*	*(1)*
[9, 25, 27]	Question 1: "Do you think you need more time to come up with an assessment?"	A	4.8	8
[20, 27, 49]	Question 2: "Have you re-examined your case upon receipt of an offer from your counterpart?"	A	5.3	6
[9, 25, 27]	Question 3: "Have you considered all the available information in deriving your current assessment?"	A, B	4.6	9
[9, 25, 27]	Question 4: "Are the current set of assumptions exhaustive?"	A, B	3.9	17
[3, 20, 51, 58]	Question 5: "Have you considered information that runs against your assumptions of the case?"	B	5.5	3
[3, 20, 51, 58]	Question 6: "Are there any facts that support your counterpart?"	B	5.9	1
Questions about attitude during mediation			*4.7*	*(2)*
[8, 22, 25, 65]	Question 7: "Are there any mutual beneficial trade-offs between you and the counterpart?"	C	5.5	4
[26, 32, 38]	Question 8: "Do you think emotion has played a part in your decision?"	C	4.4	13

(continued)

Table 7 (continued)

References	Reality testing questions	Purpose of the questions	Usefulness score	Usefulness ranking
[28, 29, 31]	Question 9: "Will you do the same if you are your counterpart?"	C	4.6	10
[28, 29, 31]	Question 10: "Let me know your understanding of the grievances of your counterpart?"	C	4.6	11
[62, 65, 68]	Question 11: "What would be the impact on future collaboration with the counterpart should the dispute is not settled?"	C	5.4	5
[46, 68, 71]	Question 12: "Do you think you are partly responsible should settlement is not achieved?"	C	4.3	14
Questions about mediation process			*4.6*	*(3)*
[13, 46]	Question 13: "What are the benefits of including new members for both of the parties at a new communication session?"	D	4.0	15
[13, 46]	Question 14: "Will you consider inviting new members to join your team for the next round of negotiation?"	D	4.0	16
[4, 25, 53]	Question 15: "How about taking a fresh look of the case and your decision before the next round of negotiation?"	A, C, D	4.9	7
[25, 47, 53]	Question 16: "Let me know if your initial needs remain unchanged?"	C, D	4.5	12

(continued)

Table 7 (continued)

References	Reality testing questions	Purpose of the questions	Usefulness score	Usefulness ranking
[25, 47, 53]	Question 17: "Are there other options that would also achieve your needs?"	A, B, C, D	5.6	2

Note (A): Remind the disputing parties to allow adequate time in decision making; (B): Remind the disputing parties to consider the opposite; (C): Remind the disputing parties to be rational; (D): Remind the disputing parties to optimize the mediation mechanism

Table 8 Profile of the experts

Person	Years of experience	The majority of disputes involved	Countries of practice
No. 1	More than 30 years	Civil engineering works	UK, HK
No. 2	More than 30 years	Building works	HK
No. 3	More than 20 years	Civil engineering works	HK
No. 4	More than 40 years	Building works	HK, Mainland China
No. 5	More than 20 years	Building works	UK, US, HK
No. 6	More than 45 years	Building works	UK, HK
No. 7	More than 20 years	Civil engineering works	HK, Macau
No. 8	More than 30 years	Civil engineering Works	HK, UK

Bias minimizing arrangements in mediation rules are not explicit or not planned for. Moreover, mediator can play a pivotal role in keeping the disputing parties away from biases and behave in a rational manner. The findings of this study suggest making de-biasing function as one of the basic skills of mediators. Seventeen reality testing questions that have the effect of de-biasing are proposed. These questions can be incorporated in mediation training. With proper reality testing, mediator could guide the disputing parties to identify the fallacies in their decisions, some may have caused by biases.

When employing these reality testing questions, timing is critical. Thus, mediators have to pick the most appropriate occasions and deliver natural communication. For example, Question 6, Question 8 and Question 12, etc. are more suitable to be asked during caucuses to save faces for the biased affect party. Some questions, such as Question 7, are more flexible and can be used in both caucuses and joint meetings. Besides, mediator should pay attention to his/her manner in asking these reality testing questions. Reality testing can only be effective when the disputing parties feel safe and being respected. Well prepared and skillful mediators would smooth the mediation process and lead to efficient communications between the disputing parties.

Summary

People are reluctant to admit being affected by biases. This study captures the invaluable expert opinions of third-party neutrals on the usefulness of bias minimizing measures. Identified from literature, four bias minimizing approaches were considered: (i) Allow adequate time and effort in making decisions; (ii) consider the opposite and question oneself; (iii) be rational and consider long-term benefit; and (iv) dispute resolution mechanism design. The third-party neutrals involved in the study agreed that these are useful ways to minimize biases. These approaches are further operationalized for case of interpretation and implementation. Mediation is the most commonly used alternative dispute resolution mechanism used in Hong Kong, two mediation rules were studied to understand if these rules include bias minimizing provisions. It was found that de-biasing is not featured. Thus, stronger reliance will be placed on the mediators' advice as far as de-biasing is concerned.

Acknowledgements The empirical work of this chapter has been reported in "Unveiling Cognitive Biases in Construction Project Dispute Resolution through the lenses of third-party neutrals" of the ASCE Journal of Construction Engineering and Management and "Embracing Debiasing in Mediator's Tactic of Reality Testing" of the ASCE Journal of Legal Affairs and Dispute Resolution in Engineering and Construction. The texts have been substantially re-written. The support of the HKSAR RGC General Research Fund Project (no. 11209118) is duly acknowledged. Special thanks go to the members of the Society of Construction Law Hong Kong (SCLHK) who had kindly provided data for the study.

References

1. Alexander FG (1980) Psychoanalytic therapy: principles and application. University of Nebraska Press
2. Anderson CA (1982) Inoculation and counterexplanation: debiasing techniques in the perseverance of social theories. Soc Cogn 1(2):126–139
3. Arkes HR (1991) Costs and benefits of judgment errors: implications for debiasing. Psychol Bull 110(3):486
4. Ashton RH, Kennedy J (2002) Eliminating recency with self-review: the case of auditors' 'going concern' judgments. J Behav Decis Mak 15(3):221–231
5. Babcock L, Loewenstein G (1997) Explaining bargaining impasse: the role of self-serving biases. J Econ Perspect 11(1):109–126
6. Baron RA (1991) Positive effects of conflict: a cognitive perspective. Emp Respons Rts J 4(1):25–36
7. Bazerman MH, Neale MA (1982) Improving negotiation effectiveness under final offer arbitration: the role of selection and training. J Appl Psychol 67(5):543
8. Bazerman MH, Curhan JR, Moore DA, Valley KL (2000) Negotiation. Annu Rev Psychol 51(1):279–314
9. Bentz BG, Williamson DA, Franks SF (2004) Debiasing of pessimistic judgments associated with anxiety. J Psychopathol Behav Assess 26(3):173–180
10. Bergman O, Ellingsen T, Johannesson M, Svensson C (2010) Anchoring and cognitive ability. Econ Lett 107(1):66–68

11. Brandon M, Robertson L (2007) Conflict and dispute resolution: a guide for practice. Oxford University Press, South Melbourne, VIC
12. Brooker P, Wilkinson S (eds) (2010) Mediation in the construction industry: an international review. Routledge
13. Burke A (2007) Neutralizing cognitive bias: an invitation to prosecutors. NYUJ LL 2:512–530
14. Chapman GB, Johnson EJ (1999) Anchoring, activation, and the construction of values. Organ Behav Hum Decis Process 79(2):115–153
15. Chapman GB, Johnson EJ (2002) Incorporating the irrelevant: anchors in judgments of belief and value. In: Gilovich T, Griffin D, Kahneman D (eds) Heuristics and biases: the psychology of intuitive judgment. Cambridge University Press, Cambridge, pp 120–138
16. Charlton R (2000) Dispute resolution guidebook. LBC Information Services, Sydney
17. Cheung SO (2010) Construction mediation landscape in the civil justice system in Hong Kong. J Legal Aff Dispute Resolut Eng Constr 2(3):169–174
18. Cheung SO (2014) The effective use of ADR processes in construction. Construction Dispute Research. Springer, Cham, pp 299–317
19. Cheung SO, Pang KHY (2013) Anatomy of construction disputes. J Constr Eng Manag. https://doi.org/10.1061/(ASCE)CO.1943-7862.0000532
20. Croskerry P, Singhal G, Mamede S (2013) Cognitive debiasing 2: impediments to and strategies for change. BMJ Qual Saf, bmjqs-2012-001713
21. Dick A, Ballantine T (2001) The art of family law: skills for successful practice. W. Green/Sweet & Maxwell, Edinburgh
22. Drolet AL, Morris MW (2000) Rapport in conflict resolution: accounting for how face-to-face contact fosters mutual cooperation in mixed-motive conflicts. J Exp Soc Psychol 36(1):26–50
23. Epley N, Gilovich T (2006) The anchoring-and-adjustment heuristic why the adjustments are insufficient. Psychol Sci 17(4):311–318
24. Fischhoff B, Beyth-Marom R (1983) Hypothesis evaluation from a Bayesian perspective. Psychol Rev 90(3):239–260
25. Fisher R, Ury WL, Patton B (2011) Getting to yes: negotiating agreement without giving in. Penguin
26. Forgas JP (1995) Mood and judgment: the affect infusion model (AIM). Psychol Bull 117(1):39
27. Galinsky AD, Mussweiler T (2001) First offers as anchors: the role of perspective-taking and negotiator focus. J Pers Soc Psychol 81(4):657
28. Galinsky AD, Ku G (2004) The effects of perspective-taking on prejudice: the moderating role of self-evaluation. Pers Soc Psychol Bull 30(5):594–604
29. Galinsky AD, Moskowitz GB (2000) Perspective-taking: decreasing stereotype expression, stereotype accessibility, and in-group favoritism. J Pers Soc Psychol 78(4):708–724
30. Goldberg SB, Sander FE, Rogers NH, Cole SR (2014) Dispute resolution: negotiation, mediation and other processes. Wolters Kluwer Law & Business
31. Hammond JS, Keeney RL, Raiffa H (1998) The hidden traps in decision making. HBR 76(5):47–58
32. Hastie R (2001) Problems for judgment and decision making. Annu Rev Psychol 52(1):653–683
33. Heiman VB (1990) Auditors' assessments of the likelihood of error explanations in analytical review. Account Rev 875–890
34. Hirt ER, Markman KD (1995) Multiple explanation: a consider-an-alternative strategy for debiasing judgments. J Pers Soc Psychol 69(6):1069–1086
35. Hoch SJ (1985) Counterfactual reasoning and accuracy in predicting personal events. J Exp Psychol Learn Mem Cogn 11(4):719
36. Holaday LC (2002) Stage development theory: a natural framework for understanding the mediation process. Negotiat J 18(3):191–210
37. Idrus A, Nuruddin MF, Rohman MA (2011) Development of project cost contingency estimation model using risk analysis and fuzzy expert system. Expert Syst Appl 38(3):1501–1508
38. Jones TS, Bodtker A (2001) Mediating with heart in mind: addressing emotion in mediation practice. Negotiation J 17(3):217–244

39. Karim A, Pegnetter R (1983) Mediator strategies and qualities and mediation effectiveness. Ind Relat 22(1):105–114
40. Kelly JB (1983) Mediation and psychotherapy: distinguishing the differences. Mediat Q 1:33
41. Kennedy J (1995) Debiasing the curse of knowledge in audit judgment. Account Rev 70(2):249–273
42. Koriat A, Lichtenstein S, Fischhoff B (1980) Reasons for confidence. J Exp Psychol Learn Mem Cogn 6(2):107–118
43. Kray LJ, Galinsky AD (2003) The debiasing effect of counterfactual mind-sets: increasing the search for disconfirmatory information in group decisions. Organ Behav Hum Decis Process 91(1):69–81
44. Landa Y, Silverstein SM, Schwartz F, Savitz A (2006) Group cognitive behavioral therapy for delusions: helping patients improve reality testing. J Contemp Psychother 36(1):9–17
45. Lande J (2002) Using dispute system design methods to promote good-faith participation in court-connected mediation programs. UCLA L Rev 50(1):69–141
46. Larrick RP (2004) Debiasing. In: Koehler DJ, Harvey N (eds) Blackwell handbook of judgment and decision making, Blackwell Publishing, pp 316–338
47. Leung HM (2014) Hong Kong mediation handbook, 2nd edn. Sweet and Maxwell, Hong Kong
48. Li K, Cheung SO (2016) The potential of bias in multi-tier construction dispute resolution processes. In: Chan PW, Neilson Edinburgh (eds) Proceedings of the 32nd annual ARCOM conference, vol 1, pp 197–205
49. Li K, Cheung SO (2019) Unveiling biases in construction project dispute resolution through the lenses of third-party neutrals. J Constr Eng Manag (in press)
50. Li K, Cheung SO (2018) Bias measurement scale for repeated dispute evaluations. J Manag Eng. https://doi.org/10.1061/(ASCE)ME.1943-5479.0000617
51. Lord CG, Lepper MR, Preston E (1984) Considering the opposite: a corrective strategy for social judgment. J Pers Soc Psychol 47(6):1231
52. Lyons C (2009) I win, you win: the essential guide to principled negotiation. Bloomsbury Publishing
53. McCorkle S, Reese MJ (2015) Mediation theory and practice, 2nd edn. SAGE, Thousand Oaks, CA
54. McLaughlin ME, Carnevale P, Lim RG (1991) Professional mediators' judgments of mediation tactics: multidimensional scaling and cluster analyses. J Appl Psychol 76(3):465
55. Menkel-Meadow and Porter-Love L (2014) Mediation: practice, policy, and ethics. Wolters Kluwer Law & Business
56. Mowen JC, Gaeth GJ (1992) The evaluation stage in marketing decision making. J Acad Mark Sci 20(2):177–187
57. Mussweiler T, Englich B, Strack F (2004) Anchoring effect. In: Pohl R (ed) Cognitive illusions—a handbook on fallacies and biases in thinking, judgment, and memory. Psychology Press, pp 183–200
58. Mussweiler T, Strack F, Pfeiffer T (2000) Overcoming the inevitable anchoring effect: considering the opposite compensates for selective accessibility. Pers Soc Psychol Bull 26(9):1142–1150
59. Posthuma RA, Dworkin JB, Swift MS (2002) Mediator tactics and sources of conflict: facilitating and inhibiting effects. Ind Relat J Econ Soc 41(1):94–109
60. Qu Y, Cheung SO (2013) Experimental evaluation of logrolling as an effective mediating tactic in construction project management. Int J Proj Manage 31(5):775–790
61. Richbell D (2007) Mediation of construction disputes. Blackwell Publishing Ltd
62. Sanna LJ, Schwarz N (2004) Integrating temporal biases. Psychol Sci 15(7):474–481
63. Schulz-Hardt S, Jochims M, Frey D (2002) Productive conflict in group decision making: Genuine and contrived dissent as strategies to counteract biased information seeking. Organ Behav Hum Decis Process 88(2):563–586
64. Schwenk CR, Cosier RA (1980) Effects of the expert, devil's advocate, and dialectical inquiry methods on prediction performance. Organ Behav Hum Perform 26(3):409–424

65. Sedikides C, Campbell WK, Reeder GD, Elliot AJ (1998) The self-serving bias in relational context. J Pers Soc Psychol 74(2):378
66. Shaheen AA, Fayek AR, AbouRizk SM (2007) Fuzzy numbers in cost range estimating. J Constr Eng Manag. https://doi.org/10.1061/(ASCE)0733-9364(2007)133:4(325)
67. Singh D, Tiong LK (2005) A fuzzy decision framework for contractor selection. J Constr Eng Manag. https://doi.org/10.1061/(ASCE)0733-9364(2005)131:1(62)
68. Soll J, Milkman K, Payne J (2014) A user's guide to debiasing
69. Tembo C, Ndekugri I, Hammond F (2010) Practice and procedure in the mediation of construction industry disputes: an exploratory study. In: W113-special track 18th CIB world building Congress May 2010 Salford, UK, pp 203–214
70. Thompson LL, Lucas BJ (2000) Judgmental biases in conflict resolution and how to overcome them. The handbook of conflict resolution: theory and practice, pp 255–282
71. Williams BJ (2017) Accountability as a debiasing strategy: does race matter?
72. Yiu TW, Cheung SO, Cheung CH (2007) Toward a typology of construction mediator tactics. Build Environ 42(6):2344–2359
73. Zadeh LA (1965) Fuzzy sets. Inf Control 8(3):338–353
74. Zhao J, Bose BK (2002) Evaluation of membership functions for fuzzy logic-controlled induction motor drive. In: IEEE 28th annual conference of the industrial electronics society, vol 1, pp 229–234

Impediments Against Settlement

Inequity and Dispute

Liuying Zhu and Sai On Cheung

Introduction

Contracts are used to govern economic exchanges whereby parties to a contract get what they intended. While the principle of freedom of contract is based on the idea that parties to a contract have exercised their free will in concluding the contract, classical contract theory assumes that contracting parties will commit to the terms of the contract that reflects their respective rights and obligations. However, Adam [5] advocated that contracting parties may enter a contract because of necessity, commercial reality and in extreme case under coercion. It is not uncommon to find one-sided contracts are used in construction projects, especially in those highly competitive construction markets where cut-throat pricing is regularly practised. In such circumstance, contractors are entering into contracts that are inequitable as far as ownership of risks and responsibilities are concerned. These disparities between the contracting parties are major departure from the notion of equal footing as assumed in classical contract theory [73]. On critical issue related to dispute is whether inequitable contract provisions would affect the contracting behaviours of the illy-treated party. In fact, Equity Theory [3] projects that whether one will honour the terms of a contract depends not only on what one gets, but also on whether the same is in parity with that to be received by their counterpart. In this study, equity gap (EG hereafter) is used as a collective term to describe such disparities. The effects of EG on contracting behaviours would surface upon commencement of physical works. Unaddressed disparities are often met with retaliatory behaviours

L. Zhu (✉) · S. O. Cheung
Construction Dispute Resolution Research Unit, City University of Hong Kong, Hong Kong, China
e-mail: liuyinzhu3-c@my.cityu.edu.hk

S. O. Cheung
e-mail: Saion.cheung@cityu.edu.hk

S. O. Cheung (ed.), *Construction Dispute Research Expanded*, Springer Tracts in Civil Engineering, https://doi.org/10.1007/978-3-030-80256-1_6

such as non-cooperation, procrastination and opportunism [4]. These responses are counterproductive and trigger disputes.

For example, with the shifting of power ex poste, it is also not surprising to find contractors practising opportunism to express their dissatisfaction over the inequitable treatment. Schieg [65] describe this as a principal-agent phenomenon whereby the agent (contractor) leveraging information advantage ex post to exploit the principal. Inequitable contractual arrangements are thus having an enduring negative impact on contracting behaviours that are counterproductive a dispute driven.

Most capital construction projects are of intermediate duration spanning a few years while mega projects may take even longer to complete. Coordinated effort of multi-disciplinary project team is the prerequisite for efficient and effective delivery of the project. Lui and Ngo [56] found that notwithstanding cooperative working among project team members is a necessary condition for successful delivery, this is not always achieved. In fact, the negative impact of non-cooperation caused by inequitable treatments will inevitably undermine project performance. The concept of asset specificity under transaction cost economics theory is central to the practice of opportunism [56, 72]. High asset specificity and interdependency asymmetry also provide the breeding ground of non-cooperative behaviour [17]. Evidently, EG is one of the sources of problematic contracting behaviours that breed disputes [15]. This study examines the elements of EG and proposes measures to address the disparities. The ultimate aim is to make contracting environment less dispute prone. This chapter reports the followings:

1. Identifying elements of EG between developer and contractor;
2. Developing a conceptual framework of EG; and
3. Suggesting measures to alleviate happening of dispute through addressing EG.

Equity Gap in Construction Contracting

This section examines the characteristic of the relationship between developer and contractor and identifies the disparities arising therefrom. These are termed as elements of equity gap that is used as a collective term doer the disparities. The potential damages of EG on contracting behaviours are then summarized.

The Relationship Between Developer and Contractor in Construction Projects

An agency relationship has arisen between two (or more) parties when one, designated as the agent, acts for the other, designated the principal [64]. In construction, the principal is the developer and the agent are the contractor. They are independent commercial entities,they would have their own interests despite joint efforts are

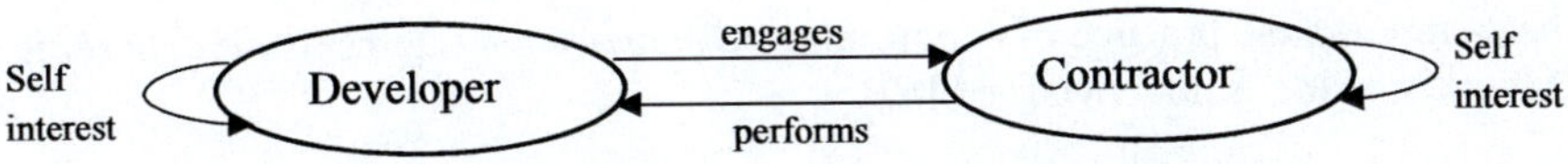

Fig. 1 The project developer-contractor relationship. Adapted from [8]

needed to develop the construction project. Figure 1 presents their interdependent relationship as principal and agent.

What significant issues arise from this interdependency? Smith and Barclay [67] claimed that cooperation is a necessary condition for effective discharge of highly interdependent construction tasks. Moreover, agency theory projects that there is a potential conflict between the principal and the agent because of their self-interest. Williamson [73] offer two important concepts, asset specificity and uncertainty, to explain the practice of opportunism resulting from self-interest.

Williamson [72] identified asset specificity (AS hereafter) as the "durable investments that are undertaken in support of particular transactions, the opportunity cost of which investments is much lower in best alternative uses or by alternative users should the original transaction be prematurely terminated." For construction projects, substantial resources are deployed progressively as the project unfolds. The investments therefore from the contracting parties are in general much higher than typical buyer–seller relationship. In extreme case, contractual determination would bring substantial loss. Thus AS makes it possible for the more flexible party to exploit the less flexible one [23] in practising opportunism [72]. Moreover, in a social context, Liu et al. [57] advocate that under social exchange theory, with relation-specific assets, multiple transactions enhance devotions from both parties and lead to more cooperative behaviour [32]. Accordingly, the practice of opportunism can be restrained when future dealings are anticipated.

Uncertainty is widely recognised as a critical design parameter of contractual governance. Uncertainty can be viewed as a state that ranges from just short of certainty to a complete lack of knowledge about a result [39, 73]. Uncertainty coupled with bounded rationality make contract inevitably incomplete. As not all the events can be predicted, the unplanned impact calls for the use of power to make sure the affected party unhurt as far as possible. Unfortunately, this usually only applies to the principal. Besides, uncertainty is also a trigger of construction disputes [16]. Construction projects are typically unique and involve meticulous coordination. The uncertainties arising from the physical environment can lead to overreactions, unnecessary interventions, second guessing, mistrust, and distorted information flows [39]. Thus, Williamson [73] concluded that the level of uncertainty dictates the type of contractual governance, the more more uncertain the environemnt, the more relational should the governance be [73]. Regretably the actual happening is the opposite.

Another distinctive feature of typical construction projects is the use of project team assembled from particiapting organisations to manage the project [23]. Project teams only function during the project duration and thus often being termed as temporary managemnt organization (TMO) [69]. As the team members are guided primarily by their respective own interest, protective behavior can be expected. Specifically for the contractor, the practice of opportunism to enhnace her interest is often observed if future dealing is not envisaged [62].

Elements of Equity Gap

The elements of the equity gaps are rooted in developer-contractor relationship. Agency Theory predicts that there is imbalance of **information** [64]and **risks** [14] in a principal-agent relationship [4]. The consequent of AS gives rise to **power asymmetry**. Besides, Transaction Cost Theory [73] also suggests that AS renders a less dependent party the power to decide how interests to be distributed, hence the **expected return**.

Information

Principal and agent are having their own goals, risk preferences and information sources [64]. Agency problem arises when (1) the principal cannot verify has the agent had behaved as contracted,and (2) the principal and agent are having different attitudes toward risks. The assumptions are about people (e.g., self-interest, bounded rationality, risk aversion) and organizations (e.g. goal conflict among members), and information (e.g., information is a commodity which can be purchased) [30].

Holmstrom [44] explains that when the principal can only observe the results, the agent may behave in a way that would jeopardise the interest of the principal while the result would appears reasonable [64]. In construction, if the developer cannot monitor the contractor or if he is not able to deduce stringently the quality of his service while work is in progress, an information imbalance in favour of the contractor will result [65]. In this regard, Xiang et al. [75] discussed the information dominance/inferiority between developer and contractor during the bidding and construction stage. Table 1 summarises their findings.

Because of information asymmetry, opportunistic behaviour happens when one party purposely not disclosing information with the belief that the other party may take advantage of the weaknesses to increase his or her own profit. Accordingly, both developer and contractor would withhold certain core information purposely. As such, the transaction becomes less and less transparent with rising suspicions and protectionism. All these would make the exchange less efficient [75]. Ratchet effect is another explanation of why contracting parties are holding back information [36]. When the agent with a high performance today would face higher future demand [52]. Disclosing performance information at is thus not preferred [52].

Table 1 Information asymmetry between developer and contractor

Participants	Bidding stage		Construction stage	
	Dominance	Inferiority	Dominance	Inferiority
Developer	• Construction purpose • Financial strength • Construction project procedures	• Contractor's qualifications • Technological and management ability, performance • Business reputation	• Financial payment capacity • Management ability • Business reputation	• Contractor's management ability • Employee talent • Business reputation • Construction technology, equipment
Contractor	• Quality: • Technology • Equipment • Management, and service	• Developer's construction purpose • Financial payment capacity • Business reputation	• Developer's talent • Construction method and technology • Management ability and instruments, material quality	• Developer's business reputation • Financial payment capacity

Adapted from [75]

Risk

Risk can be viewed as deviation from anticipation [12]. Construction projects are one-off endeavours characterized by extended duration, complicated processes, resources laden and dynamic management [66]. Risks arising from the hostile working environment may materialize at any time of the project. These risks can cause cost inflation, time delay, substandard quality and safety hazard during construction. In some cases, environmental disaster may result [79, 80].

Risks in construction projects can hardly be eliminated. Typically risks are shared or transferred among the contracting parties [1]. Risks can be broadly categorized as environmental and behavioural [77, 80].

Environmental risks refer to those often caused by the changes of the natural, economic, social, legal, and technological environments., like unforeseeable physical conditions and cost fluctuation because of the market. Theoretically, environmental risks should be distributed between and shared by the two parties [34]. Behavioural risks are resulted from the behaviours of the contracting parties [34]. Some examples include delayed payment caused by the developer or delayed information of the project reported by the contractor. Theoretically, each party should take responsibility for the risks caused by themselves respectively.

Equitable risk allocation between construction contract is pivotal in putting parties on a cooperative working platform [14]. Several principles of allocation have been advocated. The most commonly used is the set of principles proposed by Abrahamson [2]. Essentially, risk allocation should observe the principles of foreseeability, manageability and controllability [55]. When faced with uncertainties, developers typically would avoid risk as far as possible by allocating most of the risks to the contractor [14, 76]. Procurement through competitive tendering and the lowest price selection are very common for construction projects. Contractors are therefore often in an ex ante disadvantaged position with little bargaining power over the terms of the contract. However, unilateral risk shifting to contractor occurs in many contracts. Zhang et al. [77] collected data from 284 Chinese project professionals and showed that some environmental risks are indiscriminatingly shifted to the contractor. It was further found that pro-employer contractual terms would backfire when contractor choose to practice opportunism ex post.

Inequitable risk allocation in construction contracts impedes cooperative behaviour and is one of the underlying causes of dispute [13, 14]. Similarly, an investigation of the construction industry in Canada and United States found that one major cause of construction disputes is inappropriate risk allocation [76].

Return on Efforts

Williamson [73] explained that the differential of expected return comes from the dependence asymmetry. Dependence Theory [31] explains that people evaluate outcomes as gains or losses with reference to certain yardsticks. The extent of deviation from the yardstick will affect their devotion to their responsibilities. For example, if the added value is expected to be squeezed from the most contributing party, he or she would choose to take conservative actions to prevent further widening the deviation. AS also has a part to play, the party who is less dependent on the other will have the chance to direct whose interest shall take priority thereby aggregate the imbalance. When the contractor is threatening to suspend work if no extra compensation is provided. The developer would to weight which option will cost more: determine the employment of the contractor and conceding to the demand.

Power

Emerson [31] defined power as the resistance offered by participant A to overcome domination by participant B in a relationship. It can also be characterized as the ability to influence, control or restraint others. Power is an attribute of position in a network and is identified by the participants' behaviour [20]. The presence of asymmetry and imbalance of power is common in contractual relationship [23, 22]. In construction where contracting parties are highly interdependent, the issue of power differential is more acute [32]. Interdependence asymmetry occurs when one party holds power advantage over the counterpart. Inter-organisational power can be unilateral (sanction) or bilateral (bargaining) [38].

a. Sanction Power

Sanction power is used primarily to penalise nonconformance of the specified. Its use serves to influence others by the damage that could be caused by the sanction. In construction contracts, unilateral sanctions are mostly punitive and can be exercised as of right [61]. For the developer, unilateral sanction of levy like damages and contract changes are notable examples of control over the contractor [6]. Moreover, wrongful sanction can be challenged in the form of dispute. The most powerful sanction is determination of employment [18]. At the early phase of a construction project, the developer usually has power advantage as the switching cost is relatively low. With input of contractor increases, the developer becomes less flexible with the increasing transaction costs of re-tendering and the extra costs resulting from removal of a 'defaulting' contractor. As classical example of asset specificity [73] in construction, exercising determination may inflict more harm to the developer especially at the later part of the project [11].

b. Bargaining Power

Bargaining power can expressed by the extent to which one party would inflict concession by the counterpart [10]. Bargaining power is derived from ownership or control of scare resources that the other side needs [20]. In negotiation, bargaining power can be exercised by the ability of depriving the counterpart values that he possesses or by obstructing the attainment of desired values [70]. Bargaining power aims to achieve one's own benefits by exploiting the differentials between [20].

In construction, because of the competitive tendering and lowest bid selection, developer usually has greater bargaining power as compared to the contractor at the bidding stage [49]. In order to obtain the contract, the contractor is willing to compromise in the contract price negotiation. At the construction stage, bargaining power swings towards the contractor as physical work proceeds. Hold-up problem occurs and the developer becomes vulnerable in ordering changes [11]. Capitalising on the sunk cost of re-tendering [73], contactor might exploit the vulnerability of the developer. Winch [74] described this as 'opportunistic margin' for the contractor. If the opportunistic behaviour costs less than the cost of switching, the developer would concede to the demand of the contractor [9]. This situation becomes acute when the project reaches critical stage where delay will hurt the developer dearly [11, 59].

Based on the afore-stated theoretical deliberations, the elements of EG in construction contracting are summarized in Table 2.

The potential effects of EG on contracting performance can be discussed from two perspectives: (1) EG inhibits inter-organizational relationship development; and (2) EG reduces project efficiency. There is a strong preference for fairness in human interaction. People who have experienced unfairness tend to react with anger, resentment and loss of motivation [49]. Achieving fairness is also considered important in developing inter-organizational relationship [58]. If one party takes advantage of the imbalance of status and deliberately widen these gaps, mistrust between them will grow [76]. Das and Teng [24] observed that the weaker party often practices

Table 2 Theoretical bases of EG in construction contracting

	Certainty of risks	Information ownership	Expected return	Power	Key references
Agency theory	√	√			[30]
Prospect theory	√		√		[50]
Transaction cost theory	√		√		[72]
Social exchange theory				√	[32]
Power-dependence theory				√	[31]

opportunism and thereby triggers defensive reaction of the counterpart. All these behaviours would lower the overall project efficiency. Some contractors may build in high risk premium in their bids to insure against potential losses. However, this act may lower their chance of getting the contract. Instead, pursuing post-contract claims to maximise their returns is often adopted. Developers and consultants often find claim conscious contractors offensive. As such, having a harmonious working project team is unlikely. The potential damages due to EG between developer and contractor are summarized in Table 3.

Table 3 The potential damage of EG on construction contracting

Elements of EG	Consequences		References
	Inter-organizational relationship	Project efficiency	
Information	Mistrust Uncertainty of other parties' behaviour	Hamper innovation Cost wastage for information obtainment	[36, 53, 75]
Risk	Mistrust Unwilling to cooperate	Hamper innovation	[28]
		Cost wastage and disputes	[76]
Return on efforts	Less devotion to the project Opportunistic behaviour	Hamper innovation Disputes	[73]
Power	Impede commitment behaviour Withdrawal behaviour	Impede concessions and agreements Disputes	[32] [54]

Conceptualizing EG in Construction Contracting

Key Elements of EG: Empirical Findings for a Pilot Case Study

A pilot case study was first conducted to explore the existence of EG in a real project, the Hong Kong-Zhuhai-Macau Bridge (HZMB hereafter). In this regard, Zhu et al. [78] found that effective management measures by the HZMB Authority can enhance inter-organizatioanl comuunication. This had effectively narrowed information asymmetry created ex ante. Likewise, the HZMB project offereed a valued opportuntity to test the existence of the elements of EG (Table 3).

This pilot case study interviewed 20 senior construction project participants of the HZMB project. Among the 20 interviewees, half worked as developer and the remaining as contractor. A questionnaire was set to collect their viewpoints about EG. The interviewees were asked to indicate their view on the degree of EG in a Likert Scale 1–7. Using the quality risks as an example, template of the questions is listed in Table 4. Two evaluations one at commencemnt and one near the completion were done to identify the changes of asymmetries.

The results of Part one is shown in Table 5. The positive score implies that the Developer had the advantage while negative score means that the interviewees believed that contractor had advantage instead.

Table 4 Questionnaire template to measure the equity gap between developer and contractor

No.	Description	The position of your engaged party	Degree of asymmetry Low–High
A. At the commencement of the project			
A.1	The distribution of the risks		
A.1.3	Quality risks	Advantage/disadvantage/same	0 1 2 3 4 5 6 7

Table 5 The data and the changes of EG

No.	Description	Stage of the project		Ratio (%)
		Commencement	Completion	
A.1	*Risk*	*3.14*	*2.46*	−22
A.1.1	Environmental risks	1.36	1.29	*−5*
A.1.2	Behaviour risks	2.79	2.14	−30
A.2	*Power*	*4.93*	*4.18*	*−15*
A.2.1	Sanction power	5.14	4.43	−14
A.2.2	Bargaining power	4.71	3.93	−17
A.3	*Return on efforts*	2.29	*3.71*	*+63*
A.4	*Information*	*4.43*	*3.43*	−23
	Total			−37

Observations from the responses of the interviewees:

(1) It is agreed by all the interviewees that there are disparities between developer and contractor throughout the project duration. The existence of EG elements was basically confirmed.
(2) Comparatively, Developer was in better position than the contractor in terms of risk, power, expected return and information. Power and information asymmetry are more notable. This is not surprising as most construction contract are organised to provide developer these relative advantages.
(3) Sanction and bargaining power have the most notable asymmetry in level of EG through the project. At the beginning of the project, power has the highest asymmetry (4.93) while expected return has the lowest (2.29). Information (−23%) and Risk differential (−22%) present the most significant drop when comparing the evaluations between the commencement and completion stage of the project. The change of power asymmetry takes the second place (−15%). Developer has the dominating power throughout the project. Expected return in efforts is the only element that had increased as the project unfolds. The interviewees explained that it was due to the fact that the developer has dominant sanction power and thus has overwhelming influence over the final payment, thus expected return asymmetry aggregates at completion of contract.

The general view of the interviewees is that they do recognize the existence of EG. The level of disparity does change as the project unfolds. The nature of disparity decides that some had narrowed while others had widened. This phenomenon corresponds to the projection of influence of asset and process specificities under transaction cost economics theory [72]. That means the contractor became more influential with the progress of physical works.

To further analyse the elements of EG, a comprehensive literature review suggested the inclusion of identifications of the EG elements that are summarised in Table 6.

Conceptual Framework of EG

The conceptual framework of EG is then developed in Fig. 2.

Empirical Testing of the Framework

A Partial Least Square-Structural Equitation Modelling (PLS-SEM hereafter) based analysis was conducted to test the robustness of the framework. Data were collected through a questionnaire that has 2 parts. Part 1 is about the personal particulars. In Part 2, the respondents were asked to recall a project they have been involved in for at least 1 year as either Developer or Contractor. All the measurement items developed

Table 6 Identifications of EG elements

1.	Information		
1.1	Project details	In the bidding stage, Developer had more information about project details	[75]
1.2	Adverse selection	In the bidding stage, Developer had an information disadvantage about the contractor's ability	[65, 75]
1.3	Project performances	In the construction stage, Contractor had an information advantage relating to market changes	[75]
1.4	Moral Hazard	In the construction stage, Developer cannot monitor all the detailed Contractor's behaviour relating project performance	[65, 75]
2.	Risks: the allocation of the following risks in contract favour the Developer		
2.1	Environmental risks		[77]
a.	Unforeseeable physical conditions		
b.	Cost fluctuation (inflation of prices)		
c.	Adverse climatic conditions		
2.2	Behaviour risks		
d.	Defective design by owner		
e.	Delayed payment		
f.	Delayed instructions or information		
g.	Access to site		
3.	Return on efforts		
3.1	Unequal sharing of project surplus	At the beginning of the project, the returns or rewards for one party were unfair in view of his contributed resources to the project	[48] [72]
3.2		For this project, one party's profit was squeezed when there are additional profits	[10] [72]
3.3		For this project, one party beard more losses when there are unforeseeable losses	[72] [10]
4.	Power		
4.1	Sanction power		
a.	Asset specificity	The unilateral termination behaviour had more threatens to one of the parties	[11]
b.	Power for project control	For project disputes, for one party, unilateral decisions could serve as weapons against another party to achieve their own goals	[6]

(continued)

Table 6 (continued)

c.	Coordination failure	During construction procedure, one party was unwilling to cooperate for events which are critical to the other party	[37]
4.2	Bargaining power		
a.	Process specificity (ex-ante)	Comparing two parties, in bidding stage, one party used to feel more constrained and sacrificed in negotiation of contract price because of foreseeable losses	[10]
b.	Process specificity (ex-poste)	Comparing two parties, in construction stage, one party used to feel more constrained and sacrificed in renegotiation of contract price and interim payments because of foreseeable losses or disputes	[10] [47]
c.	Hold-up problem	Comparing two parties, one party used to feel pressure to agree a claim value beyond the 'true' cost of the additional change of work	[10]
d.	Time specificity	It is posited that making a compromise in a short time was needed for one party as the time pressure of switching partner and value loss at some critical moment of the project	[59] [10]

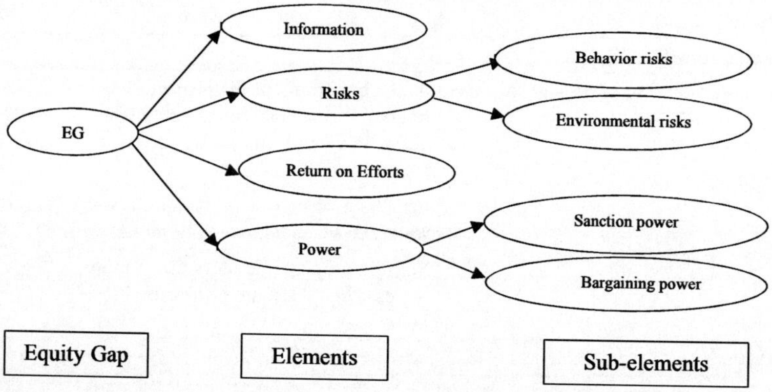

Fig. 2 A conceptual framework of EG in construction contracting

from theories were converted into questions (Table 6). They were asked to indicate in a Likert Scale of 1 (Strongly disagree) to 7 (Strongly agree). Their level of agreement of the statement represent the extent of happening of the EG identifications during the project. Confirmatory Factor Analysis (CFA hereafter) was applied to examine statistically the relationships as shown in Fig. 2.

SEM is a family of statistical models that seek to explain the relationships among multiple variables for CFA. It examines the structure of interrelationships expressed in a series of equations, similar to a series of multiple regression equations [41]. PLS-SEM is applied in this study because it is more flexible on the theoretical bases and also comparatively suitable for small sample size [42].

In view of the complex component structure of the four EG elements, applying hierarchical component model (HCM) is also necessary. Establishing higher-order models or hierarchical component models (HCMs) are usually referred to in the context of PLS-SEM. It is also important to verify the measurement framework first before analysing the relationships between different factors [51]. Testing second-order models that contain two-layer structure of constructs are often involved [71].

There are three main reasons to include HCM in a PLS path model [42]:

a. Reduce the number of relationships in the structural model;
b. Prove valuable if the first-order constructs are highly correlated;
c. Prove valuable if formative indicators exhibit high level of collinearity, and discriminant validity may not be established.

The HCMs are also divided for reflective and formative measurement models. The major difference between these two types is the contributions of the indicators in forming the construct and measures [26]. Reflective indicators can be viewed as representative sample of possible items available within the conceptual domain, which may be relevant with each other. In contrast, formative measurement is based on the assumption that all causal indicators form the construct are interdependent and considered as linear combinations [42]. Research also suggests that formative measurement is not an equal attractive alternative to reflective measurement in developing new measures or choosing among alternative existing measures [46]. In this study, the reflective measurement model is therefore selected.

Data Collection and Analysis

Personal Particulars

Over 300 questionnaires were distributed, and 106 valid responses were received. The response rate is about 30%. It is reasonably close to the median rate (35.7%) of the survey conducted in the United States for 1607 organizational academic studies [29]. The response rate of questionnaire survey is also similar to the relevant studies conducted in the construction industry studies that are usually ranged from 25 to 30% [29]. Therefore, the response rate for this study is considered acceptable. Table 7 summarizes the personal particulars of respondents.

Table 7 shows that the respondents cover the typical roles in construction projects and include management and professional staff. There are about equal number of respondents working in developer and contractor. The data is therefore useful to

Table 7 Personal particulars

No.	Description	Number	%
1.1 Your position			
1	Management staff	34	32
2	Professional staff	72	68
1.2 Working experience			
1	<5 years	22	21
2	5–10 years	28	26
3	11–20 years	36	34
4	>20 years	30	28
1.3 Your organization			
1	Developer	59	56
2	Contractor	47	44
1.4 Your counterpart			
1	Developer	47	44
2	Contractor	59	56

examine if there is any inter group differences. Q1.4 is used to countercheck if the respondents understand the study arrangements. The result shows that all these 106 responses were valid. The result shows that the feedbacks are reasonable.

PLS-SEM Analysis

a. Data Description

The relevant identification items of Table 6 were operationalized as measurement statements. Respondents need to indicate from a Likert Scale of 1 (Strongly disagree) to 7 (Strongly agree) how accurate the statement represent the happening of the project. The descriptive statistics are shown in Table 8.

Table 8 gives the general descriptive statistics of the data. It was found that most of the mean score of information, risk, return on efforts, are above 4 (out of 7). This suggests that the respondents agree in general the existence of the EG identifications in the respective project they have participated. Q2.17 has the lowest mean score (3.74) and it is lower than the neutral score of 4. It means that the respondents tend to agree that contractors are willing to cooperate in the construction period. Because the developer often has the dominant power in many contractual and management procedures, contractors would find opportunities to express their discontent and adopting a cooperative behaviour. Q2.21 has the highest mean score (4.93). Both the developer and contractor groups agree that asset specificity asymmetry is significant at some critical moment during the construction procedure. Furthermore, the K-W test result suggest there is no significant group differences between Developer and Contractor.

Table 8 Measurement statements and descriptive statistics

No	Equity gap		Min	Max	Mean	Std
	Elements/sub-elements	Identifications				
Q2.1	Information	At the bidding stage, the developer had an information advantage about the project details	3	7	4.89	1.05
Q2.2		At the bidding stage, the developer had an information disadvantage about the contractor's ability	1	7	4.08	1.15
Q2.3		At the construction stage, the contractor had an information advantage relating to market changes	3	6	4.62	0.95
Q2.4		At the construction stage, the developer could not monitor comprehensively the Contractor's behaviour relating project performance	1	7	4.05	1.39
	Risks	The allocation of the risks in the contract favoured the developer in terms of				
Q2.5	Risk (environmental risk)	Unforeseeable physical conditions	2	7	4.58	0.98
Q2.6		Cost fluctuation (inflation of prices)	1	7	4.02	1.30
Q2.7		Unforeseeable loss because of adverse climatic conditions	2	6	4.13	0.9
Q2.8	Risk (behaviour risk)	Unforeseeable loss because of defective design	1	7	4.14	1.31
Q2.9		Time for payment	2	6	4.19	0.96
Q2.10		Time for providing information/instructions	1	7	4.32	1.12
Q2.11	Return on efforts	At the bidding stage, price competition was fully leveraged to drive down contractor's profit	2	6	4.59	1.06
Q2.12		The return for one of the parties was not commensurate to his contribution in resources to the project according to the contract	2	6	4.11	0.75

(continued)

Table 8 (continued)

No	Equity gap		Min	Max	Mean	Std
	Elements/sub-elements	Identifications				
Q2.13		At the construction stage, return for changes was not commensurate to his contribution in resources to the project	2	6	4.38	0.90
Q2.14	Sanction power	At the construction stage, unilateral termination by the contractor presented greater threat than the developer	1	6	4.27	1.01
Q2.15		Unilateral decision authority over project dispute had been the major weapon used by the developer to achieve his own goals	1	6	4.24	1.13
Q2.16		At the construction stage, the developer was unwilling to cooperate for events which are critical to the contractor	2	7	4.04	0.84
Q2.17		At the construction stage, the contractor was unwilling to cooperate for events which are critical to the developer	2	6	3.74	0.91
Q2.18	Bargaining power	At the bidding stage, the contractor felt more constrained and sacrificed in negotiating contract terms in relation to compensation for foreseeable losses	3	7	4.47	1.06
Q2.19		At the construction stage, the developer felt more constrained and sacrificed in renegotiation of contract terms in relation to compensation for foreseeable losses or disputes	1	6	4.43	1.05
Q2.20		The developer felt being forced to settle claims below his entitlements for change of work	2	6	4.18	0.85

(continued)

Table 8 (continued)

No	Equity gap		Min	Max	Mean	Std
	Elements/sub-elements	Identifications				
Q2.21		Making compromise was needed for the developer in view of the time pressure in switching contractor	3	6	4.93	0.83

b. PLS-SEM Analysis

SmartPLS3 software was applied to estimate the framework. The evaluation procedure followed the guidelines of PLS-SEM analysis [42]. Applying a 5% significance level the significance of the path coefficients was all tested by bootstrapping with a 5000 subsample. The one-tail t-test was also used to test the hypotheses. For $p > 0.05$, the null hypothesis is rejected. Figure 3 shows the PLS-SEM analysis result.

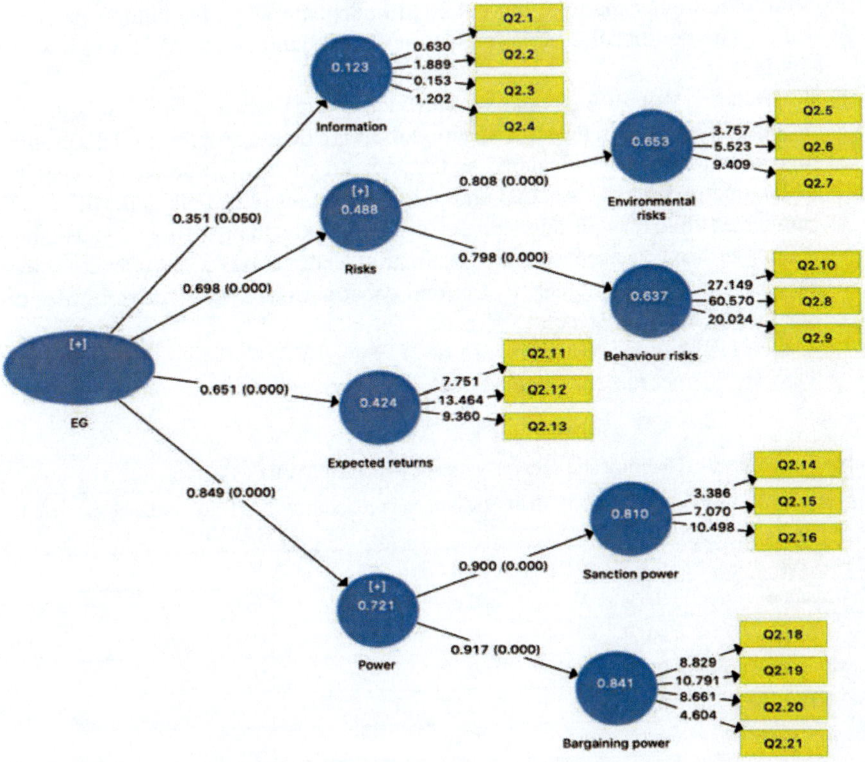

Fig. 3 PLS-SEM analysis of the EG framework

The robustness of the framework is also assessed by examining the criteria of PLS-SEM analysis [42]:

(1) Common Method Variance (CMV). As the measuring constructs were all measured in a one-time questionnaire, common method variance (CMV) problem may exist which could affect the hypothesised relationships in the structural model. It is suggested that CMV exists if a significant factor is found to explain over 50% of the variance for all variables in factor analysis [41]. The results indicate that the factor only explained 21.01% of the variance. The results suggest that CMV unlikely affects model evaluation.

(2) Internal consistency reliability (Composite Reliability) and Average Variance Extracted (AVE). Cronbach's alpha (α) is widely used to assess internal consistency of the construct. Threshold value of 0.7 is suggested by Davcik [25]. For PLS analysis, composite reliability is a more appropriate measure for internal consistency [42]. Composite reliability of all constructs should satisfy the threshold of 0.70 [42] and 0.60 is acceptable [25]. To establish convergent validity, researchers consider the outer loadings of the indicators, as well as the Average Variance Extracted (AVE) [40]. The general acceptable AVE should be higher than 0.5 but 0.4 is still adequate when the composite reliability is higher than 0.7 [35]. The Composite Reliability and AVE are shown in Table 9.

(3) R^2 Value, f^2 Value and predictive relevance Q^2 are the mostly used measures to evaluate the fitness of the structure model. R^2 is a measure of the model's predictive accuracy and is calculated as the squared correlation between a specific endogenous construct's actual and predicted values [42]. Falk and Miller [33] suggested that R^2 and adjusted R^2 values greater than 0.10 are acceptable. Effective size f^2 is used to further evaluate all endogenous constructs. Its objective is to evaluate whether the omitted construct has a substantive impact on the endogenous constructs.

(4) Cohen [19] has suggested the use of Q^2 value of 0.02, 0.15 and 0.35 to indicate weak, medium or strong effects, respectively. Also, to further evaluate

Table 9 Composite reliability and average variance extracted (AVE)

	Cronbach's Alpha	Composite reliability	Average variance extracted (AVE)
Equity gap	0.78	0.82	0.42
Information	0.63	0.63	0.47
risks	0.67	0.79	0.54
Environmental risks	0.59	0.79	0.56
Behaviour risks	0.86	0.91	0.78
Return on efforts	0.81	0.89	0.73
Power	0.75	0.82	0.43
Bargaining power	0.60	0.77	0.46
Sanction power	0.53	0.76	0.53

the magnitude of the R^2 values, the Stone-Geisser's Q^2 value is examined [42]. This value is obtained by means of the blindfolding procedure, estimates the model parameters, and predicts the omitted part by using the previously computed estimates. The smaller the difference between the predicted and the original values, the greater the Q^2 value [63]. Since Q^2 values greater than 0 indicate predictive relevance for a certain endogenous construct, the relevance is considered as small, medium, and when the respective values are 0.02, 0.15 and 0.35. Table 11 shows the predictive relevance Q^2 of the constructs used in this study. The Q^2 value are all acceptable on a reasonable level. Table 10 shows all the analysis results.

(5) Heterogeneity. Heterogeneity occurs when different groups of data show significant differences in terms of model parameters. For example, the Developer and Contractor may hold different view towards different questions. A Multi-Group Analysis (MGA hereafter) is developed to investigate the differences between different observed groups. For the non-parametric data, PLS-MGA is applied [43]. Group analysis between Developer and Contractor was conducted. The results are presented in Table 11. There are no significant differences (p value under 0.05) detected between these two groups.

Discussions and Recommendations

The proposed EG framework (Fig. 2) is supported by the PLS-SEM analysis statistically. As such, the elements of EG are considered well placed. Power (0.849) is the most notable element. Bargaining power (0.917) is having a higher contribution to EG than sanction power (0.900). For bargaining power, the most relevant is Q2.18: "At the bidding stage, the contractor felt more constrained and sacrificed in negotiating contract terms in relation to compensation for foreseeable losses". It shows that at the biding stage, contractor is at an inferior position. Q2.16 is the most influential factor in sanction power: "At the construction stage, the developer was

Table 10 R^2 value, effect size f^2 and blindfolding results

	R^2	R^2 Adjusted	f^2	SSO	SSE	Q^2 (=1 – SSE/SSO)
EG				2120	2120	
Information	0.123	0.115	0.141	424	417.61	0.12
Risks	0.488	0.483	0.952	530	426.96	0.19
Environmental risks	0.653	0.65	1.885	318	212.31	0.33
Behaviour risks	0.637	0.633	1.752	318	168.46	0.47
Return on efforts	0.424	0.418	0.735	318	228.82	0.28
Power	0.721	0.718	2.584	742	546.67	0.26
Sanction power	0.841	0.84	4.275	318	189.24	0.41
Bargaining power	0.81	0.809	5.308	424	268.90	0.37

Table 11 The MGA result between developer and contractor

	Path coefficients-diff (Developer – contractor)	t-value (lDeveloper vs contractorl)	p-value (Developer vs contractor)
EG → Expected returns	−0.011	0.058	0.954
Power → Sanction power	−0.008	0.151	0.881
EG → Risks	−0.097	0.182	0.856
Risks → Environmental risks	−0.04	0.303	0.763
EG → Power	−0.102	0.314	0.755
EG → Information	−0.129	0.52	0.606
Risks → Behaviour risks	0.2	1.241	0.221
Power → Bargaining power	0.063	1.358	0.181

unwilling to cooperate for events which are critical to the contractor". Comparatively, the contractor gradually takes the dominant position at the construction stage.

Return on efforts and risks are also relevant when addressing EG between Developer and Contractor. The unequal distribution of returns is reflected by the responses. For risks, greater influences are found for the differential of environmental (0.808) and behavioural risks (0.798). Comparatively, information asymmetry has the lowest contribution to EG among the four elements.

The PLS-MGA result shows no significant difference between the developer and contractor group. Both groups thus share similar view about the existence of EG elements. As both groups are commercial based, opportunistic behaviours may happen to maximise their own interest should circumstances allow. With EG being one of the contextual enablers, unrestrained opportunistic behavers would lead to serious dispute [16].

The findings of this study also suggested that EG should be addressed to alleviate the chance of happening of dispute during construction. For example, at the contract planning stage, identifying the extent of EG is the first step to devise strategies to install a perception of fairness. Corresponding strategies should be put forward based on different project nature and characteristics. The following possible quasi-contractual arrangements are suggested:

(1) Setting relational incentive to balance power differential

Equity concerns and emergent interpersonal commitments would constrain and impede the use of power [20]. Status recognition was proposed as an effective strategy by Power Dependence theory [31]. To alleviate EG, the party with power advantage should make higher motivational and relational investments towards the party with less power [31]. The recognition including aligned goal commitments [12]; shared relational attitudes [68]; offer mutual support and developing mutual trust [60]. For example, developer can set incentive schemes at the contract planning stage as additional payment to compensate the additional 'risks' the contractor is taking to reach a more balanced risk ownership. Instead of exercising power to suppress potential retaliation, recognising the risks taken by the contractor and with reasonable return on running the risks would prevent further deterioration of their relationship [21].

Furthermore, relational investment aims to let the weaker party to be better recognized and increase the sense of engagement in this relationship. Similarly, offering status cognition is also suggested by Stewardship Theory [27]. The differences between Agency theory and Stewardship Theory is the degree of participation of the agent in decision-making. Stewardship Theory further suggests giving more support and freedom for the agent to manage effectively the transaction [27]. The improved bargaining power should have positive impacts on collaborative working and trust building.

(2) Allowing reallocation of risk and return as deemed necessary and appropriate

Under classical contract theory, contract terms cannot be adjusted unless the parties agree to establish supplemental agreement for the proposed changes. Moreover, when a project is facing great uncertainties, frequent use of supplemental agreement is not efficient. Alternatively three instruments are suggested: establishing common targets, reducing information asymmetries, and reallocating risks [65]. *Ex post* revision in profit sharing is found as an inducer for contractors to align their goals with those of the developer. Allowing reallocation of risks would also change the risk attitude of project participants. In terms of crest for long-term improvement, realignment of innovation risks can be the turning twist to promote creativity [7].

(3) Enhancing tasks programmability for ease of monitoring and evaluation

For information asymmetry, improve observability is one way to reduce moral hazard [30]. Based on Agency theory, tasks should be detailly programmed to facilitate the observability of the agent's compliance or otherwise [44]. It is because the behaviour of agents engaged in more programmed jobs is easier to be observed and thus evaluated. Therefore, the more programmed the tasks, the more attractive are behaviour-based contracting and information about the agent's behaviour is readily determined [30]. Highly programmed tasks can also reduce indeterminacy [45] and increase the accuracy of outcome evaluation [72]. In construction, using more tangible objective targets and milestones are instrumental. In fact, the strategy is commonly used in setting targets of incentive schemes. Hopefully the aligned common interest would encourage the contractor to share progress information. The information exchange is therefore embodied and beneficial for the communication of two parties. Nonetheless, it is admitted that specifying objective standard remains a challenge.

Summary

Contractors often enter into contracts that are notably inequitable in terms of risks and responsibilities. Retaliatory responses are common ex post and, in many instances, have led to dispute [16]. In this study, equity gap (EG) is used to describe the differentials created ex ante between the developer and the contractor. This study contributes to construction dispute research by analysing the roles of EG in cultivating disputes. The objectives of the study include: (i) identifying elements of EG between developer and contractor; (ii) developing a conceptual framework of EG; and (iii) suggesting measures to address EG.

Through literature review, four EG elements in construction projects are identified: information, risk, power and return on efforts. To develop the conceptual framework of EG, a pilot case study of the HZMB project was first conducted. Through structured interview with 20 project participants it was found that developers are in general having the upper hand position for all the four EG elements. The interviewees agreed that all these elements did exist in the HZMB project. The most notable element is power and is represented by sanction and bargaining power. The conceptualisation of EG was systemised by arranging the elements into a framework. This representation enabled statistical testing of the framework. Data were collected from 106 senior project professionals working for developer or contractor. The conceptual framework was validated through PLS-SEM analysis. No significant differences were detected between the developer and contractor group. The strength of the relationship links of the framework inform the respective level of contribution towards EG. With these, the following recommendations are put forward: (a) setting relational incentive to balance power differential; (b) allowing reallocation of risk and return as deemed necessary and appropriate; and (c) enhancing tasks programmability for ease of monitoring and evaluation. Collectively, it is anticipated that by addressing EG, the chance of having disputes arising from retaliatory behaviours of contractors would be reduced.

Acknowledgements The empirical work of this chapter has been reported in a paper entitled "Equity Gap in Construction Contracting: identification and ramifications" of the Journal of Engineering, Construction and Architectural Management. The funding support of a CityU SRG (project no. 7005557) is duly acknowledged.

References

1. ANDI (2006) The importance and allocation of risks in Indonesian construction projects T. Constr Manag Econ 24:69–80. https://doi.org/10.1080/01446190500310338
2. Abrahamson MW (1982) Risk management. In: International construction law conference
3. Adams JS (1963) Toward an understanding of inequity. J Abnorm Soc Psychol 67(5):422–436. https://doi.org/10.1093/past/96.1.22
4. Adams JS (1965) Inequity in social exchange. Adv Exp Soc Psychol 2:267–299

5. Adams JS (1965) Inequity in social exchange. Adv Exp Soc Psychol 2(C):267–299. https://doi.org/10.1016/S0065-2601(08)60108-2
6. Back WE, Grau D, Mejia-Aguilar G (2013) Effectiveness evaluation of contract incentives on project performance. Int J Constr Educ Res 9(4):288–306. https://doi.org/10.1080/15578771.2012.729551
7. Bower D et al (2002) Incentive mechanisms for project success. J Manage Eng
8. Ceric A (2016) Trust in construction projects. Talor& Francis Group
9. Chan APC, Yeong CM (1995) A comparison of strategies for reducing variations. Constr Manag Econ 13(6):467–473. https://doi.org/10.1080/01446199500000054
10. Chang CY et al (2007) Reversal of bargaining power in construction projects: meaning, existence and implications. Constr Manag Econ 25(8):845–855. https://doi.org/10.1080/01446190601164113
11. Chang C, Ive G (2007) The hold-up problem in the management of construction projects: A case study of the channel tunnel. Int J Project Manage 25:394–404. https://doi.org/10.1016/j.ijproman.2007.01.001
12. Chapman C, Ward S (2008) Developing and implementing a balanced incentive and risk sharing contract. Constr Manag Econ 26(6):659–669. https://doi.org/10.1080/01446190802014760
13. Cheung SO (1997) Risk allocation: an essential tool for construction project management. J Constr Procurement 3:16–27
14. Cheung SO et al (2014) Construction Dispute Research. Springer
15. Cheung SO, Pang KHY (2013) Anatomy of construction disputes. J Constr Eng Manag 139(January):15–23. https://doi.org/10.1061/(ASCE)CO.1943-7862.0000532
16. Cheung SO, Yiu TW (2006) Are construction disputes inevitable? IEEE Trans Eng Manage 53(3):456–470
17. Cheung SO, Zhu L, Lee KW (2018) Incentivization and interdependency in construction contracting. J Manag Eng 34(3):1–13. https://doi.org/10.1061/(ASCE)ME.1943-5479.0000601
18. Chow PT et al (2005) Impact of trust and satisfaction on the commitment-withdrawal relationship, 31(Gounaris), pp 1–8. https://doi.org/10.1061/(ASCE)ME.1943-5479.0000331
19. Cohen J (1988) Statistical power analysis for the behavioral sciences, 2nd edn. Lawrence Erlbaum associates, Hillsdale, New Jersey
20. Cook KS, Emerson RM (1978) Power, equity and commitment in exchange. Am Sociol Rev 43(5):721–739. https://doi.org/10.3102/0034654311405999
21. Cox A, Ireland P (2002) Managing construction supply chains: the common sense approach. Eng Constr Arch Manage (Wiley-Blackwell) 9(5/6):409–418. https://doi.org/10.1046/j.1365-232X.2002.00273.x
22. Cropper S et al (2008) The oxford handbook of inter-organizational relations. Oxford University Press, New York
23. Cropper S et al (2008a) Introducing inter-organizational relations. In: Oxford handbooks online introducing, pp 1–23. https://doi.org/10.1093/oxfordhb/9780199282944.003.0001
24. Das TK, Teng B-S (1998) Between trust and control: developing confidence in partner cooperation in alliances. Acad Manag Rev 23(3):491–512
25. Davcik NS (2014) The use and misuse of structural equation modeling in management research. J Adv Manage Res 11(1):47–81. https://doi.org/10.1108/JAMR-07-2013-0043
26. Diamantopoulos A, Riefler P, Roth KP (2008) Advancing formative measurement models. J Bus Res Elsevier Inc. 61(12):1203–1218. https://doi.org/10.1016/j.jbusres.2008.01.009
27. Donaldson L, Davis JH (1991) Stewardship theory or agency theory. Aust J Manage 16:49–66. https://doi.org/10.1177/031289629101600103
28. Dulaimi MF, Ling FYY, Bajracharya A (2003) Organizational motivation and inter-organizational interaction in construction innovation in Singapore. Constr. Manage Econ, 21(3):307–318. https://doi.org/10.1080/0144619032000056144
29. Easterby-Smith M, Thorpe R, Lowe A (1991) Management research: an introduction. SAGE Publications, Inc., London, p 2455
30. Eisenhardt KM (1989) Agency theory: an assessment and review. Acad Manag Rev 14(1):57–74

31. Emerson RM (1962) Power-dependence relations. Am Sociol Rev 27(1):31–41
32. Emerson RM (1976) Social exchange theory. Ann Rev Sociol 2(1976):335–362
33. Falk RF, Miller NB (1992) A primer for soft modelling. University of Akron Press, Akron
34. Fang D et al (2004) Risks in Chinese construction market—contractors perspective. J Constr Eng Manag 130(2):235–244. https://doi.org/10.1061/(ASCE)0733-9364(2004)130
35. Fornell C, Laecker DF (1981) Structural equation models with unobservable variables and measurement error:algebra and statistics
36. Freixas X, Guesnerie R, Tirole J (1985) Planning under incomplete information and the Ratchet Effect. Rev Econ Stud 52(2):173–191. https://doi.org/10.2307/2297615
37. Fu Y et al (2015) Promoting cooperation in construction projects: an integrated approach of contractual incentive and trust. Constr Manage Econ Routledge 33(8):653–670. https://doi.org/10.1080/01446193.2015.1087646
38. Goldhamer H, Shils EA (1939) Types of power and status. Am J Sociol 45(2):171–182. https://doi.org/10.1086/218263
39. Gosling J, Naim M, Towill D (2013) Identifying and categorizing the sources of uncertainty in construction supply chains. J Constr Eng Manag 139(1):102–110. https://doi.org/10.1061/(ASCE)CO.1943-7862.0000574
40. Hair JF et al (2012) An assessment of the use of partial least squares structural equation modeling in marketing research. J Acad Mark Sci 40(3):414–433. https://doi.org/10.1007/s11747-011-0261-6
41. Hair JF et al (2010) Multivariate data analysis, 7th edn. Pearson Education. https://doi.org/10.1016/j.ijpharm.2011.02.019
42. Hair JF et al (2014) A primer on partial least squares structural equation modeling (PLS-SEM). SAGE Publications, Inc., p 2455
43. Henseler J, Ringle CM, Sinkovics RR (2009) The use of partial least squares path modeling in international marketing. Adv Int Mark 20:277–319. https://doi.org/10.1108/S1474-7979(2009)0000020014
44. Holmstrom B (1979) Moral hazard and observability. Bell J Econ 10(1):74. https://doi.org/10.2307/3003320
45. Hosseinian SM (2016) An optimal time incentive/disincentive-based compensation in contracts with multiple agents. Constr Econ Build 16(4):35–53
46. Howell RD, Breivik E, Wilcox JB (2007) Reconsidering formative measurement. Psychol Methods 12(2):205–218. https://doi.org/10.1037/1082-989X.12.2.205
47. Hughes W, Yohannes I, Hillig J-B (2007) Incentives in construction contracts: should we pay for performance?. In: CIB world building congress, pp 2272–2283. Available at: https://www.irbnet.de/daten/iconda/CIB5038.pdf
48. Ibbs CW (1991) Innovative contract incentive features for construction. Constr Manag Econ 9:157–169
49. Kadefors A (2007) Fairness in interorganizational project relations : norms and strategies Fairness in interorganizational project relations : norms and strategies. Constr Manage Econ 6193. https://doi.org/10.1080/01446190500184238
50. Kahneman D, Tversky A (1979) Prospect theory: an analysis of decision under risk by. https://doi.org/10.1016/B978-0-12-388437-4.00006-5
51. Kuppelwieser VG, Sarstedt M (2014) Applying the future time perspective scale to advertising research. Int J Advert 33(1):113–136. https://doi.org/10.2501/IJA-33-1-113-136
52. Laffont J-J, Tirole J (1988) The dynamics of incentive contracts. Econometrica J Econometric Soc 56(5):1153–1175
53. Lampel J, Millert R, Floricelt S (1996) Information asymmetries and technological innovation in large engineering construction projects. R&D Manage 26(4):357–368
54. Lawler EJ, Yoon J (1993) Power and the emergence of commitment behavior in negotiated exchange. Am Sociol Rev 58(4):465–481
55. Lloyd H (1996) Prevalent philosophies of risk allocation—an overview. Int Constr Law Rev 13(4):502–548

56. Lui SS, Ngo HY (2004) The role of trust and contractual safeguards on cooperation in non-equity alliances. J Manage 30(4): 471–485. https://doi.org/10.1016/j.jm.2004.02.002
57. Lui SS, Wong Y, Liu W (2009) Asset specificity roles in interfirm cooperation: Reducing opportunistic behavior or increasing cooperative behavior? J Bus Res Elsevier Inc. 62(11):1214–1219. https://doi.org/10.1016/j.jbusres.2008.08.003
58. Luo Y (2007) The independent and interactive roles of procedural, distributive, and interactional justice in strategic alliances. Acad Manag J 50(3):644–664. https://doi.org/10.5465/AMJ.2007.25526452
59. Masten SE et al (1991) The costs of organization. J Law Econ Organ 7(1):1–25. https://doi.org/10.1093/oxfordjournals.jleo.a037001
60. De Melo RSS, Granja AD, Ballard G (2013) Collaboration to extend target costing to non-multi-party contracted projects: evidence from literature. In: 21st Annual conference of the international group for lean construction 2013, IGLC 2013, pp 21–30
61. Oliver P (1980) Rewards and punishments as selective incentives for collective action: theoretical investigations. Univ Chicago Press J 85(6):1356–1375
62. Pang HY et al (2015) Opportunism in construction contracting: minefield and manifestation. Int J Proj Organ Manage 7(1). https://doi.org/10.1504/IJPOM.2015.068004
63. Ringle CM et al (2018) Partial least squares structural equation modeling in HRM research. Int J Hum Resour Manage, pp 1–27. https://doi.org/10.1080/09585192.2017.1416655. Routledge
64. Ross SA (1973) The economic theory of agency: the principal's problem. Am Econ Rev 63(2):134–139
65. Schieg M (2008) Strategies for avoiding asymmetric information in construction project management. J Bus Econ Manag 9(1):47–51. https://doi.org/10.3846/1611-1699.2008.9.47-51
66. Shen LY, Wu GWC, Ng CSK (2001) Risk assessment for construction joint ventures in China. J Constr Eng Manage 127(1):76–81. https://doi.org/10.1016/j.molcel.2015.01.036
67. Smith JB, Barclay DW (1997) The effects of organizational differences and trust on the effectiveness of selling partner relationships. J Mark 61(1):3–21. https://doi.org/10.2307/1252186
68. Suprapto M et al (2015) Sorting out the essence of owner-contractor collaboration in capital project delivery. Int J Proj Manage 33(3):664–683. https://doi.org/10.1016/j.ijproman.2014.05.001. Elsevier Ltd and International Project Management Association
69. Suprapto M et al (2016) How do contract types and incentives matter to project performance? Int J Proj Manage 34(6):1071–1087. https://doi.org/10.1016/j.ijproman.2015.08.003. Elsevier Ltd and Association for Project Management and the International Project Management Association
70. Thibaut J, Gruder CL (1969) Formation of contractual agreements between parties of unequal power. J Pers Soc Psychol 11(1):59–65. https://doi.org/10.1037/h0027033
71. Wetzels M, van Oppen GO-SC (2009) Using PLS path modeling for assessing hierarchical construct models: guidelines and empirical illustration. MIS Q 33(1):177–195
72. Williamson, O. E. (1985) *The Economic Institutions of Capitalism., China social sciences publishing house chengcheng books ltd.* doi: https://doi.org/10.5465/AMR.1987.4308003.
73. Williamson OE (1979) Transaction-cost economics: the governance of contractual relations. J Law Econ 22(2)
74. Winch G (1998) Zephyrs of creative destruction: Understanding the management of innovation in construction. Build Res Inf 26(5):268–279. https://doi.org/10.1080/096132198369751
75. Xiang P et al (2012) Study on the use of industrialised building system in Malaysian private construction projects. J Constr Eng Manag 138(11):1303–1311. https://doi.org/10.1061/(ASCE)CO
76. Zaghloul R, Hartman F (2003) Construction contracts: the cost of mistrust. Int J Project Manage 21(6):419–424. https://doi.org/10.1016/S0263-7863(02)00082-0
77. Zhang S et al (2016) Contractual governance: effects of risk allocation on contractors cooperative behavior in construction projects. J Constr Eng Manag 142(6):1–11. https://doi.org/10.1061/(ASCE)CO.1943-7862.0001111

78. Zhu L et al (2020) Success DNA of a record-breaking megaproject. J Constr Eng Manag 146(8):05020009. https://doi.org/10.1061/(asce)co.1943-7862.0001878
79. Zou PXW, Zhang G (2009) Managing risks in construction projects: Life cycle and stakeholder perspectives. Int J Constr Manag 9(1):61–77. https://doi.org/10.1080/15623599.2009.10773122
80. Zou PXW, Zhang G, Wang J (2007) Understanding the key risks in construction projects in China. Int J Project Manage 25(6):601–614. https://doi.org/10.1016/j.ijproman.2007.03.001

Inter-organisational Relationship and Conflict Resolution

Liuying Zhu and Sai On Cheung

Introduction

Interorganizational relationships describe the pattern of bonding among organizations. When independent and autonomous organisations have to work together like construction projects, effective bonding among them is essential [12]. Most of the theories on motivation are about individual behaviours. Moreover, in construction contracting, individual team members are agents of their respective organizations and do not participate merely as individuals [6]. Thus these motivation theories may not be directly applicable. In this regard, Guitot [24] clarified that organizations involve role relationships rather than interpersonal relationship. It has been argued that the ways in which individuals make sense of others' intentions and behaviours are subject to what role they are playing. This suggests that behaviour may change when individuals are performing in a role context. An individual may be willing to work with his counterpart in his "qua persona" relationships, and he may not be able to do the same in the capacity of a member of his organization. Studies of motivation in construction contracting should thus be carried out at organization level. In construction project, managing inter-organizational relationships (IOR hereafter) is a delicate but critical management function [37]. It is challenging to coordinate mega project team for involvement of large number of member intercommoned with a my mind of contractual network [57]. It has been commented that construction project team members work together in a temporary that has certain objectives. Moreover members also have their own interests too [6]. It is not uncommon to find these members are acting for the interest of their own organisation. Disputes and

L. Zhu (✉) · S. O. Cheung
Construction Dispute Resolution Research Unit, City University of Hong Kong, Hong Kong, China
e-mail: liuyinzhu3-c@my.cityu.edu.hk

S. O. Cheung
e-mail: Saion.cheung@cityu.edu.hk

S. O. Cheung (ed.), *Construction Dispute Research Expanded*, Springer Tracts in Civil Engineering, https://doi.org/10.1007/978-3-030-80256-1_7

conflicts would surface when team members seek to maximize their own benefits at the expense of others [42]. Typically, their attitude would be defensive if not opportunistic. Major change in this type of working style is rigidly needed. Based on the analysis of four construction projects in Australia, Rose and Manley [52] found that seamless relationships underpin efficient project management. The linkage between IOR maintenance and project performance improvement is the focus of this study [39]. Four working objectives are listed:

(1) To conceptualize inter-organizational relationships in construction contracting;
(2) To examine the relationship between IORs and conflict resolution;
(3) To empirically test the relationship developed in (2); and
(4) To suggest ways to improve IORs.

Conceptualizing IOR in Construction

Oliver [49] developed six elements of IOR, which include (i) necessity (relationships are formulated because of legal requirements); (ii) stability (an adaptive response to environmental uncertainty); and (iii) legitimacy (organizations are motivated to interconnect for social reputation because of environmental pressure [49]. The other three IOR elements are efficiency, asymmetry and reciprocity. These elements are organization specific and may be established through contracts. In these regards, these three elements are used to identify IOR in this study. The following subsections articulate these elements in a construction context.

Efficiency

Transaction cost economics theory [58] illustrates that the formation of IORs can be invaluable to assist organisations to minimise costs of the transaction [49]. Cooperation among different organizations has been proved to be instrumental to raise efficiency. Organizations seeking to enhance inter-organizational cooperation are taking proactive steps to preserve valuable resources [2]. In fact, project monitoring systems are quite commonly used to control the use of resources. Attainment of specification is the minimum. Raising efficiency means achieving standard higher than the baseline requirements.

Asymmetry

Asymmetry between organizations can be expressed by the power or control one organization has over the other [49]. Whilst enhancing efficiency drives organizations to cooperate, asymmetry may result in organizations attempting to exert power, influence, or control over other units, especially those having scarce resources. Contract governance is therefore used to shape, influence or control others' behaviour in a contractual relationship. Information differential may well be the most crucial form of asymmetry away contracting parties [26]. Principal-agent theory suggests that if a principal (developer) cannot observe the agent's (contractor's) behaviour properly, moral hazard will arise. In these circumstances, the agent will only maximize his own benefit even at the expense of the interest of the principal [53]. In response, the principal may seek to exercise greater control through the use of contractual power. Eisenhardt [17] proposed that observability through incentives or disincentives is a useful way to balance the information differential [17]. For construction projects, self-reporting and right to inspect are typically installed as part of the project management system to make the performance of contractors more observable [1, 11].

Reciprocity

The human instinct of altruism can be a powerful force to bring self-interested individuals to cooperate [19]. Effective multiparty coordination and equitable exchange characterise admirable IOR [49]. Exchange theory projected that cooperation can be reciprocal to the level of interdependency [18, 49]. Cheung et al. [10] did find that interdependency underpins IOR building. It is further advocated that by aligning parties' objectives, a more proactive, cooperative relationship among organizations can be resulted. Nonetheless, major cultural shifts from the self-interest focused mindset is needed [5]. Project members are then more likely to cooperate to achieve common goals when reciprocity can be expected. When this cycle becomes initialised, cooperative working will be more enduring.

The Relationship Between IOR and Conflict Resolution

Simon [55] pointed out that individuals are "passionate economists" when making decisions because their rationality is somewhat bounded. They may settle with "satisfying" instead of "optimal" outcomes. People are sensitive to the identified patterns of the relation exchange. For example, an individual with low status is more dependent on an individual with high status and is therefore more sensitive to relationship issues [20]. Similar sentiment is also featured in inter-organizational relationships. For example, people may feel angry about unfair transactions [22]. Without the possibility of restoring equity, distress may inhibit the development of IOR. Moreover, in the case of a lacking of mutual trust, cooperation is less likely. The natural consequence is ineffective communication between them.

Their business relationship will be worsen. They will become more noncooperative with the state of distrust aggregates, When this happens, instead of working together to face crisis or problems, opportunistic exploitation is the more likely scenario [41].

Effective contractual governance is therefore needed to manage IOR. Cheung et al. [10] illustrated that establishing interdependency can improve IOR. Sophisticated project management strategies can be devised to promote the smooth running of mega projects. Some classical examples includes the contractual use of joint risk management [47], partnerships [3], or information sharing platforms [54].

IOR is one of the primary contextual factors for effective project monitoring. Through analysing 113 capital projects, Suprapto et al. [57] suggested that a partnering/alliance contract with a positive relational attitude and good team working quality is likely to perform better than conventional contractual arrangements. It is further found that cooperative construction organizations help minimize transaction costs for projects of high uncertainty and complexity [39]. Relational contracting has been advocated as an effective means to improve performance and profit margins in construction projects [36]. Summarizing the abovementioned IOR elements, an IOR-project performance relationship framework is proposed and presented in Fig. 1.

Case Study: The PRES in the HZMB Project

To examine the proactively of the framework in practice, the Hong Kong-Zhuhai-Macao Bridge (HZMB) project is studied.

Project Particulars

The HZMB project is located at the Pearl River Estuary adjacent to the Hong Kong International Airport. This project was jointly initiated by the Guangdong Government, the Hong Kong Government and the Macao Government (the three governments hereafter). The HZMB has been planned to be one of the landmark infrastructures in China. The Hong Kong portion of the HZMB is also one of 10 major infrastructure projects initiated by the HKSAR Government [27, 29]. The HZMB used over 400,000 tons of steel that is enough to build 60 Eiffel Towers. It is recognized as the world's longest across channel bridge and the steel structure with the highest tonnage.

At the initial stage of the HZMB project, some basic principles in terms of finance, construction and operation were agreed upon by the three governments. In 2010, the Hong Kong Zhuhai Macao Bridge Authority (HZMBA hereafter) was established. The HZMBA is responsible to manage the construction, operation and maintenance of the HZMB project. The HZMBA has the following pledges [30]: (1) build a world-class channel; (2) provide high-quality service to users and (3) deliver a landmark bridge in China. In addition to these pledges, four major challenges were identified [30]: coordination, construction, technical difficulties and environmental protection.

(1) Coordination: The project was under the management of the three governments with different regulations, management systems and communication styles [61]. Complicated power relations and multilevel governance preserved extra challenges in a tripartite project [60].
(2) Construction: The three places had different construction standards and there had been no previous reference project to follow. The sophisticated construction tasks also require innovations. Many new techniques were needed. The collective effort of all parties involved was the only way possible to deal with the challenges.
(3) Technical difficulties: The project faced many unprecedented technical challenges. The changing state of the ocean current affected the formation of the artificial islands. There were also difficulties regarding sedimentation prevention in the immersed tunnel. Moreover, according to the construction specifications, advanced undersea waterproof materials needed to achieve a 120-year design life span.
(4) Environmental protection: The bridge is located across the Chinese White Dolphin National Nature Reserve District, comprehensive environmental measures were required to protect the marine ecology.

The case study for this study was based on the most critical and challenging portion of the HZMB and involves two artificial islands connected by a 6.7 km immersed tunnel (ATA hereafter) [30]. ATA is a new type of construction. First, the two artificial islands are surrounded by 120 steel drums as island walls. The eastern island contains 61 drums, and the western island contains 59 drums. Second, the immersed tunnel is the longest (5.7 km) and deepest (48 m) immersed tube tunnel in the world. The immersed tube tunnel is connected by 33 pipe joints, of which 29 are 180 m long. Four other immersed tubes connecting two artificial islands, which are 112.5 m long, consist of 5 segments [32]. Third, each standard immersed tube weighs 74 thousand tons, making these tubes the heaviest that have ever been fabricated. The technical challenges made the project exceptionally risky. The uncertainties were enormous. To capitalise on the expertise of the contractor, design-and-build procurement was used. The ATA project is therefore a super mega project characterized by features identified by Flyvbjerg [21]: (1) large-scale, (2) complex, (3) high value, (4) long period, and (5) having significant social impacts. Table 1 summarizes the particulars of the ATA project.

Flyvbjerg [21] found that most mega projects ended with cost overruns, delays and disputes. Nonetheless, the ATA project appears to be an exception. After eight years of exceptional efforts, the HZMB project was opened on time for use in October 2018. This was facilitated by many engineering and management innovations that include 450 patents that made the project completion meeting the planned schedule [30]. As an example, the new technology used for rapid island formation shortened the construction period by more than two years. All of the technical accomplishments together with the management experience will benefit similar cross-channel bridges and immersed tunnel mega projects in the future.

Table 1 Particulars of the ATA project

Project particulars		Description
1	Scope of work	1. Permanent works a. Overall design of the artificial islands, tunnel b. Construction work of artificial islands, tunnel, etc. 2. Temporary works a. All temporary work for the required permanent project, such as planning, dismantling and restoring the original form b. Construction and dismantling of relevant camps
2	Procurement method	Design and build
3	Contracting arrangement	Joint venture
4	Project contract value	The estimated cost of the whole project is about 13.1 billion RMB
5	Form of contract	FIDIC (with modifications)
6	Payment period	Quarterly
7	Contract duration	Maximum 75 months
8	Commencement date	January 2011
9	Time for completion	83 months
10	Date of completion	December 2017
11	Defect liability period	5 years

In view of the expectations of the governments and people of the three places, embracing the concerted efforts of the project team members was top priority for the project management team. A project reputation evaluation system (PRES) was devised. PRES serves as a project management tool. In addition, the disincentivisation arrangements heled in creating state of interdependency between developer and contractor. Given the success of the ATA project, insightful lessons can be learned from understanding the operation and value of the PRES. The analysis of PRES was carried out from two perspectives:

(1) The use of the PRES as a project control tool in the ATA project and
(2) Anecdotal evidence of the PRES in promoting IORs and reducing construction conflicts.

The Use of the PRES as a Project Control Tool in the ATA Project

PRES was designed to assist the HZMBA to manage this challenging project. Given the enormous difficulties and expectations, the HZMBA realized that working closely with the contractors was necessary [62] so that problems can be tacked on the spot.

At the same time, the effort of the contributors cannot be ignored. On these bases, the HZMBA decided to include these anticipations in the contract governance through a project reputation evaluation system (PRES) [8]. To start the case study, a pilot study was first conducted with 10 senior project team members to have an overview of the project management strategies. The findings from the pilot study are summarized as follows:

(1) The PRES was a primally a useful **project control tool**. As the initiator of the PRES, the HZMBA confirmed that the design of the PRES aimed to cover all key targets of the project. Therefore, the PRES scores thus can indicate the overall performance of the contractor. The PRES seeks to detail all forms of non-compliance behaviours so that compliance or otherwise can be assessed on a quarterly basis. The scores were used to pinpoint existing or upcoming problems.
(2) The PRES was also instrumental in building **IORs**. All the interviewees agreed that the detailed reporting required by the PRES had indirectly forced the stakeholders to communicate openly because they were all inter-connected. opined that the PRES improved interorganizational communication, which was vital in speeding up decision making. At the beginning of the project, the project team members were not familiar with each other's management styles. Moreover, the PRES systemised somewhat what and when responses were needed. Thus following a system avoided surprises. In fact, the PRES facilitated communication and exchange of observations. When the parties became more familiar with each other, the number of non-compliance behaviours dropped notably. Communication and the sense of involvement were further enhanced through the opportunity of discussing the quarterly scores by the contractor with the evaluation offline.

The PRES therefore worked like a special project management tool for the ATA project. The PRES has 148 clauses. To motivate the contractors' performance, disincentivisation arrangements were also installed to work with the PRES.

The PRES had four principle functions:

1. **Goal commitment**

All the project pledges and key milestones were operationalised in the PRES as 6 project goals:

(1) Quality management: Ensure that the project had a 120-year life span and met all the required standards with a 100% acceptance rate.
(2) Health, safety and environment (HSE) management: Pursue zero injury, zero pollution and zero accidents in this project. Reach the advanced level of construction project requirements in health, safety and environmental management. Protection of white dolphins was specifically stated.
(3) Procedure management: (1) Completion on time. (2) Maximum utilisation of resources. (3) Total management plan to cover works, quality and budget control.

(4) Cost Control: (1) Cost efficiency is considered from a life-cycle perspective. The cost management of the project referred to the life cycle cost efficiency and value management. All life cycle cost measures (such as agency costs (ACs), user costs (UCs), and environmental costs [51]) were developed based on the "Measures for the Preparation of Investment Estimation for Highway Engineering Capital Construction Projects" [46]. (2) The preliminary design estimate should be maintained with a tolerance of ±10%. (3) The total cost move over should be the estimate for approved preliminary design.
(5) Information management: Establish an information sharing system to support the holistic management of the project due regard to regulations of the these governments.
(6) Innovation: Cultivate scientific innovations in technology and management to promote project efficiency.

2. **Reward/Responsibility reallocation**

Two percent of the total contract value was set aside to support the operation of the PRES. Performance thresholds were set based on previous project performance records collected from the evaluation committee (EC) of the HZMBA. The thresholds were used as reward attainment targets.

3. **Monitoring method**

Comprehensive evaluation was carried out quarterly by the EC, whose members were the heads of different departments of the HZMBA. The responsibilities of the EC include:

(1) Setting detailed standards according to the contents of each assessment;
(2) Organizing quarterly meetings to report and discuss performance evaluations;
(3) Examining the final assessment scores; and
(4) Evaluating the creditability of the contractors.

4. **Performance Assessment**

The EC conducted independent performance evaluation on contractors according to the grading guidelines. The maximum score was 100, and the score distribution according to different project goals is shown in Table 2.

The evaluation was carried out by way of mark deduction according to a predetermined scale included in the tender document. When project inspectors from the EC observed non-compliance behaviours, the points would be deducted accordingly. If a contractor made major errors or deviated from the provisions stipulated in the

Table 2 Percentage weightings of different targets

Item	Quality	HSE	Procedure	Cost	Information	Innovation
Ratio of score (%)	35	35	15	5	5	5

Table 3 The performance level and respective payment ratio of the PRES

Comprehensive evaluation score: L	Performance assessment grade	Payment adjustment ratio
$L \geq 90$	AA	100%
$85 \leq L < 90$	A	90%
$80 \leq L < 85$	B	70%
$75 \leq L < 80$	C	50%
$L < 75$ or the qualification is cancelled	D	0

contract, the assessment would be 0 points. Cost penalties would be enforced. In addition, other penalisations like down quality of credentials was also possible. The calculation of the project performance score in each quarter is shown in Eqs. (1) and (2):

$$Deduction_i = \sum (D_1 + D_2 + \ldots + D_n) \quad (1)$$

$$PERSC_i = SC - Deduction_i \quad (2)$$

where $Deduction_i$ = The overall deductions for non-compliance behaviours in quarter i. According to the project duration, there were 28 quarters in this project.

$D_1 \ldots D_n$ = The recorded deduction for the specific n non-compliance behaviours in quarter i.

$PERSC_i$ = The performance score in quarter i.

SC = The total performance score (typical value = 100).

Performance was directly linked to the payment adjustment system in the PRES [50, 33]. Table 3 summarises the payment ratios according to the performance assessment results.

A contractor that received two consecutive "D" grade was considered to have breached the contract, and the HZMBA could terminate the contract. Deductions were implemented according to the performance score of each quarter if the EC observed any non-compliance behaviours according to the PRES requirements. All the scores, together with the rankings of all the contractors, were announced each quarter. The specifications of the PRES are given in Table 4.

The Incentivizing Functions of the PRES

For project performance, the PRES functioned as a contractual project control tool. The primary design of the PRES was to formalise working targets and rewards. The clear requirements were established through consultation and served as the catalyst in

Table 4 The Specification of the PRES

Item	Project reputation evaluation system
Main objective	Improve this project's performance based on different goals
Goals and distribution of scores	Quality (35%); HSE (35%); Procedure (15%); Cost (5%); Information management (5%); Innovation (5%)
Bonus/penalty ratio	2% of contract value
Nature of Incentivisation	Disincentivisation
Payment method	The payment of the bonus was integrated into every interim payment
Evaluation method	Quarterly
Assessment method	The calculation was mainly by deductions. A form of classified payment was set. (Scores over 90 -100%, descending in turn)
Future chance for bidding	The performance was noted as a reference for future tendering. Contractors with high scores would be given priority
Feedback from the contractors	The deduction was too harsh in the first quarter. Negotiations were conducted to enhance flexibility after the first quarter

bring about the utmost efforts of both the developer and the contractor [7]. Eisenhardt [16] and Hosseinian [31] suggested that for monitoring highly programmable tasks, setting clear rewards/punishments is the prerequisite. Clear targets minimize the chance of misunderstanding the expectations. Furthermore, breaking down tasks into more discrete packages can make the targets more manageable [31, 58]. For construction projects, clear targets capitalise expectations [31]. Agency theory also postulates that the criteria and measurability of performance should be prepared by the principal [16, 17]. Outcome-based behaviours have only a partial control effect, while behaviour monitoring helps synchronize multiple goals [13, 15]. In the ATA project, the PRES offered both project monitoring and performance incentivising function [35].

The review of the PRES also pinpointed the importance of promoting IOR and enhancing project performance [31, 45]. In the ATA project, this was mainly reflected as the incentivising effect. Two percent of the total contract value was designated as the 'reward' of the PRES. Incentivisation and disincentivisation are regularly used as project control measures to alleviate opportunism in construction projects [7]. Reward will be given when performance exceeds agreed targets while underperformance will be penalised [9, 44]. Many studies have analysed the use of incentive schemes in reducing disputes and nurturing innovation [33]. Oliver [48] found that disincentives can raise unanimous cooperation. Disincentives are costless when compared with a positive financial bonus. In an ideal situation, if everyone cooperates, the cost of disincentives is only the design and implementation efforts [48].

The pilot interviews provide guidance for understanding the effect of the PRES. Many interviewees considered that the PRES was valuable system to manage IOR. As a result, they were willing to cooperate [31, 45]. Table 5 shows the IOR elements and the devices under the PRES.

Table 5 The IOR elements as vehicles under the PRES

No.	IOR element	Intention	Devices through project monitoring:	Manifestations in the PRES	Key references
1	Efficiency	An organization's attempt to improve its input/output ratio	Setting tasks as highly detailed contractual specifications to achieve specific project objectives	"The Authority set six major project targets with plans to monitor and assess the project performance" "Project team members with better assessment results should be given full-line notification and praises, while those with worse assessment results should be reported and criticized. Rewards or penalties should be released based on the assessment of the performance"	[23, 58]
2	Asymmetry	The potential to exercise additional power or control over another organization and/or its resources	Improving observability through setting programmable tasks	"In the process of project construction, the participating units strictly comply with the requirements of objectives, norms, contracts and documents. The Authority should supervise and inspect the implementation of relevant requirements. Assessment should be conducted in each quarter to ensure that the quality of project construction is always under control"	[4, 15]

(continued)

Table 5 (continued)

No.	IOR element	Intention	Devices through project monitoring:	Manifestations in the PRES	Key references
3	Reciprocity	Emphasizes cooperation, collaboration, and coordination among organizations, rather than domination, power, and control	Aligning the objectives of all project members and promoting cooperation	"All the project team members shall follow the outline for the HZMB project, with the aims to (1) build a world-class oversea channel, (2) provide high-quality services for users, and (3) become a landmark bridge in China" "On the basis of the strict implementation of contracts, this project promotes the concept of a win–win partnership and integrated management for all project team members"	[18, 49]

Adopting the abovementioned IOR framework (Fig. 1), the framework for the ATA project is presented in Fig. 2.

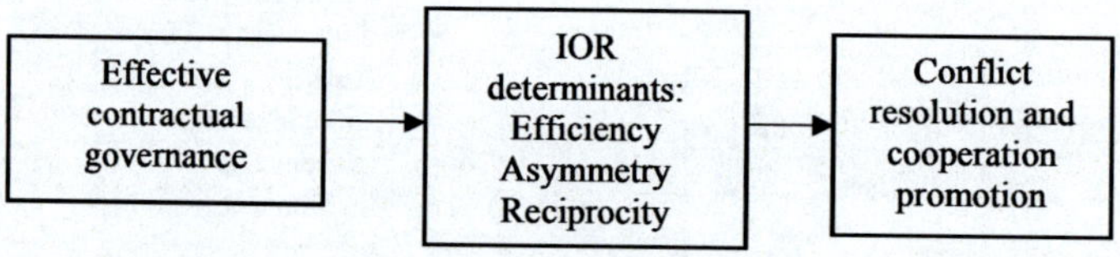

Fig. 1 An IOR project performance relationship framework

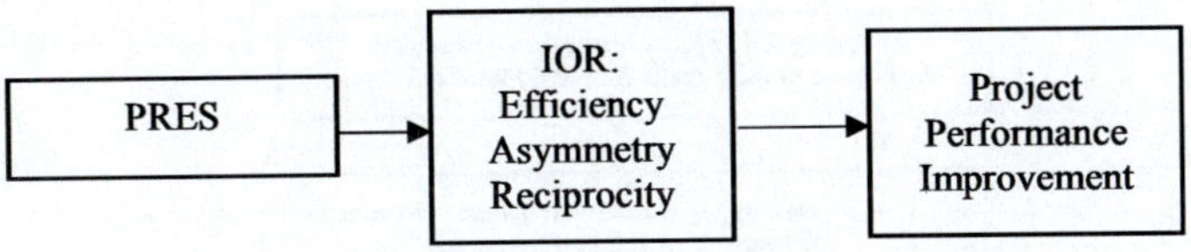

Fig. 2 An IOR-project performance relationship framework

Anecdotal Evidence of the PRES in Promoting IOR

Figure 2 hypothesized the potential linkages between the PRES, IOR and project performance improvement. Further evidence was sought through the following:

(1) Trend Analysis

Trend line analysis was performed in order to obtain the pattern of change overtime [25]. Such analysis can track the changes in project performance as well as the IOR elements during the project duration. There are two parts of data analysis:

(i) Part A: Investigate the evidence of IOR improvement respective to project goals.

To accomplish this objective, 28 sets of quarterly PRES evaluation scores covering the overall 8-year construction period were examined. The data include all the non-compliance behaviours of the main contractor. 197 non-compliance behaviours were recorded with deductions of performance scores. These non-compliance behaviours are first classified with reference to the six goals (Quality, HSE, Schedule, Cost, Innovation, Information management). To further investigate the intention of making all these deductions, the non-compliance behaviours were further grouped with reference to the IOR elements. Trend line were plotted by the log value of the scores. Trend line analysis was considered the best fit when the rate of change in the data increased or decreased quickly and then levelled out [14]. As hypothesized, the deductions related to the three IOR elements showed a gradual drop.

(ii) Part B: Investigate the influence of different IOR elements on general project performance improvement.

The second part of data concerned the analysis examines the relationship between the three IOR elements and the overall project performance. The trend of the scores may show if the PRES had helped achieving the project outcomes. Thus, this part of data analysis aim to examine (a) whether the contractor had achieved the specific performance level as expected; and (b) in what ways the IOR elements contributed to the achievement of the project objectives.

A second round of discussion was then conducted with key project team members to comment on the observations of the trend analysis. Figure 3 summarises the flow of the data analysis.

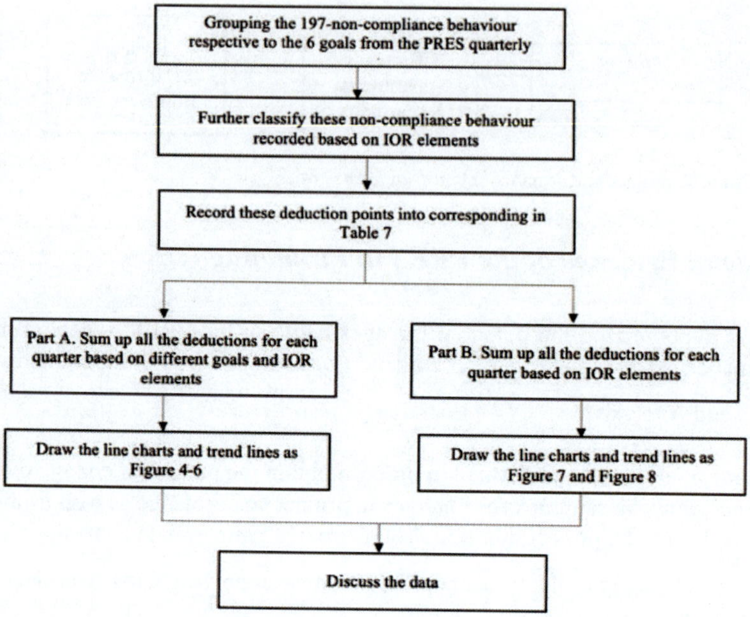

Fig. 3 Flowchart of the empirical study

Part A. The evidence of IOR enhancement

As mentioned in the previous section, to evaluate how the IOR concept was developed in this project, all non-compliance behaviours were grouped respective to the six project targets. To further evaluate the development of IOR in the ATA, all these non-compliance behaviours were then grouped respective to the three IOR elements. Table 6 shows how the data were grouped.

Table 7 shows the distribution of the deductions based on different project objectives and IOR elements.

With reference to Table 7, there had been major deductions for non-compliance of quality and HSE targets. However, no deductions were found for the cost target and only a few for procedure, information and innovation. Ever with deductions, the deduction ratio was relatively small and did not have significant effects on payment. Thus, quality and HSE were having the most influence on the overall score. For the IOR elements, the differential was also seen in different project targets. Asymmetry and efficiency represented similar portions of deductions on quality. However, for HSE, most of the deductions were related to efficiency. The general decreasing trend of deductions for quality and HSE are noted in Fig. 4.

The deductions on quality gradually reduced. The deduction dropped from 8 to 4 for the quality scores. However, the performance score for HSE fluctuated. Comparatively, from Q1 to Q28, the deductions of HSE fluctuated around 2. The relatively larger drop in quality non-compliance behaviour suggests that the PRES

Table 6 Sample for the data analysis

No.	Assessment period	Non-compliance behaviours	Quarterly deduction	Category	
				Goal	IOR elements
1	Q1	Overall quality plan not submitted in time	1.0	Quality	Asymmetry
2	Q1	Poor quality plastic drainage board for soft foundation treatment found in construction camp	2.0	Quality	Efficiency
3	Q2	Poor coordination between the constructor and the designer	1.0	Quality	Reciprocity

were having effect. The trend analyses for the other IOR elements are shown in Figs. 5 and 6.

These two figures present the changes in the three IOR elements. A clear decreasing trend is shown for asymmetry in both quality and HSE. However, for efficiency, there is an opposite trend for these two targets. A decreasing trend of efficiency was detected in Fig. 5, while a slight increasing trend was noted for HSE (Fig. 6). There was basically no change for reciprocity in either quality or HSE.

Part B. The roles of IOR in project performance improvement

To obtain an overall view of how these three factors influenced project performance, a trend analysis of the project performance scores achieved by the design and build contractor was prepared (Fig. 7).

Figure 7 shows that except for the first quarter, the performance scores were consistently above 90 points. This means that the contractor obtained 100% of the expected payment except in the first quarter (90% for that quarter). For the general trend, there was a slight increase at the beginning, after which very slight increase was noted. The average score fluctuated around 93 representing a 3% higher than the baseline of grade AA (Table 7). This result seems resonating the optimizing behaviour of contractor identified by Wong et al. [59] who found that most contractors were "optimizers" and adjusted their resources to sustain their performance at a level that would not jeopardise their future work opportunities. There were no serious attempt to maximise their performance.

To further investigate the contribution of IOR elements in this project, analysis of the total deductions was also conducted. Figures 8 shows the trend of the total points deducted the three IOR elements.

Figure 8 shows that there was basically no change in reciprocity related deductions. In contrast, the deductions regarding asymmetry and efficiency decreased after

Table 7 The distribution of the deductions based on different project objectives and IOR elements

Quarter (i)	Quality 35%			HSE 35%			Procedure 15%			Information 5%			Innovation 5%			Cost 5%	PERSC
	As	Re	Eff	As	Re	Eff	As	Re	Eff	As	Re	Eff	As	Re	Eff		
1	4.5	0	3.5	0	0	3	0.5	0	0	0	0	3	0	0	0	0	85.5
2	3.5	0	0	1	0	2	0	0	2	0	0	0.5	1	0	0	0	90
3	3.5	1	4	0	0	0	0	0	0.5	0	0	0	0	0	0	0	91
4	0	0	4	0	0	0	0.5	0	0	0	0	0	0	0	0	0	95.5
5	3	0	2	0	0	3.5	0	0	0	0	0	0	0	0	1	0	90.5
6	2.5	0	1	0	0	3	0	0	0	0	0	0	0	0	0.8	0	92.7
7	2.4	0.5	2	2	0	0.5	0	0	0	0	0	0	0	0	1	0	91.6
8	2	1.5	1.5	0	0	1.5	0	0	0	0	0	0	0	0	0	0	93.5
9	0	1	3.5	0	0	1	0	0	1	0	0	0	0	0	0	0	93.5
10	0.5	1.5	0	0	1.5	0.5	0	0	1	0	0	0	0	0	0	0	95
11	0.5	0	3.5	0	0	2.5	0	0	0	0	0	0	0	0	0	0	93.5
12	1	1	1	0	0	1.5	0	0	0	0	0	0	0	0	0	0	95.5
13	2.5	0	2	0.5	0	0	0	0	0	0	0	0	0	0	0	0	95
14	0	1	5	1	0	2	0	0	0	0	0	0	0	0	0	0	91
15	3	0	1	0	0	3	0	0	0	0	0	0	0	0	0	0	93
16	1.5	2	1	0	0	1	0	0	0	0	0	0	0	0	0	0	94.5
17	3	1	0	0	0	4	0	0	0	0	0	0	0	0	0	0	92
18	3	1	0	0	0	4	0	0	0	0	0	0	0	0	0	0	92
19	5	0	1	0	0	1	0	0	0	0	0	0	0	0	0	0	93
20	0.5	0	1.5	0	0	4	1	0	0	0	0	0	0	0	0	0	93

(continued)

Table 7 (continued)

Quarter (i)	Quality 35%			HSE 35%			Procedure 15%			Information 5%			Innovation 5%			Cost 5%	PERSC
	As	Re	Eff	As	Re	Eff	As	Re	Eff	As	Re	Eff	As	Re	Eff		
21	1	0	2.5	0.5	0	2.5	0	0	0	0	0	0	0	0	0	0	93.5
22	2.5	3	0.5	0	0	0	0	0	0	0	0	0	0	0	0	0	94
23	0	0	3	0	0	0.5	0	0	1	0	0	0	0	0	0	0	95.5
24	1	0	0	0	0	5	0	0	3.5	0	0	0	0	0	0	0	90.5
25	0	0	3	0	0	3	1	0	0	0	0	0	0	0	0	0	93
26	1	1	4.5	0	0	2	0	0	0	0	0	0	0	0	0	0	91.5
27	0	0	1	0	0	2	0	0	0	0	0	0	0	0	0	0	97
28	0	0	0	0	0	2	0	0	0	0	0	0	0	0	0	0	98
Sum	**47.4**	**15.5**	**52**	**5**	1.5	**55**	3	0	9	0	0	3.5	1	0	2.8	0	85.5
%	**41%**	**13%**	**45%**	**8%**	2%	**89%**	25%	0%	75%	0%	0%	100%	26%	0%	74%	–	

Note As = Asymmetry; Re = Reciprocity; Eff = Efficiency; PERSC = Performance score

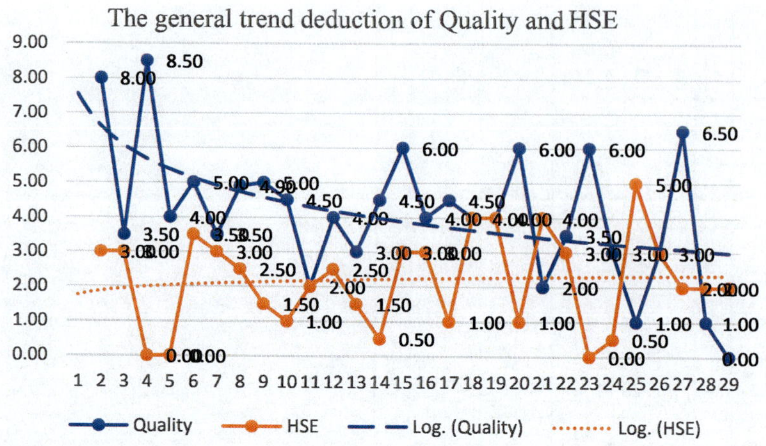

Fig. 4 The general trend of deduction of Quality and HSE

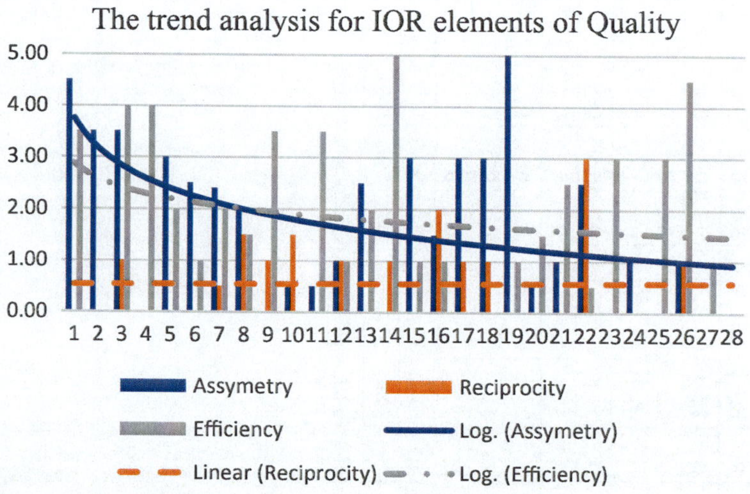

Fig. 5 The trend analysis for IOR elements of Quality

the first quarter. These echo the comments by the interviewees that a more understanding and reasonable approach to evaluate performance had been adopted after the learning from the first quarter. After the first few quarters, the decreasing trend of efficiency became very marginal. Conversely, there was a significant drop in asymmetry. At the final stages of the project, in Q24, Q26, Q27 and Q28, there were no deductions relating to asymmetry.

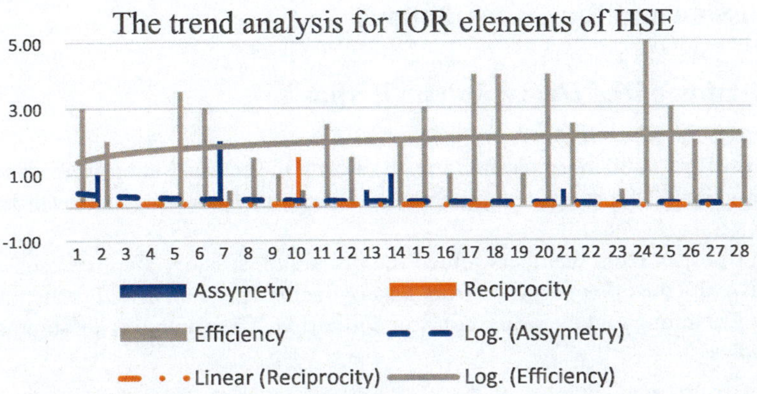

Fig. 6 The trend analysis for IOR elements of HSE

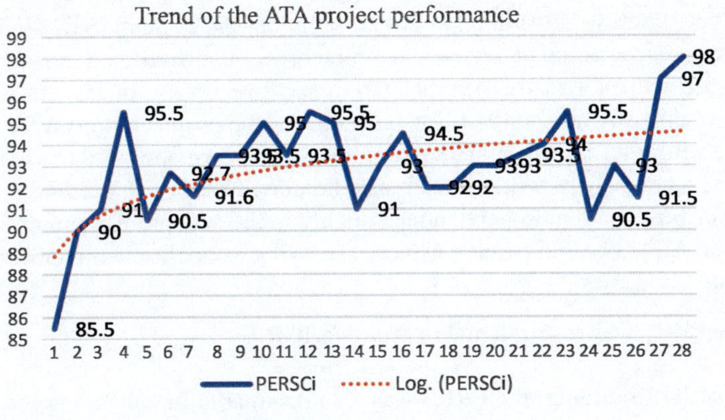

Fig. 7 Trend of the ATA project performance

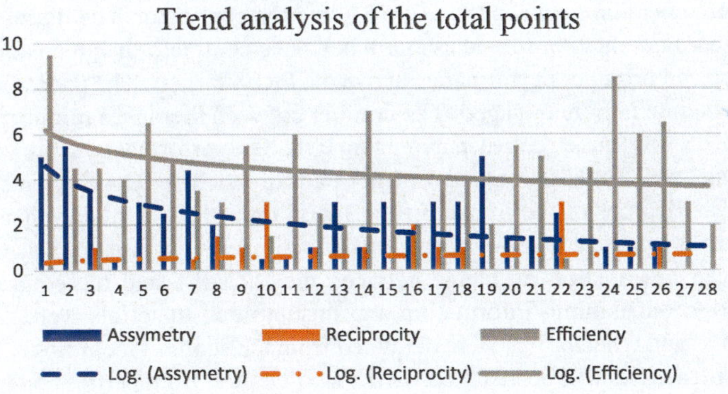

Fig. 8 Trend analysis of the total points

Discussion and Recommendations

Discussion of the Data Analysis Results

As part of the validation, the trend analyses and the interpretation were presented to the major ATA project team members. This part of the work is to validate the observations.

Four project team members joined the discussion session. Two represented the HZMBA and the other two were from the contractor. All of them had participated in the project throughout the whole construction period. Their opinions are summarized as follows:

(1) The PRES had been instrumental and most effective to manage quality target.

Both the contractor and the developer argued that the PRES was useful in monitoring the contracting behaviours of all parties under the evaluation by the PRES. In essence, non-compliance behaviours were recorded and discussed during the evaluation process. This had the benefit of retrospective review of ten the result was released. In this way less conflict was resulted. All the project targets were achieved at the end of the ATA project. The rise in project performance and the lowering of deductions were very pleasing for all stakeholders. Notably, upholding the quality targets was most challenging and indeed quality issues was more common in a mega project like ATA. The end result was very satisfying as quality was attained through many innovate efforts.

(2) The PRES was instrumental in building IOR.

In terms of IOR building, the PRES was also invaluable In enhancing the communication among the parties. At the beginning of the project, the organizations had yet known each other well enough. Deductions levied by the HZMBA had led to open communication between the HZMBA and the contractor. The discussion after the first quarter deduction served as alarm belt. The contractor had then adjusted her work force and raise her performance above the PRES targets. The parties since then had become much more engaged. The conflict between them was minimised.

The PRES was thus effective in managing IOR. The performance score was kept at a satisfying level. The different movement trends of the three IOR elements reflected the different degrees of the impact of the PRES on IOR. The major changes were effected on asymmetry. The sharp drop in deductions for asymmetry was likely due to enhanced information exchange between the HZMBA and the contractor. For each quarter, outstanding information and unsubmitted materials were raised and discussed as and when points were deducted from the scores. Accordingly, remedial or follow up actions will be installed for the next quarter. Both parties confirmed that exchanges of useful information were achieved during the whole project process.

The following summarizes the findings of the trend analyses and the views of the interviewees.

(i) The importance of setting performance targets

Among the six project goals, significant impacts on performance as primarily on quality and HSE. No deductions were recorded for cost. In fact, the contractor had very little relation with the project cost because the ATA project was a lump sum contract. In general, the HZMBA was less concerned about the cost than the contractor. For information and innovation, the contributing ratio are relatively small and in deed very few points were deducted. The low deductions support that showed the effective communication between the two parties was quite effective after then goals were aligned and crystalised through PRES.

(ii) Trend Analysis

- There was significant project performance improvement during the project. The improvement in quality was most notable.
- The improvement in asymmetry was more apparent than efficiency and reciprocity.

The PRES appeared to be able to balance information asymmetry and enhance inter-organizational communication during the project execution stage. A substantial drop in the project efficiency score was recorded for the first quarter. After Q23, the reduction in project efficiency presented basically no change. The trend of asymmetry deductions generally reached to zero. Suggesting that their communication had become seamless.

The deductions also show that imperfect information sharing existed at the beginning of the project. Although design and build procurement is considered a good way to integrate design and construction, research also shows that project delivery methods do not differ significantly in soliciting cooperation [38, 40, 54]. Some researchers have suggested that the source of motivation is not at the individual level but rather at the organizational level [6]. Effective contractual management and appropriate incentives can enhance communication and prevent opportunistic behaviours. The PRES helped improve project performance directly by enhancing interdependency and fostering cooperation [10]. The interviewees commented that despite the observation of no significant improvement in the three IOR elements, the PRES as a whole served as a means to bring the organizations together in terms of language, communication mode and performance goals.

Recommendations for Project Management

This study harvested valuable insight on construction project management:

(1) IOR enhancement as an integral part of contract planning and implementation

It has been well reported that amicable IOR could bring about conflict avoidance and performance improvement. Building IOR is therefore very well worthy. Relationship investment should therefore be an integrate part of project management [6]. If people enter into arrangements with the belief that their counterparts will present self-serving behaviours, they will adopt a defensive attitude. To guard against this, relational contracting [56] and invest on relationship can be a good strategy to cultivate IOR, To this ends, establishing a spirit of mutual trust and cooperation was considered absolutely necessary at the beginning of the project. For the ATA project, fostering IORs was on the top of the agenda. IORs aligned the project objectives of all the project members and aimed to promote cooperation.

(2) The use of incentivisation to reduce opportunistic behaviours

Both financial and nonfinancial rewards have been regularly used in construction projects to engender extra efforts. Can this be extended to cultivate IOR? The empirical evidence provided by the HZMB ATA project provided an affirmative answer. When parties' interest are aligned by the common goals like the reward targets of an incentive scheme, their attention would be directed to cooperate for the reward instead of raising dispute. Incentive schemes also bridge the asymmetry between the parties. As a result their relationship would be improved. The PRES of the ATA project was an exemplar. Several features of the PRES are worthy to be mentioned have. The publication of evaluation scores offered unintended motivator on the participating contractors as none of them would like to be seen as the 'black sheep'. Nonetheless, the PRES also has its contributor as most participating units were satisfying rather than maximising. Nevertheless, the PRES created the platform whereby the parties could stay away from the conventional opportunistic game plan.

(3) Close project monitoring as an effective way to reduce project conflicts

Based on principle-agent theory [17], opportunistic behaviour may occur when the principal (developer) cannot properly monitor the behaviour of the agent (contractor). Extra-contractual tools to reduce information asymmetry would be useful from the perspective of the principal. In the ATA project, the PRES provided the deadly needed platform for open information exchanges. The PRES addressed all major project objectives and guided innovative quarterly performance reports. The developer took a proactive role in setting performance standards and was committed to implementing the disincentivisation arrangements. More importantly, when necessary and appropriate, the working targets were adjusted to reflect the practical situation [43]. Because of that, many problems faced during the projects were brought up in the quarterly meetings. The scores reflected indicators of the performance level with scores lower than the agreed target indicated underperformance. The open performance reporting system helped reduce the likelihood of disputes and opportunistic behaviours. For mega projects, therefore, there is no substitute for well-planned project targets and dedicated project monitoring and control. By addressing information asymmetry and raising efficiency, the chances of the principal and the agent working cooperatively is not an impossibility.

Summary

Inter-organizational relationships (IORs) in construction contracting provides the bonding construction project team members. It is widely accepted that engendering motivation should be pitched at organizational level for construction projects [6]. In fact, managing IORs among contracting parties in mega-construction projects is particularly challenging. Non-cooperative attitudes and opportunistic behaviours are quite common in these complex mega projects. Effective contract management is thus essential to regulate contracting behaviours and promote cooperative working. The relationship between IOR maintenance and conflict resolution was investigated in this study that has four objectives: (1) to identify IOR concepts in construction; (2) to discuss the relationship between IOR and conflict resolution; (3) to verify the relationship through empirical study; and (4) to suggest ways to improve IORs. Efficiency, asymmetry and reciprocity were identified as the key constructs to conceptualize IORs in construction contracting. An IOR-project performance relationship framework was proposed. The Hong Kong-Zhuhai-Macao Bridge (HZMB) project offered valuable empirical evidence to support the proposed relationship framework. The HZMB project demonstrated exemplary use of disincentivisation arrangements in the forms of a project reputation evaluation system (PRES). The PRES was found to act as both a project monitoring tool and an incentivising agent. To further investigate the effects of the PRES in building inter-organizational relationships, 28 quarterly evaluation scores under the PRES were analysed. Analysing from the results of the trend analyses and two rounds of focus group discussions with key team members of the project, it was found that (1) IOR enhancement should be an integral part of contract planning and implementation; (2) incentivisation should be used to reduce opportunistic behaviour; and (3) IOR enhancement is an effective way to reduce conflict and hence disputes.

Acknowledgements The empirical work of this chapter has been reported in a paper entitled "Success DNA of a Record-Breaking Mega Project" of the Journal of Construction and Engineering Management. Special thanks to Mr. Gao Xinglin, Dr. Li Qian and Dr. Liu Gang in helping with the empirical study. Thanks for the support from the Hong Kong Zhuhai Macao Bridge Authority as well.

References

1. Abu-Hijleh SF, Ibbs CW (1989) Schedule-based construction incentives. J Constr Eng Manag 115(3):430–443. https://doi.org/10.1061/(ASCE)0733-9364(1989)115:3(430)
2. Barringer BR (2000) Walking a tightrope: creating value through interorganizational relationships. J Manag 26(3):367–403. https://doi.org/10.1177/014920630002600302
3. Bayliss R, Cheung SO, Suen HC, Wong SP (2004) Effective partnering tools in construction: a case study on MTRC TKE contract 604 in Hong Kong. Int J Project Manage 22(3):253–263. https://doi.org/10.1016/S0263-7863(03)00069-3

4. Benson JK (1975) The interorganizationa network as a political economy. Adm Sci Q 20(2):229–249
5. Bower D, Ashby G, Gerald K, Smyk W (2002) Incentive mechanisms for project success. J Manage Eng
6. Bresnen M, Marshall N (2000) Motivation, commitment and the use of incentives in partnerships and alliances. Constr Manag Econ 18(5):587–598. https://doi.org/10.1080/014461900407392
7. Bubshait AA (2003) Incentive/disincentive contracts and its effects on industrial projects. Int J Project Manage 21(21):63–70
8. Burati JL, Weed RM, Hughes CS, Hill HS (2003) Optimal procudures for quality assurance specifications.
9. Chan DW, Lam PT, Chan AP, Wong JM (2010) Achieving better performance through target cost contracts—the tale of an underground railway station modification project. Perf Meas Manage Facil Manage 28(5/6):261–277
10. Cheung SO, Zhu L, Lee KW (2018) Incentivization and interdependency in construction contracting. J Manag Eng 34(3):1–13. https://doi.org/10.1061/(ASCE)ME.1943-5479.0000601
11. Construction Industry Council (2012) Guidelines on the adoption of the pay for safety scheme, pp 1–20. Available at: www.hkcic.org
12. Cropper S, Ebers M, Huxham C, Ring PS (2008) Introducing Inter-organizational relations. In: Oxford handbooks online introducing, pp 1–23. doi: https://doi.org/10.1093/oxfordhb/9780199282944.003.0001
13. Dekker HC (2004) Control of inter-organizational relationships: Evidence on appropriation concerns and coordination requirements. Acc Organ Soc 29(1):27–49. https://doi.org/10.1016/S0361-3682(02)00056-9
14. Deshpande B (2014) 6 trend analyses to consider prior to time series forecasting. Available at: http://www.simafore.com/blog/bid/205162/6-trend-analyses-to-consider-prior-to-time-series-forecasting. Accessed: 26 July 2019
15. Eisenhardt KM (1985) Control: organizational and economic approaches. Manage Sci 31(2):134–149
16. Eisenhardt KM (1988) Agency and institutional theory explanations: the case of retail sales compensation. Acad Manag J 31(3):488–511
17. Eisenhardt KM (1989) Agency theory: an assessment and review. Acad Manag Rev 14(1):57–74
18. Emerson RM (1962) Power-dependence relations. Am Sociol Rev 27(1):31–41
19. Fehr E, Fischbacher U (2003) The nature of human altruism—proximate and evolutionary origins. Nature 425(October):785–791
20. Flynn FJ, Reagans RE (2006) Helping one's way to the top: Self- monitors achieve status by helping others and knowing who helps whom. J Pers Soc Psychol 91 (6): 1123–1137. https://doi.org/10.1037/0022-3514.91.6.1123
21. Flyvbjerg B (2017) The Oxford handbook of megaproject management. Oxford University Press. https://doi.org/10.1084/jem.20050882
22. Fu Y, Chen Y, Zhang S, Wang W (2015) Promoting cooperation in construction projects: an integrated approach of contractual incentive and trust. Constr Manage Econ Routledge 33(8):653–670. https://doi.org/10.1080/01446193.2015.1087646
23. Golicic SL, Foggin JH, Mentzer JT (2003) Relationship magnitude and its role in interorganizational relationship structure. J Bus Logist 24(1):57–75. https://doi.org/10.1002/j.2158-1592.2003.tb00032.x
24. Guitot M (1977) Attribution and identity construction: some comments. Am Sociol Rev 42:692–704
25. Hojem MA, Ottenbacher KJ (1988) Empirical investigation of visual-inspection versus trend-line analysis of single-subject data. Phys Ther 68(6):983–988. https://doi.org/10.1093/ptj/68.6.983
26. Holmstrom B (1979) Moral hazard and observability. Bell J Econ 10(1):74. https://doi.org/10.2307/3003320

27. Hong Kong Special Administrative Region Government (2007) The 2007–08 policy address Hong Kong
28. Hong Kong Special Administrative Region Government (2010) Ten major infrastructure projects, 2010–11 policy address. Available at: www.policyaddress.gov.hk/10-11/eng/pdf/projects.pdf
29. Hong Kong Zhuhai Macau Bridge Authority (2009) Introduction of HZMB project. Available at: http://www.hzmb.org/cn/bencandy.asp?id=2
30. Hong Kong Zhuhai Macau Bridge Authority (2017) The research challenges of Hong Kong-Zhuhai-Macau Bridge
31. Hosseinian SM (2016) An optimal time incentive/disincentive-based compensation in contracts with multiple agents. Constr Econ Build 16(4):35–53
32. Hu ZN, Xie YL, Wang J (2015) Challenges and strategies involved in designing and constructing a 6km immersed tunnel: a case study of the Hong Kong-Zhuhai-Macao Bridge. Tunn Undergr Space Technol 50:171–177. https://doi.org/10.1016/j.tust.2015.07.011. Elsevier Ltd
33. Hughes D, Williams T, Ren Z (2012) Is incentivisation significant in ensuring successful partnered projects? Eng Constr Archit Manag 19(3):306–319. https://doi.org/10.1108/09699981211219625
34. Hughes CS, Moulthrop JS, Tayabji S, Weed RM, Burati JL (2011) Guidelines for quality related pay adjustment factors for pavements
35. Hughes W, Yohannes I, Hillig J-B (2007) Incentives in construction contracts: should we pay for performance ? In: CIB world building congress, (2000), pp 2272–2283. Available at: https://www.irbnet.de/daten/iconda/CIB5038.pdf
36. Jelodar MB, Yiu TW, Wilkinson S (2016) A conceptualisation of relationship quality in construction procurement. Int J Proj Manage 34(6):997–1011. https://doi.org/10.1016/j.ijproman.2016.03.005. Elsevier Ltd and Association for Project Management and the International Project Management Association
37. Kim, K. K., Park, S., Ryoo, S. and Park, S. K. (2010) 'Inter-organizational cooperation in buyer-supplier relationships: Both perspectives', *Journal of Business Research*. Elsevier Inc., 63(8), pp. 863–869. doi: https://doi.org/10.1016/j.jbusres.2009.04.028.
38. Koolwijk JSJ, van Oel CJ, Wamelink JWF, Vrijhoef R (2018) Collaboration and Integration in Project-Based Supply Chains in the Construction Industry. J Manag Eng 34(3):1–14. https://doi.org/10.1061/(ASCE)ME.1943-5479.0000592
39. Kumaraswamy MM, Anvuur AM (2008) Selecting sustainable teams for PPP projects. Build Environ 43(6):999–1009. https://doi.org/10.1016/j.buildenv.2007.02.001
40. Laffont J-J, Tirole J (1988) The dynamics of incentive contracts 56(5):1153–1175
41. Lawler EJ, Yoon J (1993) Power and the emergence of commitment behavior in negotiated exchange. Am Sociol Rev 58(4):465–481
42. Lui SS, Ngo HY (2004) The role of trust and contractual safeguards on cooperation in non-equity alliances. J Manage 30(4):471–485. https://doi.org/10.1016/j.jm.2004.02.002
43. Luthans F (1997) A meta-analysis of the effects of organizational behaviour modification on task performance. Acad Manag J 40(5):1122–1149
44. Meng X (2015) Incentive mechanisms and their impact on project performance. In: Handbook on project management and scheduling. https://doi.org/10.1007/978-3-319-05915-0_17
45. Meng X, Gallagher B (2012) The impact of incentive mechanisms on project performance. Int J Proj Manage 30(3):352–362. https://doi.org/10.1016/j.ijproman.2011.08.006. Elsevier Ltd and IPMA
46. Ministry of Transport of the People's Republic of China (2018) Measures for the prepartation of investment estimation for highway engineering capital construction projects'
47. Motiar Rahman M, Kumaraswamy MM (2002) Joint risk management through transactionally efficient relational contracting. Constr Manag Econ 20(1):45–54. https://doi.org/10.1080/01446190110089682
48. Oliver P (1980) Rewards and punishments as selective incentives for collective action: theoretical investigations. Univ Chicago Press J 85(6):1356–1375

49. Oliver C (1990) Determinants of interorganizational relationships: integration and future directions published by: academy of management linked references are available on JSTOR for this article: determinants of interorganizational relationships: integration and futu. Acad Manag Rev 15(2):241–265
50. Praticò FG (2007) Quality and timeliness in highway construction contracts: a new acceptance model based on both mechanical and surface performance of flexible pavements. Constr Manag Econ 25(3):305–313. https://doi.org/10.1080/01446190601042426
51. Praticò F, Saride S, Puppala A (2011) Comprehensive life-cycle cost analysis for selection of stabilization alternatives for better performance of low-volume roads. Transp Res Rec 2204:120–129. https://doi.org/10.3141/2204-16
52. Rose T, Manley K (2011) Motivation toward financial incentive goals on construction projects. J Bus Res 64(7):765–773. https://doi.org/10.1016/j.jbusres.2010.07.003. Elsevier Inc.
53. Ross SA (1973) The economic theory of agency: the principal's problem. Am Econ Rev 63(2):134–139
54. Schieg M (2008) Strategies for avoiding asymmetric information in construction project management. J Bus Econ Manag 9(1):47–51. https://doi.org/10.3846/1611-1699.2008.9.47-51
55. Simon H (1972) Theories of bounded rationality, decision and organization. North-Holland Publishing Company. http://innovbfa.viabloga.com/files/Herbert_Simon___theories_of_bounded_rationality___1972.pdf
56. Smyth H, Edkins A (2007) Relationship management in the management of PFI/PPP projects in the UK. Int J Project Manage 25(3):232–240. https://doi.org/10.1016/j.ijproman.2006.08.003
57. Suprapto M, Bakker HL, Mooi HG, Hertogh MJ (2016) How do contract types and incentives matter to project performance? Int J Proj Manage 34(6):1071–1087. https://doi.org/10.1016/j.ijproman.2015.08.003. Elsevier Ltd and Association for Project Management and the International Project Management Association
58. Williamson OE (1985) The economic institutions of capitalism. China social sciences publishing house chengcheng books ltd. https://doi.org/10.5465/AMR.1987.4308003
59. Wong PSP, Cheung SO, Wu RTH (2010) Learning from project monitoring feedback: a case of optimizing behavior of contractors. Int J Proj Manage, pp 469–481. https://doi.org/10.1016/j.ijproman.2009.07.003
60. Yang C (2006) The geopolitics of cross-boundary governance in the Greater Pearl River Delta, China: A case study of the proposed Hong Kong-Zhuhai-Macao Bridge. Polit Geogr 25(7):817–835. https://doi.org/10.1016/j.polgeo.2006.08.006
61. Yu J, Huo J, Shi Q, Sheng G, Zhang H (2013) An exploration on the organizational structure design of the Hong Kong-Zhuhai-Macao Bridge Authority. In: The 19th International Conference on Industrial Engineering and Engineering Management, pp 975–985. https://doi.org/10.1007/978-3-642-38433-2
62. Zeng W, Zhang J, Wang H, Zhou H (2018) Supplier development and its incentives in infrastructure mega-projects: a case study on Hong Kong–Zhuhai–Macao Bridge. Front Eng Manage 5(1). https://doi.org/10.15302/j-fem-2018077

A Note on Intention to Settle

Sen Lin and Sai On Cheung

Introduction

Negotiations are commonly used to make deals or resolve differences. This applies to daily matters, family issues, business transactions and political affairs. The mixed-motive nature of negotiation can be portrayed with three "**I**'s"—an **i**nteraction between two or more **i**nterdependent parties who run into **i**ncompatible goals. Negotiation can be an art, given the inherent differences between the negotiating parties. Both parties rely on each other to fulfil their own needs and at the same time satisfy their counterparts' interests. Thus, negotiators are required to cooperate and compete simultaneously. Construction dispute negotiation (CDN) is further complicated by the existence of contracts that may limit the settlement options. The paradox of bridging the differences while maintaining their rights can undermine prospective settlement opportunities. In most construction contracts, if negotiation fails, a dispute will be subjected to a multitiered procedure that includes alternative dispute resolution and arbitration [1, 2]. To avoid costly proceedings and the loss of business reputations, there should be a strong incentive to end disputes through negotiation. However, as construction projects are nondiscrete transactions with high risk and uncertainty, overcoming the differences and reaching a mutually acceptable settlement is not easy [3]. Dispute decision-makers are subjected to the information available, their own personalities or skills, and the unexpected and changeable negotiation environment [4, 5]. Construction practitioners are usually required to react spontaneously to different types of unanticipated events. It is worth noting that some disputes can be settled peacefully and successfully, while others cannot reach an

S. Lin (✉) · S. O. Cheung
Construction Dispute Resolution Research Unit, City University of Hong Kong, Hong Kong, China
e-mail: slin24-c@my.cityu.edu.hk

S. O. Cheung
e-mail: Saion.cheung@cityu.edu.hk

S. O. Cheung (ed.), *Construction Dispute Research Expanded*, Springer Tracts in Civil Engineering, https://doi.org/10.1007/978-3-030-80256-1_8

acceptable agreement even with serious attempts and efforts [6]. Yiu et al. [6] found that the occurrence of CDN failure is highly probable, and the reasons include stringent contract governance and inadequate preparation. Even though these notable causes of negotiation failure may have been well known, there is still no guarantee that a negotiated settlement can be reached [7]. In some cases, negotiators may even ignore the notable resolution chances on the table and fail to reach a consensus. This suggests that the human factor may be a crucial settlement decider.

In this regard, Cheung and Yiu [8] suggested that project participants are one of the leading causes of conflict. Meng and Boyd [9] also highlighted the critical role of managers in dispute management. The social cognitive psychology research has documented that people's reactions to a specific event depend on how they perceive the event rather than the event itself [10]. More specifically, different people may have differing views on a particular event, thus, they will make different decisions regarding the same event. To move towards a negotiated settlement, it would be necessary to attach direct importance to the negotiators' subjective perceptions of the negotiation and thereby raise the intention to settle. Intention represents people's aspiration for a particular outcome in an event and will encourage people to engage in a given target behaviour [11, 12]. The theory of reasoned action and the theory of planned behaviour [13] state that intention is a strong indicator of the practice of the related actions. Therefore, we argue that obtaining the negotiators' intention to settle is a necessary condition for a negotiated settlement [14].

Intention to settle is a prerequisite of settlement in CDN. Despite the extensive negotiation literature, there has been very little consideration of an essential state during the negotiation process, i.e., the negotiators' willingness to actually be engaged in and put an end to the negotiation, especially in CDNs, is nearly uncharted. If one or both negotiating parties express no intention to settle their disagreement, other factors will matter very little. Thus, the pillars of a settlement are the elements that facilitate the negotiators' intention to settle. As such, the primary interest in this study is the root causes of the negotiators' intention to settle. As CDN is usually protracted, the negotiators' intention can be affected by different factors stemming from the overall negotiation process, such as the bargaining situation, negotiating parties' changeable positions, interests, priorities, and alternatives. Notably, efficiency will be enhanced if favourable conditions can be ensured. In this regard, this study consolidated efforts to summarize a list of elements that affect negotiators' intention to settle in CDNs and formed a framework of these specific variables to serve as a guide for effective negotiation management and dispute resolution.

Anatomy of Intention to Settle

Few studies have focused on intention to settle, let alone defining it. Nonetheless, there is a similar term named "willingness to negotiate", which has varying definitions depending on the applied settings or fields [15]. For example, in the study of labour conflict, willingness to negotiate refers to the intention to negotiate in contract concessions; in law, it means resolving a conflict out of the court. Drawing from Stein [16], who defined willingness to negotiate in the setting of public relations as "the state of being favourably disposed to meet with other parties to a conflict, discuss issues of common concern, and exchange proposals for resolving the conflict". Willingness to negotiate can be viewed as a trigger to start the communication process, leading to the decision of whether to be involved in a formal negotiation. Willingness to negotiate occurs at the prenegotiation stage. However, in CDNs, as they have common goals to finish the project, there is usually no reason for both parties to refuse to enter the negotiation; moreover, whether there is an intention to settle is worth considering. In our study, intention to settle embraces willingness to negotiate. However, it is not limited to a particular stage of the negotiation, and it represents the state of being favourably disposed to participate and actively put forth proposals to resolve the negotiation.

One difficulty in studying the intention to settle is the lack of an operational definition of the term. There are virtually no negotiation studies in which dimensions of elements affecting the intention to settle are explicitly identified. However, studies on negotiated settlement and negotiation outcomes can provide some inspiration and evidence for what influences negotiators' intentions. Lumineau and Henderson [17] emphasized the importance of contractual governance in buyer–supplier disputes. Chebet et al. [18] focused on negotiation skills that can help break down barriers and achieve excellence in business negotiation. Some studies highlight the importance of the relationship between the negotiating parties. For example, Christen [15] proved that perceived power and trust can affect negotiators' decisions towards settlement. Kteily et al. [19] found that imbalanced power would influence negotiators' willingness to come to the table. Macfarlane [20] summarized the following three essential factors that encourage a conflict to be settled: meeting disputants' expectations, feeling equally treated and a rational risk assessment. Curhan et al. [21] reached a similar conclusion and believed that negotiators' feelings and perceptions about settlement are essential to negotiation outcomes. Thompson [22] summarized the empirical evidence and found that the relationship between negotiation behaviour and outcome is nominal. However, negotiators' motivational and cognitive models significantly influence the negotiation outcome. It can be observed that many factors would be involved when considering negotiators' intention to settle and negotiated settlement. In this regard, a systematic literature review was conducted to identify the elements for the conceptualization of negotiated settlement in CDNs.

Several major steps were taken. Step 1: Determine the key terms related to the research aim (e.g., "negotiation", "construction dispute negotiation", "intention to settle", "negotiated settlement", "negotiation outcome", "negotiation success", and "influencing factor"). Step 2: Conduct an extensive literature review; a wide range of engineering and management databases have been chosen, and research journals, books, newspapers, magazines, and websites on related topics have been searched. Sufficient resources are available to provide a wide range of information from the psychological and behavioural studies on negotiation. Step 3: Evaluate the quality of each study; choose the relevant research that satisfies our research aim. As there

Table 1 Summary of elements for negotiated settlement

Elements for negotiated settlement	References
(1) Task versus relation; (2) emotion versus rationality; (3) competition versus cooperation	[23]
(1) Alleviate obstacles; (2) problem-solving ability	[3]
(1) Take adequate preparation about significant negotiation elements, which are what, who, where, when and how; (2) determine your reservation price; (3) present yourself well; (4) watch your language and attitude; (5) follow the rules and play fair	[24]
(1) Cognitive ability; (2) aspiration-level; (3) dual-concern; (4) risk attitude; (5) judgment accuracy	[22]
(1) Relationship between negotiators; (2) negotiator goals; (3) expected cooperation; (4) cooperation behaviour	[25]
Negotiation skills: (1) preparation and planning; (2) adopt a win–win approach; (3) practice communication competence; (4) maintain personal integrity and build solid relationships	[18]
(1) Experience; (2) preparation; (3) power; (4) relationship	[26]
(1) Meeting disputant's expectations; (2) feel of being fairly treated; (3) appraise risk rationally	[20]

are few studies that focus on "intention to settle", especially in the context of CDN, the key criterion for selecting literature is to provide information about the factors influencing the negotiated settlement. Table 1 shows the summary of the elements for negotiation derived from the literature review. Step 4: Synthesize and form a framework of the findings in the specific context of CDNs.

Towards a Framework of Intention to Settle in CDN

The negotiated settlement elements summarized in Table 1 are reorganized to present a conceptual framework for the intention to settle in CDN. The elements of settlement are operationalized in light of a negotiation process. Accordingly, several issues are taken into account, including the negotiation situations, negotiating parties involved, own issues and interests, counterpart's issues and interests, and relationships between parties. These are directly related to how negotiators perceive and deal with the negotiation. Four fundamental groups of root causes for negotiators' intention to settle are proposed. These are preparation, negotiation skill, relationship, and the self, as presented in Fig. 1.

The previous research revealed that preparation (conceptualized as "preparation") could significantly influence the negotiators' intention and negotiation outcomes. Preparation occurs at the prenegotiation stage, so the factors are more project-specific, including contract governance, trade-off zone, and senior management involvement.

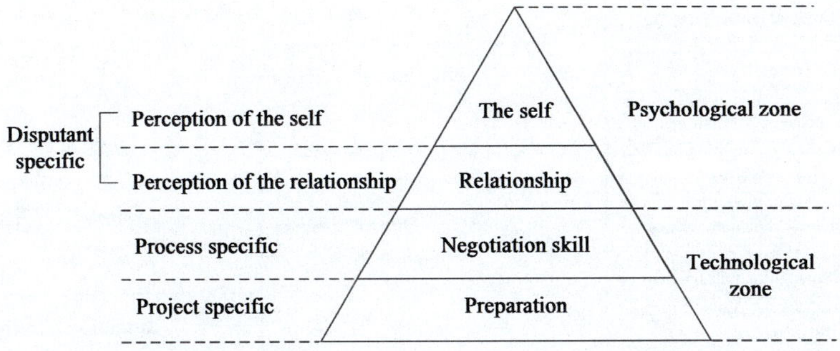

Fig. 1 Root causes of intention to settle

At the negotiation table, negotiators' ability (conceptualized as "negotiation skill") is another critical root cause of intention. It includes negotiating styles, communication competence, mindful emotion management, and judgement accuracy. The level of negotiation competence may affect the negotiators' confidence during the negotiation process, thus influencing their intention to settle. In addition, as negotiators are the leading factor in CDNs, the following two categories are disputant specific. The third is summarized from the perspective of the negotiators' perception of the negotiation relationship (conceptualized as "relationship"), including trust, justice, and balanced power. The last category is the negotiators' perceptions about the self (conceptualized as "the self"), which includes commitment to negotiation, self-efficacy, matching expectations, and loss aversion. The factors from the first two categories are more technical, while the factors in the last two categories are psychological and disputant specific. These four classifications are broadly and fundamentally adequate to be operationalized as the critical root causes of intention to settle in CDN. The research framework is shown in Fig. 2.

Preparation

Construction projects are commonly one-off, long-duration, and complex [27]. CDNs are often information asymmetric and challenging due to the complicated internal and external environments. Adequate preparation cannot be emphasized strongly enough. Entering a negotiation without performing enough homework will result in a disaster. Due to the high stakes, it is essential to carefully prepare and plan before the negotiation [18]. As the proverb says, "good preparation is half the battle". Before initiating a negotiation, the following three preparatory works should be realised: contract governance, trade-off zone, and senior management involvement. These are necessary works for an efficient negotiation. Good preparation is believed to promote the negotiators' intention to settle in CDNs [6].

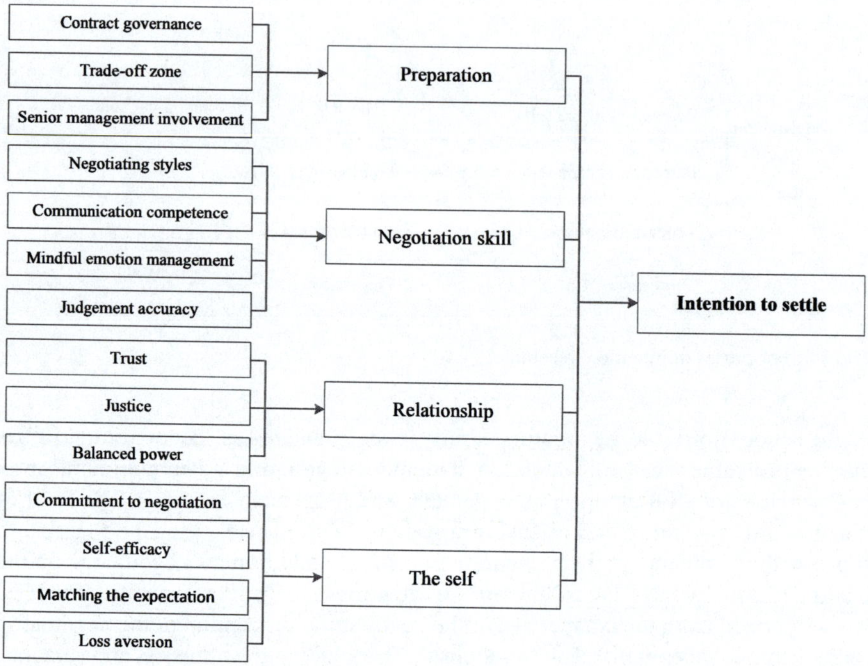

Fig. 2 A hierarchical framework of intention to settle

Contract Governance

Construction activities are complex, with a set of interrelated relationships assembled by the contractual network [28]. Construction projects are usually completed by a temporary project team. Team members with different professional backgrounds generally have different goals and needs. To achieve project success, the parties' obligations and rights are clearly regulated in the contract conditions. Formal contracts among team members are essential to prevent opportunistic behaviours. The constraints imposed by the construction contracts are called the contract governance [28]. Construction contracts contain various matters for which compensation may be sought. Provisions for instructing variations, acceleration, and postponement together with the corresponding time and monetary adjustments should be clearly stated in the construction contracts [3]. Additionally, dispute resolution is an indispensable part of construction contracts. Claims that occur among contracting parties focus on additional time or payment. The initial negotiation inevitably has to be conducted within the contract framework. Cheung and Pang [29] classified construction disputes into the following two types: contractual and speculative disputes. Contract incompleteness underpinned both types of construction disputes, indicating the critical role of contracts in dispute management. Yiu et al. [6] also summarized the procedural requirements, the burden of proof, and ambiguous provisions as the key contract governance factors that could be obstacles to effective dispute negotiation. Nonetheless, construction contracts are the main base under which disputes are negotiated. It is not only the primary reference for negotiation settlement regarding the reasonableness of the offer but also provides the framework for a negotiation to be conducted [27].

Based on the constraints and provisions of construction contracts, negotiators should prepare a compelling document to support their demands. A proper claim document should consist of at least the following lists: (i) a breakdown of issues to be recovered; (ii) relevant support of project correspondence and contract specifications; and (iii) legal and evidential proof. Detailed information collection and well-prepared documents would enable a negotiator to have the necessary background to conduct rational analysis. The effort needed for a thorough preparation signifies the intention to settle the dispute.

Trade-Off Zone

When preparing for a negotiation, negotiators should understand and identify where there might be conflict and plan for it accordingly. At this stage, negotiators will evaluate the settlement options, set their priorities based on interests, and select the initial demands according to their aspiration levels [30]. For example, there should be a reservation price for which a negotiator may accept the offer [31]. The reservation price is the lowest outcome that a negotiator is willing to accept, when the price is breached, he or she will not settle [32]. Apart from the reservation price, another reference point is the aspiration price, which is the most desired and best outcome a negotiator can reasonably expect [31]. Between the two reference points lies the negotiators' trade-off zone. Figure 3 shows the two parties' trade-off zone.

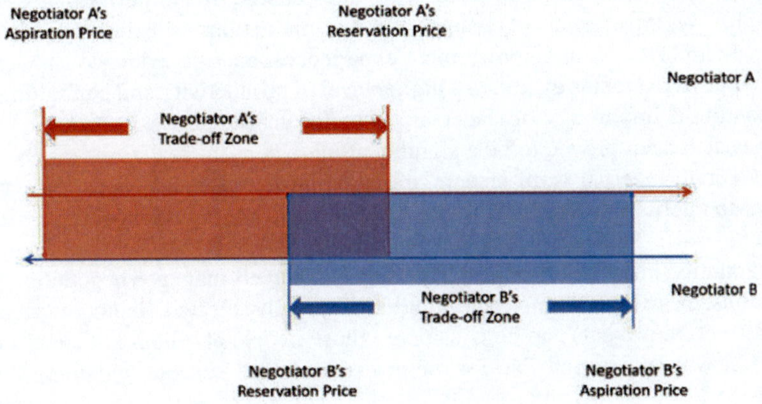

Fig. 3 Two negotiating parties' trade-off zone

Negotiators should figure out the zone clearly. In other words, many proposals can be proposed as long as these are developed within the trade-off zone.

CDNs typically involve multiple issues; while not all the needs presented are of equal importance, it is possible to make a proposal better than one expected if appropriate trade-offs can be capitalized [33, 34]. A trade-off occurs when one party lowers its demands regarding some negotiation issues and simultaneously demands more on others [35], for example, paying a higher price for an earlier delivery date or vice versa. Negotiators are very cautious regarding the core issues and will not easily put their cards on the table. Furthermore, negotiators' decisions may change with the emergence of new information and through their interaction with their counterparts. In this case, the principle of trade-offs remains essential to the settlement of negotiations. The efficiency of bargaining is related to how negotiators make trade-offs. To ensure that they are better prepared for the negotiation, negotiators need to decide (i) what to concede and what to insist on, (ii) when to concede, (iii) for which party and (iv) how much should be conceded [36]. With a well-defined trade-off zone, negotiators can concede on low priority issues in exchange for concessions on issues of higher priority to them, which will assist with obtaining integrative outcomes.

Senior Management Involvement

Effective negotiation team building is necessary for negotiations [37]. As part of this process, team members' resources, skills, and roles should be fully considered, with the aim to match members' abilities with specific needs. Among the members, the manager's role is vital. Leadership has been proven to be a significant antecedent for team effectiveness and successful project execution in many studies [38, 39]. Mathieu et al. [40] further commented that the key functions of leaders are facilitating teamwork and promoting the completion of the taskwork. External leadership represents the influence of a leader who takes charge of the performance of the team [40]. Kirkman et al. [41] suggested that the actions of external leaders can enhance employee team empowerment experiences, and the evidence showed that more empowered teams exhibited a higher level of productivity and commitment. In addition to the importance of leadership to performance and task outcomes, leadership has also been proven to be a significant ingredient for realizing team bonding [42]. Overall, external team leaders are essential because they not only serve as liaisons to internal and external organizations but also as guides for setting the team's vision [43, 44].

Negotiation in construction projects also requires well-managed negotiation teamwork; thus, the external team leader should be carefully selected. Senior management in construction projects can be considered those people at a home office or corporate level who are responsible for the project's overall success, including dispute resolution [45]. Their responsibility extends beyond completion of the project to establishing a long-term business relationship and engendering customer satisfaction. Senior management's attitude towards the other party can influence how negotiators act [3]. The involvement of senior management can reflect the planning and proactive efforts of the negotiating parties, while their participation can enhance negotiators' cohesion and boost morale. It has been proven that the early and heavy involvement of senior management could bring positive effects towards dispute resolution [46]. Notably, the leadership provided by senior managers can imply some level of strategic importance, which, in turn, drives negotiators to settle and achieve a better outcome [47].

Negotiation Skill

Negotiation skills have attracted significant attention because they are a vital contributor to the negotiation process and outcomes [48]. It is especially important in CDNs due to the changeable negotiation environment and complicated interests. Thus, the need for practical negotiation skills in CDNs is obvious. A central proposition in this part is that a higher level of competence in negotiation may affect the negotiators' confidence, and more task-specific confidence will enhance the negotiators' intention to engage in the negotiation process [49]. Considering the interaction process between the negotiation parties, we summarized the following four types of negotiation skills: negotiating styles, communication competence, mindful emotion management, and judgement accuracy.

Negotiating Styles

The negotiating style was first summarized by Follett [50]. He suggested the following five approaches to deal with conflicts: domination, compromise, integration, avoidance, and suppression. Black [51] proposed a conceptual framework of interpersonal conflict management that included forcing, withdrawing, smoothing, compromising, and problem-solving. Along with the development of the study of negotiating styles [52, 53], the following five conflict handling styles were proposed: integrating, obliging, compromising, dominating, and avoiding. Ruble and Thomas [53] differentiated the five conflict handling styles by the following two primary dimensions: concern for self (the high or low degree to which a person attempts to satisfy his own concern) and concern for others (the high or low degree to which a person wants to satisfy the concern of others). It is worth noting that the two dimensions show the motivational orientations of an individual in the face of conflict [54]. Dual-concern models suggest that "concern for self" and "concern for others" can motivate negotiators to choose conflict-handling strategies that facilitate dispute resolution [55]. Stevens (1963) pointed out that acknowledging the other party's needs and interests and understanding the other party's perspective better can assist in the development of negotiation. Rahim [56] asserted that a severe concern for both personal goals and relationships could result in cooperation. Figure 4 shows the combination of the dual-concern model with the conflict handling styles, which

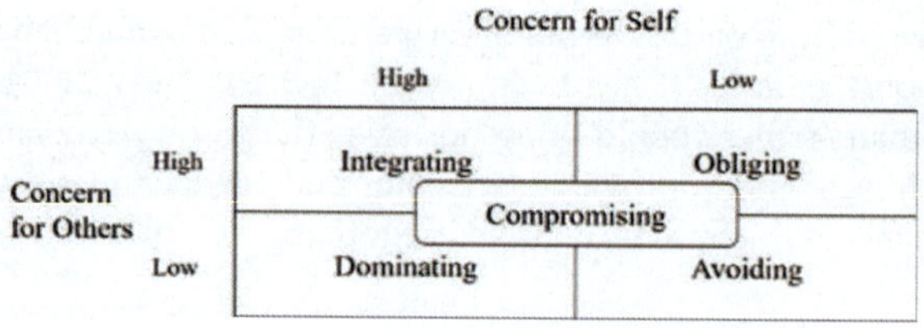

Fig. 4 The dual concern model of handling interpersonal conflict [56]

is called the dual concern model of handling interpersonal conflict [56]. The five conflict handling styles have been commonly adopted as negotiating styles [54, 57].

With the integrating style, collaboration between the parties will occur, including openness, exchanging of information about goals and priorities, and actively examining the differences in the proposals or alternatives to meet both parties' decision criteria [58]. Follett [50] proposed that facing the real issue, uncovering the dispute, and openly communicating are the cardinal rules to reach a "win–win" resolution. As Cheung and Yiu [8] indicated, there are two elements that exist during integration, i.e., confrontation and problem solving. Confrontation can assist negotiators in communicating openly and directly, which will then lead to problem solving. For ongoing projects, open problem solving can also help to sustain positive relationships [58]. With a deep concern for both the self and their counterpart, the integrating style is considered the most effective approach. Because it provides the chance to compare and address both parties' interests and goals, mutually acceptable solutions that are unique to the problem may be found [59].

The other four styles reduce the concern for the self or others to varying degrees. Obliging is a type of self-sacrificing style that involves placing others above oneself. Negotiators who use obliging style will try to reduce divergences and emphasize commonalities to satisfy their counterparts' needs and concerns [8]. Compromising has moderate levels of concern for the self and the counterpart. It is a neutral strategy that tends to neither compete nor cooperate [60]. The negotiators in this style prefer a middle ground be reached between the two parties. A dominating style is forceful and competitive [61]. With conflicting interests, a dominating negotiator will put forth as much effort as possible to pursue his or her goals and ignore the counterpart's expectations. Avoiding is the most destructive and ineffective style and has low concern for the self and others. Avoiding behaviours include withdrawal, buck-passing, or sidestepping [56].

Negotiating styles reflect the relative importance that individuals place on their own versus joint outcomes in negotiation interactions [62, 63]. Negotiators who are mainly concerned about their own interests and outcomes are competitors, and those who are concerned about both their own and their counterparts' outcomes are cooperators [64]. It is generally believed that negotiators as cooperators are more willing to settle and can create higher negotiation outcomes than competitors, whose only concern is how to maximize their own interests [65]. Furthermore, cooperatively and socially motivated negotiators can help with short-term collaboration and long-term trust development [66]. Most researchers believe that the integrating style is the most appropriate way to manage conflict [51, 54]. In addition, compromising is also regarded as a practical approach. Cheung et al. [54] found that it is the style most often used by negotiators in CDNs. This was supported by Rahim [67], who believed that some conflict in compromising negotiation could be beneficial to the projects. However, this does not mean that other negotiating styles are useless. One style may be more appropriate than another depending upon the situation, thus, it is crucial to adopt a negotiating style that fits the circumstances [53, 68].

Communication Competence

Negotiation is carried out via both verbal and nonverbal communication. Negotiators communicate face to face, electronically, or through mail. They exchange or add information that is beneficial to their party or to both parties. Essentially, CDN is both interorganizational and interpersonal communication. The aim of the negotiating parties is to sell their proposals about overcoming conflicts to reach consensus [24].

The negotiating parties make private interpretations of the counterparts' motives, degree of good faith, expectations, and competency during the interaction [20]. It is evident that appropriate communication competence is essential to deliver the correct information. Chebet et al. [18] suggested that good communication can lead to a brilliant resolution, however, poor communication can complicate matters. Meiners and Miller [69] proposed that a more formal communication environment can result in a more detailed elaboration and concessions during a negotiation process. A personalized and friendly relational tone will lead to open and direct negotiation and improve the efficiency of reaching an agreement. Chebet et al. [18] recommended that to achieve a better outcome, professional negotiators should increase their communication competence. Negotiators should ask both open- and closed-ended questions to ensure that the message is well received and communicated accurately. They should also learn communication skills, such as when to withhold or reveal information to their counterparts, the negotiation sequence of different issues, and how to create a friendly business atmosphere to relieve tensions [18]. Huczynski and Buchanan [70] stated that negotiators should listen with both their eyes and ears. Nonverbal communication, which is the sending and receiving of wordless messages, may be more significant than what they said. In general, negotiators in CDNs should not only master professional knowledge but also understand how to communicate well during the negotiation process.

Mindful Emotion Management

Emotion is central and plays a critical role during the negotiation process. A growing number of studies have begun to highlight the role of emotion in resolving organizational conflicts [71–73]. Specific emotions can influence negotiation-related cognition and behaviour [74]. Some psychology researchers have found that a positive mood is highly related to cooperative behaviour and intention to settle, including the following: (i) the desire for active conflict resolution by Baron [75], (ii) increasing the propensity to concede in conflicts [76], and (iii) fostering the intention to cooperate during conflict resolution [77]. Negotiators with positive moods tend to be more concerned about their counterparts, thus promoting their motivations and behaviours to find integrative gains [78].

However, negotiators who experience dejection-related emotions will have harmful effects on negotiations and make the negotiations more competitive [75, 76]. Allred et al. [79] found that angry negotiators with low compassion for others would (i) have less desire to work with each other in the future and (ii) achieve fewer joint gains. Angry negotiators are more self-centred regarding their preferences and are less accurate in evaluating their counterparts' interests [80]. People with negative emotions tend to focus on the negative side of the situation and are more likely to reject reasonable offers. They treat their counterparts as opponents rather than partners, which can be an obstacle to settling the negotiation [73, 81, 82]. Negative emotions may quickly bring a negotiation to an impasse. Reilly [83] summarized several ways of managing emotion, as follows: (i) be open to both pleasant and unpleasant emotions; (ii) pay attention to, monitor and reflect on emotions anytime; (iii) master the ability to engage, prolong, or withdraw from an emotional state; (iv) adjust emotions in oneself and others. Kopelman et al. [84] also suggested that both positive and negative emotions can be used to align with different negotiation strategies to help achieve the desired outcome.

Judgement Accuracy

Negotiating parties usually do not directly express their goals or expectations; their interests will be conveyed by their behaviours, emotions, or statements [85]. It is critical that each side understand what the counterpart thinks and feels about the negotiation, such as how conflict arises, what is at stake, and the relative importance of issues. These internal judgements are the basis for negotiators to make their decisions. Judgement accuracy can be defined as "the participant's ability to correctly discern the relative importance of issues to his or her opponent" [85]. Correct judgement about the negotiation situation is the starting point before suggesting proper offers or persuading counterparts to change their minds [20]. If negotiating parties realize that they rank the issues to be negotiated differently, an integrative solution is possible. However, if negotiators assume that they have the same priorities regarding negotiation issues, a fixed-pie error will arise [22]. A fixed-pie error means that negotiators may fail to accurately understand the negotiation process and their counterparts' preferences, judging that their own interests are intrinsically opposed to those of the other side [86]. Studies have proven that the fixed-pie error can partially explain why some negotiations fail to reach a negotiated settlement or integrated agreements [87, 88]. When negotiators make incorrect judgement regarding the negotiation situation or misperceive their counterparts, this may lead to missed opportunities to settle or suboptimal negotiation outcomes [22, 86, 89]. The critical role played by judgement accuracy on negotiation outcomes has been reported. For example, Kim et al. [85] suggested that when negotiators have a better understanding of the two sides' interests and priorities, they have a better chance of creating more value. Accurate judgement enables negotiators to identify "logrolling" solutions to fit each party's interests, allowing negotiators to propose more mutually acceptable solutions and formulate more generous settlements [90].

Furthermore, as CDNs may go through several rounds before reaching a consensus, negotiators' perceptions will constantly change [20]. Both parties will shift their goals, expectations, preferences, and offers during the negotiation process. Negotiators mainly seem to begin the negotiation with accurate perceptions but become less accurate due to the negotiation dynamic, thus requiring a continuous interpretation and judgement of the negotiation situation [86].

Relationship

Negotiators' relationships can be viewed as a consistent state of the ongoing negotiation process among individuals or negotiating parties [25]. The significance of relationships in negotiation has been widely proven in many studies [17, 91, 92]. Drawing from interdependency theory, relationship-specific motives can be the key to dispute resolution [80]. Relationships are usually developed with the existence of empathy, trust, long-term focus, cohesiveness, and better communication. Negotiators in a good relationship will view the negotiation in a positive way; they will collect information about their counterpart and put themselves in their counterparts' shoes, which can decrease negotiators' cognitive biases and facilitate coordination [93]. Negotiators committed to maintaining long-term interactions will engage in less opportunistic behaviour; instead, they are more willing to reciprocate constructive solutions [94]. Greenhalgh and Gilkey [90] opined that if future interactions are expected, negotiators will show more kindness to each other and prefer to settle in a peaceful manner. Halpert et al. [25] suggested that negotiators in a positive relationship are more likely to cooperatively come to the table. Sondak and Moore [95] highlighted that with a positive relationship, negotiators tend to cooperate rather than compete. Against this background, we summarize three critical relationship factors in CDNs, i.e., trust, justice, and balanced power, to explore their influence on the intention to settle.

Trust

The sources and dimensions of trust tend to be diverse and vary according to the domain. In the strategic alliance research, trust is classified as fragile or resilient, with partnership as the focus. Child et al. [96] proposed more generic categories, as follows: knowledge-based, affect-based, and calculation-based trust. Zaghloul and Hartman [97] put trust in the construction perspective and introduced a similar conceptualization, i.e., competence trust, integrity trust, and intuitive trust. Wong and Cheung [98] focused on partnering projects and suggested a four-factor framework (i.e., partners' performance, partners' permeability, relational bonding, and system-based trust) to operationalize trust in construction contracting. Zhang et al. [92] defined trust in the context of a construction project as the willingness of a party to be vulnerable to another party's action. They suggested a two-dimensional framework to measure trust that included goodwill trust, also referred to as affective trust, and competence trust, also referred to as cognitive trust. This categorization has been widely adopted in many trust studies [99, 100]. Lewicki et al. [101] summarized three contributors to the level of trust one negotiator may have for another as follows: the individual's chronic disposition towards trust, situation factors (e.g., the opportunity for the parties to communicate with each other adequately); and the history of the relationship between the parties.

Regardless of the resources or dimensions of trust, the current trend observed in construction projects indicates that the establishment of trust is the most critical factor in fostering the spirit of partnering [102]. Black et al. [103] added that trust is one of the crucial facilitators of interorganizational relationships. However, in the construction industry, due to the high level of risks, the divergent objectives of the contracting parties, and the disjoint working patterns, there is usually a lack of trust [104]. When conflicts arise, the confrontation between the disputing parties becomes prominent. Even though trust building is difficult, researchers have reached a consensus that trust among negotiating parties contributes to cooperation, willingness to settle, and optimal outcomes [15, 92]. Trust can be regarded as the glue between negotiating parties that fosters the negotiation process; in this regard, trust development is significant for CDN.

Justice

Justice is also a major element of interorganizational relationships. The importance of justice has been verified in construction management, for example, in tendering [105], evaluating construction contracts [106], contract implementation [107], and construction dispute negotiation [27]. Colquitt et al. [108] summarized three dimensions of justice measurement, i.e., distributive justice, procedural justice and interactional justice. Distributive justice is derived from social exchange theory, and it emphasizes the fairness of the distribution of positive acquisitions or negative burdens. In the CDN context, distributive justice can be regarded as how negotiators feel about the fairness of their counterparts' offers [27]. Procedural justice mainly refers to the negotiators' feelings and whether the procedures and criteria adopted by their counterparts during the negotiation process are fair enough [109]. People tend to believe that only a fair procedure can ensure a fair outcome. Regarding interactional justice, Bies and Shapiro [110] portrayed it as the "quality of interpersonal treatment received during the execution of a procedure". It can be expressed during social interactions in terms of honesty, respect, and politeness. In CDNs, interactional justice means how negotiators perceive fairness regarding interpersonal treatment and information exchange during settlement.

Lu et al. [27] proved that the three dimensions of justice are positively related to cooperative behaviours in CDNs. Aibinu et al. [111] came to a similar conclusion, indicating that negotiators' perceptions of their interactions will finally influence their dispute resolution ability. If negotiators feel a hostile atmosphere and treat each other unfairly, they are more likely to compete and bargain at every step. Kadefors [105] showed that an unfair experience would lead to anger, resentment, and loss of motivation, thus impeding the negotiators' settlement intention and decision making process. Macfarlane [20] implied that only when negotiators feel fairly treated and believe that their offers are given serious consideration will they come to the table to discuss the disputes. The significance of justice is that even if all other conditions for settlement can be met, a voluntary resolution will not occur if negotiators perceive unfair treatment. Given the above, justice is a reasonable concept to evaluate negotiators' perceptions regarding the CDN process and relationship quality. Justice should be formed and stabilized along with negotiating development, as it is critical in promoting settlement.

Balanced Power

Fisher [112] defined individual power as the ability to influence others' decisions. However, when determining the power of the other party, a comparison of their dependence on each other will take place. Accordingly, a more welcomed definition of power is derived from social exchange theory [113], which assesses power from both parties' perspectives as how perceptions of relative dependency lead to relative influence within the parties [114]. If the negotiating parties are equally dependent on each other for the exchanged resources, then their relative power is balanced [113]. Power imbalances are inevitable in different types of relationships. These power imbalances, which stem from asymmetry in dependence, will, in turn, lead to asymmetry in their influence. For example, if Party A has power over Party B, then A has more influence and voice in this relationship for goal achievement or dispute resolution [115]. This situation is more acute in markets with only a few service providers [54] (Shen and Cheung 2016).

According to social exchange and resource dependency theories, negotiation is essentially shaped and influenced by the parties' mutual influence. In CDNs, to reach an agreement, each party attempts to persuade its counterpart to compromise. With imbalanced power, the party with the relatively higher power is likely to address its own interests; however, it overlooks the demands of the party with the lower power

[116]. Christen [15] suggested that negotiators' intention to settle will decrease as perceived power increases. For the low-power party, considering its lack of resources and the impetus to improve the relatively disadvantaged position, it should be motivated to negotiate and argue for its interests [19]. However, for the more powerful party, even though they may also benefit from quelling dissent, there is still a high probability of damage occurring to their advantageous situation and resources. Thus, the high-power party will tend not to cooperate and will be less motivated to make concessions or deal with issues [15]. As such, negotiating parties with imbalanced power will hold different attitudes and perceptions towards negotiation, which will be a barrier for settlement. In light of bilateral deterrence theory, negotiating parties in a balanced power relationship will treat their counterpart as equally strong and be concerned about the consequences if negotiation fails, thus, they will be inclined to cooperate and solve the problems [117]. However, when there is a significant power gap between the parties, the party with more power will treat their counterpart as weaker and have little fear of applying a dominating style or coercive tactics, whereas the weaker party will be sceptical and vigilant when dealing with issues [118]. Thus, power asymmetry creates an atmosphere of hostility, which is detrimental to a negotiated settlement. Wolfe and McGinn [116] believed that compared to parties with a greater power difference, parties who perceived a smaller difference in relative power can reach a more integrative agreement. Both social exchange and bilateral deterrence theory suggest that balanced power can produce less hostility and more cooperative behaviour than unequal-power relations.

The Self

Iklé [119] summarized three possible options for negotiators when facing the task of making critical decisions, as follows: (i) accept the currently available offers from counterparts; (ii) continue negotiating to seek better solutions; and (iii) end the negotiation with no intention of resuming it in the short run. Negotiators' self-cognition or personal interpretation during negotiation will affect their decision about whether to settle or not [80]. Negotiators make decisions based on their inferences and judgements. Negotiated settlement will occur when the negotiators themselves feel that the conditions and status are ready for settlement. Considering the negotiators' self-perception and assessment during the negotiation process, we summarized the following four critical factors about "the self": commitment to negotiation, self-efficacy, matching expectations, and loss aversion.

Commitment to Negotiation

The concept of commitment has attracted a great deal of interest in organizational management and psychology studies. Commitment has been found to influence individual attitudes and actions, levels of endeavour, satisfaction, turnover, and organizational performance [120–122]. Moreover, low commitment is a signal of the reluctance to work and is manifested by distancing their responsibility and roles within their organization [123].

Meyer and Allen [124] suggested the following three dimensions of organizational commitment: (i) affective commitment, which means emotional attachment to the working team; (ii) normative commitment, indicating an employee's obligation to remain in the organization; and (iii) continuance commitment, which is related to an employee's economic ties to the organization and the employee's perceived costs for leaving the organization. In the context of CDN, Chow et al. [125] defined commitment as "a negotiator's continuing effort to invest in a relationship and an acceptance of joint goals and values". It can also be manifested through the three dimensions. Affected commitment reflects the negotiators' involvement in the project team and their desire to maintain membership. Higher affective commitment can foster negotiation, as negotiators are more willing to devote their time or effort to deal with conflicts. Normative commitment represents the negotiators' feeling of being obliged to continue. It is the internally normative pressure for negotiators to achieve their party's desires, goals, and interests. Continual commitment is related to the negotiators' perceived costs if they leave the negotiation table. This can help negotiators make rational appraisals, when there is a high cost of leaving and termination, negotiators have to continue with the negotiation [125]. Therefore, a high level of commitment will assist negotiators in cooperating and deriving mutual outcomes [126]. In contrast, low commitment demonstrates a low value to continuing the negotiation with the other party, leading to withdrawal behaviours. In this regard, commitment is the contributor to a negotiated settlement, as it can be achieved only when negotiators are committed to dealing with it.

Self-efficacy

Bandura [127] developed the self-efficacy theory, which states that individuals have varying levels of confidence in their ability to perform the tasks that they take up. It is a social-cognitive approach to portray a person's task-specific self-confidence and how confidence affects their behaviours or decision making [49]. Bandura [128] described four principal sources of self-efficacy, as follows: (i) mastery experiences; (ii) social modelling; (iii) social persuasion; and (iv) maintaining a good physical and mental state. Previous studies have proven that self-efficacy is a reliable variable that positively affects people's behaviour and task performance [129]. People's beliefs about their efficacy influence how they predict future scenarios of their tasks [128]. People with confidence in their task-related abilities will visualize success scenarios, focus their thoughts on achieving, and actively search for a way out when facing difficulties, which ultimately enhance their chances of obtaining good outcomes [130, 131]. Nevertheless, a low level of self-efficacy will lead to a negative, downward self-fulfilling prophecy that undermines performance as they scale back their efforts or even give up. The stronger the perceived self-efficacy, the higher the goals, and the more effective people are in their analytic thinking and in driving towards the specific action [132]. It has also been suggested that self-efficacy can affect individual selection processes [128]. People with low self-efficacy will shy away from difficult tasks that they view as personal threats, however, they prefer more controllable activities. Unless people have enough confidence to handle the issues, they have little intention to conduct the specific activities.

In CDNs, Yiu et al. [49] defined negotiation efficacy, which is derived from self-efficacy, as "the confidence of negotiators in their own ability to successfully use tactics to achieve desired outcomes", and they believed that negotiation efficacy can positively influence negotiation outcomes. Sullivan et al. [131] found that negotiators' self-efficacy towards distributive and integrative strategies will affect negotiators' initial choice of tactics, guide the negotiation course, and eventually influence the negotiation outcomes. Marilyn et al. [133] suggested that those with higher self-efficacy are more likely to succeed in pursuing their desired outcomes. However, negotiators with low self-efficacy are more likely to spiral into negative emotions and predict a frustrating negotiation, which will result in passive cooperation and negotiating styles [134]. Typically, high levels of self-efficacy to achieve a negotiated settlement will help negotiators overcome constraints and make more proper decisions. Nonetheless, it is worth noting that too much self-efficacy can also have a dark side, making negotiators overconfident so that they do not walk away from losing situations or seek outside help from a mediator or third party even if it may be in their best interests to do so [134].

Matching the Expectations

Expectancy theory constructs a framework for understanding motivation [135]. The expectation in negotiation can be regarded as a negotiator's drive for achievement and the utility for which the negotiator is striving [136]. Vroom [135] believed that people would subjectively determine the value of the desired outcome, as such negotiators will have varying preferences for the negotiation outcome. For example, negotiators with future negotiation interaction expectations tend to lower their aspiration levels, be more cooperative, and prefer the problem-solving negotiating style (Patton and Balakrishnan 2010). Zetik and Stuhlmacher [137] revealed that negotiators with higher expectations usually achieved better performance than negotiators with suboptimal expectations. Even though expectation is the fundamental motivation to start a negotiation [138], negotiators' irrational expectations or insensitive expectations can be sources of disputes [8]. Expectations impact both the rising and falling of negotiations. Inaccurate or incomplete assumptions about expectations can exacerbate negotiations. The sources of expectations are both complex and elusive. In CDNs, an experienced negotiator assesses the potential outcomes based on their experience and professional knowledge before initiating a negotiation. Furthermore, their expectations are also affected by pressure from their leaders or organizations [20]. It is critical that negotiators analyse and understand the basis of their own expectations and, as far as possible, those of their counterparts [20]. Expectations of the negotiation outcome range from the "best case" to the "acceptable" scenario. It is usually difficult to reach the best case, but the acceptable scenario, which is closer to the bottom line, is more likely to be achieved in most cases. Of course, reaching the best case is optimal for negotiating parties, but it is still essential for negotiators to understand that making concessions and adjusting their expectations are necessary steps to reach a settlement.

Ren et al. [139] described the negotiation phase in three stages, as follows: first, negotiation starts from the point where both negotiating parties try to maximize their own interests, which is their initial expectation, second, along with the negotiation development, they exchange offers and information and seek to induce or persuade their counterparts to compromise; at the same time, they will also modify their own expectations and requirements to pursue a mutual ground; finally, at an equilibrium point where the opposing interests are balanced and expectations are met, the negotiating parties reach an agreement. Similarly, Ben-Yoav and Pruitt [94] summarized three types or states before moving forward to settle a negotiation, which are as follows: yielding (i.e., lowering one's own expectation), competitive (i.e., lowering the counterpart's expectation), and problem-solving (i.e., an effort to satisfy both parties' expectations). It seems that expectations are dynamic and develop as negotiation unfolds. Negotiators are willing to engage in dispute resolution when their expectations are likely to be satisfied.

Loss Aversion

Loss aversion is a type of cognitive preference that has attracted substantial attention in the study of psychology and economics. Kahneman and Tversky [140] proved that people usually evaluate losses relative to a reference point more intensely than gains of the same magnitude, or it can be said, "losses loom larger than gains". Experimental works have shown that people are more motivated to minimize loss than maximize gain [141]. Loss aversion has been tested as one of the basic principles during the decision-making process, such as bias, the endowment effect, and the act of compromising [142, 143].

In regard to negotiation, loss aversion acts as a cognitive barrier to conflict resolution [144, 145]. If negotiators view making a concession as a certain loss, they will be reluctant to offer a concession, even though the continuous disputes at each step may lead to greater losses [146]. Negotiators usually code their outcomes as gains or losses according to the reference point, and the reference point can be manipulated by the choice of reference outcome, such as the non-agreement outcome or their different levels of aspiration [144, 146]. In the gain frame, negotiators' reference outcome is lower than their prospective outcomes, so they perceive the outcomes as positive and evaluate their concessions as a decrease in their gains. In contrast, negotiators with a loss frame will use a reference outcome that is higher than the potential outcome, thus, they will view the outcome as negative and evaluate concessions as increasing their losses [147]. Accordingly, negotiators manifest varying perceptions and reactions based on their frame adoption, which will affect their risk propensity and behaviour [86]. Changing the frame of the negotiation situation will lead to a very different prediction regarding the negotiation outcomes. When a positive frame exists, people tend to be risk averse, and a negotiated settlement is predicted [148]. Neale et al. [149] suggested that loss-framed negotiators are more aversive to making concessions than gain-framed negotiators. Further research found that gain-framed negotiators tend to be more cooperative, make larger concessions, and are more likely to settle than negatively framed negotiators [148, 149]. Shalev [141] found that increasing negotiators' loss aversion will result in worse outcomes for that negotiator in bargaining situations. Bazerman et al. [150] revealed that gain-framed negotiators completed more transactions than loss-framed negotiators. According to these studies, it seems that negotiators become risk seekers when the negotiation outcome is related to the loss (loss frame),in contrast, they tend to be risk averse with respect to negotiation gains (gain frame), which will increase the negotiators' intention to settle and the likelihood of a negotiated settlement [86].

Summary

In the construction industry, negotiation has been recognised as the most effective way of resolving disputes or claims [49]. However, negotiation failure is not uncommon. This study proposes that having an 'intention to settle' would provide the impetus for successful CDN. Negotiators' unwillingness to settle would make negotiation much more difficult. Delayed resolution or reverting to arbitration or litigation to settle are not desirable alternatives. In this regard, understanding what factors would enhance negotiators' intention can improve the chance of success. Based on a literature review on negotiation process, four success factor groups are identified: (i) preparation; (ii) negotiation skill; (iii) relationship; and (iv) the self. Comprehensive review of these elements can help formulating dispute negotiation strategies and provide insights for negotiators to prepare themselves to strive for settlement.

References

1. Chapman S (2010) Multi-tiered dispute resolution clauses: enforcing obligations to negotiate in good faith. J Int Arbitr 27(1):89–98
2. Cheung SO, Li K (2019) Biases in construction project dispute resolution. Eng Constr Archit Manage 26(2):321–348. https://doi.org/10.1108/ECAM-03-2018-0109
3. Mitropoulos P, Howell G (2001) Model for understanding, preventing and resolving project disputes. J Const Eng Manag 127(3):223–231

4. Loosemore M (1999) A grounded theory of construction crisis management. Constr Manage Econ 17(1):9–19. https://doi.org/10.1080/014461999371781
5. Simon HA (1957) Models of man; social and rational
6. Yiu TW, Cheung SO, Lok CL (2015) A fuzzy fault tree framework of construction dispute negotiation failure. IEEE Trans Eng Manage 62(2):171–183. https://doi.org/10.1109/TEM.2015.2407369
7. Bazerman MH, Chugh D (2005) Bounded awareness: focusing failures in negotiation. Frontieres of social psychology: negotiation, vol 04–062, pp 7–26
8. Cheung SO, Yiu TW (2006) Are construction disputes inevitable? IEEE Trans Eng Manage 53(3):456–470. https://doi.org/10.1109/TEM.2006.877445
9. Meng X, Boyd P (2017) The role of the project manager in relationship management. Int J Project Manage 35(5):717–728. https://doi.org/10.1016/j.ijproman.2017.03.001
10. Bandura, A. (1993). Perceived self-efficacy in cognitive development and functioning. Educ Psychol 28:117–148. Retrieved from https://www.tandfonline.com/doi/abs/10.1207/s15326985ep2802_3
11. Bagozzi RP (1992) The self-regulation of attitudes, intentions, and behavior. Soc Psychol 55(2):178–204. https://doi.org/10.1007/sl21
12. Hagger MS, Chatzisarantis NLD, Biddle SJH (2002) A meta-analytic review of the theories of reasoned action and planned behavior in physical activity: an examination of predictive validity and the contribution of additional variables. J Sport Exercise Psy 24(1):3–32
13. Ajzen I (1985) From intentions to actions: a theory of planned behavior. Action control, Springer, pp 11–39
14. Johari S, Jha KN (2020) Interrelationship among belief, intention, attitude, behavior, and performance of construction workers. J Manage Eng 36(6):1–14. https://doi.org/10.1061/(ASCE)ME.1943-5479.0000851
15. Christen CT (2004) Predicting willingness to negotiate: the effects of perceived power and trustworthiness in a model of strategic public relations. J Public Relat Res 16(3):243–267. https://doi.org/10.1080/1532-754X.2004.11925129
16. Stein JG (1989) Getting to the table: the triggers, stages, functions, and consequences of prenegotiation. Int J: Canada's J Global Policy Anal 44(2):475–504. https://doi.org/10.1177/002070208904400209
17. Lumineau F, Henderson JE (2012) The influence of relational experience and contractual governance on the negotiation strategy in buyer–supplier disputes. J Oper Manage 30(5):382–395. https://doi.org/10.1016/j.jom.2012.03.005
18. Chebet WT, Phil M, Rotich JK, Kurgat A (2015) Negotiation skills: keys to business excellence in the 21st century? European J Res Reflection Manage Sci 3(3):23–31
19. Kteily N, Saguy T, Sidanius J, Taylor DM (2013) Negotiating power: agenda ordering and the willingness to negotiate in asymmetric intergroup conflicts. J Pers Soc Psychol 105(6):978–995. https://doi.org/10.1037/a0034095
20. Macfarlane J (2001) Why do people settle? McGill Law J 46(3):663–711
21. Curhan JR, Elfenbein HA, Xu H (2006) What do people value when they negotiate? Mapping the domain of subjective value in negotiation. J Pers Soc Psychol 91(3):493–512. https://doi.org/10.1037/0022-3514.91.3.493
22. Thompson L (1990) Negotiation behavior and outcomes: empirical evidence and theoretical issues. Psychol Bull 108(3):515–532. https://doi.org/10.1037/0033-2909.108.3.515
23. Cheung SO, Chow PT (2011) Withdrawal in construction project dispute negotiation. J Constr Eng Manage 137(12):1071–1079. https://doi.org/10.1061/(ASCE)CO.1943-7862.0000388
24. Whitney FW (1983) Winning: the art of successful negotiation. Occup Health Nurs 31(5):31–34. https://doi.org/10.1177/216507998303100505
25. Halpert JA, Stuhlmacher AF, Crenshaw JL, Litcher CD, Bortel R (2010) Paths to negotiation success. Negot Confl Manage Res 3(2):91–116. https://doi.org/10.1111/j.1750-4716.2010.00051.x
26. Lenarčič B, Franc B (2013) Analysis of influences on buyer-supplier negotiation. Innov Issue Approaches Soc Sci 7(2):81–98

27. Lu W, Li Z, Wang S (2017) The role of justice for cooperation and contract's moderating effect in construction dispute negotiation. Eng Constr Archit Manage 24(1):133–153. https://doi.org/10.1108/ECAM-01-2015-0002
28. Cheung SO, Wong WK, Yiu TW, Kwok TW (2008) Exploring the influence of contract governance on construction dispute negotiation (October):391–399
29. Cheung SO, Pang KHY (2013) Anatomy of construction disputes. J Constr Eng Manage 139(1):15–23. https://doi.org/10.1061/(ASCE)CO.1943-7862.0000532
30. Kuon B, Uhlich GR (1993) The negotiation agreement area: an experimental analysis of two-person characteristic function games. Group Decis Negot 2(4):323–345. https://doi.org/10.1007/BF01384487
31. Van Poucke D, Buelens M (2002) Predicting the outcome of a two-party price negotiation: contribution of reservation price, aspiration price and opening offer. J Econ Psychol 23(1):67–76. https://doi.org/10.1016/S0167-4870(01)00068-X
32. Schaerer M, Loschelder DD, Swaab RI (2016) Bargaining zone distortion in negotiations: the elusive power of multiple alternatives. Organ Behav Hum Decis Process 137:156–171. https://doi.org/10.1016/j.obhdp.2016.09.001
33. Tajima M, Fraser NM (2001) Logrolling procedure for multi-issue negotiation. Group Decis Negot 10(3):217–235. https://doi.org/10.1023/A:1011262625052
34. Gupta S (1989) Modeling integrative, multiple issue bargaining. Manag Sci 35(7):788–806
35. Faratin P, Mary Q, College W, London E (2000) Using similarity criteria to make negotiation trade-offs (0)
36. Qu Y, Cheung SO (2012) Logrolling "win-win" settlement in construction dispute mediation. Autom Constr 24:61–71. https://doi.org/10.1016/j.autcon.2012.02.010
37. Fapohunda TM (2013) Towards effective team building in the workplace. Int J Educ Res 1:1–12
38. Salvato C, Corbetta G (2013) Transitional leadership of advisors as a facilitator of successors' leadership construction. Fam Bus Rev 26:235–255
39. Tabassi AA, Roufechaei KM, Ramli M, Bakar AHA, Ismail R, Pakir AHK (2016) Leadership competences of sustainable construction project managers. J Cleaner Prod 124:339–349
40. Mathieu J, Maynard TM, Rapp T, Gilson L (2008) Team effectiveness 1997–2007: a review of recent advancements and a glimpse into the future. J Manage 34(3):410–476. https://doi.org/10.1177/0149206308316061
41. Kirkman BL, Rosen B (1999) Beyond self-management: antecedents and consequences of team empowerment. Acad Manage J 42:58–74
42. Foels R, Driskell JE, Mullen B, Salas E (2000) The effects of democratic leadership on group member satisfaction: an integration. Small Group Res 31:676–701
43. Balkundi P, Harrison DA (2006) Ties, leaders, and time in teams: strong inference about network structure's effects on team viability and performance. Acad Manage J 49:49–68
44. Morgeson FP (2005) The external leadership of self-managing teams: intervening in the context of novel and disruptive events. J Applied Psychol 90
45. Diekmann JE, Girard MJ (1995) Are contract disputes predictable? J Constr Eng Manage 121(4):355–363. https://doi.org/10.1061/(asce)0733-9364(1995)121:4(355)
46. Cheung SO, Tam CM, Ndekugri I, Harris FC (2000) Factors affecting clients' project dispute resolution satisfaction in Hong Kong. Constr Manage Econ 18(3):281–294. https://doi.org/10.1080/014461900370654
47. Sharland A (n.d.) The negotiation process as a predictor of relationship outcomes in international buyer—supplier arrangements 559(2001):551–559
48. Chapman E, Miles EW, Maurer T (2017) A proposed model for effective negotiation skill development. J Manage Dev 36(7):940–958. https://doi.org/10.1108/JMD-01-2016-0002
49. Yiu TW, Cheung SO, Siu LY (2012) Application of Bandura's self-efficacy theory to examining the choice of tactics in construction dispute negotiation. J Constr Eng Manage 138(3):331–340. https://doi.org/10.1061/(ASCE)CO.1943-7862.0000403
50. Follett MP (2011) Constructive conflict. Sociology Organ: Struct Relationships 417
51. Black RR (1970) The fifth achievement

52. van de Vliert E, Kabanoff B (1990) Toward theory-based measures of conflict management 33(1):199–209
53. Ruble TL, Thomas KW (1976) Support for a two-dimensional model of conflict behavior. Organ Behav Hum Perform 16(1):143–155. https://doi.org/10.1016/0030-5073(76)90010-6
54. Cheung SO, Yiu TW, Yeung SF (2006) A study of styles and outcomes in construction dispute negotiation. J Constr Eng Manage 132(August):871–881. https://doi.org/10.1061/(ASCE)0733-9364(2006)132
55. Morse EA, Savage GT (1999) A test of the motivations underlying choice of conflict strategies in the dual-concern model (December). https://doi.org/10.1108/eb022817
56. Rahim MA (1983) A measure of styles of handling interpersonal conflict. Acad J Manage 26(2):368–376
57. Montes C, Rodríguez D, Serrano G (2012) Affective choice of conflict management styles. Int J Confl Manage 23(1):6–18. https://doi.org/10.1108/10444061211199304
58. Lin X, Miller SJ (2003) Negotiation approaches: direct and indirect effect of national culture. Int Mark Rev 20(3):286–303+227. https://doi.org/10.1108/02651330310477602
59. Tutzauer F, Roloff ME (1988) Communication processes leading to integrative agreements: three paths to joint benefits[J]. Commun Res 15(4):360–380
60. Sutschek LB (2002) Conflict resolution style and experience in management: moderating the effects of gender. UW-L research grant paper (1991), pp 69–78. https://doi.org/10.1080/03637759109376219
61. Pivotto C, Antunes D, Lima T, Hwang J, Rodrigues S, De Souza JM (2011) Detection of negotiation profile and guidance to more collaborative approaches through negotiation games
62. Iedema J, Poppe M (1994) The effect of self-presentation on social value orientation. J Soc Psychol 134(6):771–782. https://doi.org/10.1080/00224545.1994.9923012
63. Olekalns M, Smith PL (1999) Social value orientations and strategy choices in competitive negotiations. Pers Soc Psychol Bull 25(6):657–668. https://doi.org/10.1177/0146167299025006002
64. Messick DM, McClintock CG (1968) Motivational bases of choice in experimental games. J Exp Soc Psychol 4(1):1–25. https://doi.org/10.1016/0022-1031(68)90046-2
65. De Dreu CKW, Weingart LR, Kwon S (2000) Influence of social motives on integrative negotiation: a meta-analytic review and test of two theories. J Pers Soc Psychol 78(5):547–558. https://doi.org/10.1037//0022-35I4.78.5.889
66. De Dreu CKW, Boles TL (1998) Share and share alike or winner take all? The influence for social value orientation on the choice and recall of heuristics in negotiation. Organ Behav Hum Decis Process 76(3):253–276
67. Rahim MA (2005) Toward a theory of managing organizational conflict. SSRN Electron J (October 2003). https://doi.org/10.2139/ssrn.437684
68. Thomas KW (1977) Toward multi-dimensional values in teaching: the example of conflict behaviors. Acad Manage Rev 2(3):484–490. https://doi.org/10.5465/amr.1977.4281851
69. Meiners EB, Miller VD (2004) The effect of formality and relational tone on supervisor/subordinate negotiation episodes. West J Commun 68(3):302–321. https://doi.org/10.1080/10570310409374803
70. Huczynski A, Buchanan D (2013) Organizationa behaviour. Retrieved from http://library.mpib-berlin.mpg.de/toc/ze_2003_855.pdf
71. Barry B, Oliver RL (1996) Affect in dyadic negotiation: a model and propositions. Organ Behav Hum Decis Process 67(2):127–143. https://doi.org/10.1006/obhd.1996.0069
72. Carnevale PJ (2008) Positive affect and decision frame in negotiation. Group Decis Negot 17(1):51–63. https://doi.org/10.1007/s10726-007-9090-x
73. Rhoades JA, Arnold J, Clifford J (2001) The role of affective traits and affective states in disputants' motivation and behavior during episodes of organizational conflict. J Organ Behav 22(3):329–345. https://doi.org/10.1002/job.72
74. Morris MW, Keltner D (2000) How emotions work: the social functions of emotional expression in negotiations. Res Organ Behav 22. https://doi.org/10.1016/s0191-3085(00)22002-9

75. Baron RA (1984) Reducing organizational conflict: an incompatible response approach. J Appl Psychol 69(2):272–279. https://doi.org/10.1037/0021-9010.69.2.272
76. Van Kleef GA, De Dreu CKW, Manstead ASR (2004) The interpersonal effects of emotions in negotiations: a motivated information processing approach. J Pers Soc Psychol 87(4):510–528. https://doi.org/10.1037/0022-3514.87.4.510
77. Forgas JP (1998) On feeling good and getting your way: mood effects on negotiator cognition and bargaining strategies. J Pers Soc Psychol 74(3):565–577. https://doi.org/10.1037/0022-3514.74.3.565
78. Carnevale PJD, Isen AM (1986) The influence of positive affect and visual access on the discovery of integrative solutions in bilateral negotiation. Organ Behav Hum Decis Process 37(1):1–13. https://doi.org/10.1016/0749-5978(86)90041-5
79. Allred KG, Mallozzi JS, Matsui F, Raia CP (1997) The influence of anger and compassion on negotiation performance. Organ Behav Hum Decis Process 70(3):175–187. https://doi.org/10.1006/obhd.1997.2705
80. Bazerman MH, Curhan JR, Moore DA, Valley KL (2000) Negotiation. Soc Psychol 279–314.
81. Gutnik LA, Hakimzada AF, Yoskowitz NA, Patel VL (2006) The role of emotion in decision-making: a cognitive neuroeconomic approach towards understanding sexual risk behavior. J Biomed Inform 39(6):720–736. https://doi.org/10.1016/j.jbi.2006.03.002
82. Shepherd DA, Patzelt H, Wolfe M (2014) Moving forward from project failure: negative emotions, affective commitment, and learning from the experience. A psychological approach to entrepreneurship: selected essays of Dean A. Shepherd, vol 54, no 6, pp 546–576. https://doi.org/10.4337/9781783479801.00039
83. Reilly P (2014) Mindfulness, emotions, and mental models: theory that leads to more effective dispute resolution 201–204
84. Kopelman S, Rosette AS, Thompson L (2006) The three faces of eve: strategic displays of positive, negative, and neutral emotions in negotiations. Organ Behav Hum Decis Process 99(1):81–101. https://doi.org/10.1016/j.obhdp.2005.08.003
85. Kim K, Cundiff NL, Choi SB (2015) Emotional intelligence and negotiation outcomes: mediating effects of rapport, negotiation strategy, and judgment accuracy. Group Decis Negot 24(3):477–493. https://doi.org/10.1007/s10726-014-9399-1
86. Caputo A (2013) A literature review of cognitive biases in negotiation processes. Int J Confl Manage 24(4):374–398. https://doi.org/10.1108/IJCMA-08-2012-0064
87. Bazerman MH, Moore, DA (2012) Judgment in managerial decision making. Wiley
88. Raiffa H (2007) Negotiation analysis: the science and art of collaborative decision making. Harvard University Press
89. Mumpower JL, Sheffield J, Darling TA, Milter RG (2004) The accuracy of post-negotiation estimates of the other negotiator's payoff. Group Decis Negot 13(3):259–290. https://doi.org/10.1023/B:GRUP.0000031089.91654.26
90. Greenhalgh L, Gilkey RW (1993) The effect of relationship orientation on negotiators' cognitions and tactics. Group Decis Negot 2(2):167–183. https://doi.org/10.1007/BF01884770
91. Cheung SO, Yiu TW, Chiu OK (2009) The aggressive-cooperative drivers of construction contracting. Int J Project Manage 27(7):727–735. https://doi.org/10.1016/j.ijproman.2008.09.001
92. Zhang SB, Fu YF, Gao Y, Zheng XD (2016) Influence of trust and contract on dispute negotiation behavioral strategy in construction subcontracting. J Manage Eng 32(4):04016001. https://doi.org/10.1061/(asce)me.1943-5479.0000427
93. Galinsky AD, Moskowitz GB (2000) Perspective-taking: decreasing stereotype expression, stereotype accessibility, and in-group favoritism. J Pers Soc Psychol 78(4):708–724. https://doi.org/10.1037/0022-3514.78.4.708
94. Ben-Yoav O, Pruitt DG (1984) Resistance to yielding and the expectation of cooperative future interaction in negotiation. J Exp Soc Psychol 20(4):323–335. https://doi.org/10.1016/0022-1031(84)90029-5

95. Sondak H, Moore MC (1993) Relationship frames and cooperation. Group Decis Negot 2(2):103–118. https://doi.org/10.1007/BF01884766
96. Child J, Faulkner D (1998) Strategies of cooperation: managing alliances, networks, and joint ventures. Oxford University
97. Zaghloul R, Hartman F (2003) Construction contracts: the cost of mistrust. Int J Project Manage 21(6):419–424. https://doi.org/10.1016/S0263-7863(02)00082-0
98. Wong PSP, Cheung SO (2004) Trust in construction partnering: views from parties of the partnering dance. Int J Project Manage 22(6):437–446. https://doi.org/10.1016/j.ijproman.2004.01.001
99. Dirks KT, Ferrin DL (2002) Trust in leadership: meta-analytic findings and implications for research and practice. J Appl Psychol 87(4):611–628. https://doi.org/10.1037/0021-9010.87.4.611
100. Patzelt H, Shepherd DA (2016) The decision to persist with underperforming alliances: the role of trust and control. Decision making in entrepreneurship: selected joint papers of Dean A. Shepherd (November), pp 386–412. https://doi.org/10.4337/9781784716042
101. Lewicki RJ, Saunders DM, Minton JW, Roy J, Lewicki N (2011) Essentials of negotiation. McGraw-Hill/Irwin Boston, MA
102. Wong PSP, Cheung SO (2005) Structural equation model of trust and partnering success. J Manage Eng 21(2):70–80. https://doi.org/10.1061/(asce)0742-597x(2005)21:2(70)
103. Black C, Akintoye A, Fitzgerald E (2000) Analysis of success factors and benefits of partnering in construction. Int J Project Manage 18(6):423–434. https://doi.org/10.1016/S0263-7863(99)00046-0
104. Xue X, Shen Q, Ren Z (2010) Critical review of collaborative working in construction projects: business environment and human behaviors. J Manage Eng 26(4):196–208. https://doi.org/10.1061/(asce)me.1943-5479.0000025
105. Kadefors A (2005) Fairness in interorganizational project relations: norms and strategies. Constr Manage Econ 23(8):871–878. https://doi.org/10.1080/01446190500184238
106. Bubshait AA, Almohawis SA (1994) Evaluating the general conditions of a construction contract. Int J Project Manage 12(3):133–136. https://doi.org/10.1016/0263-7863(94)90027-2
107. Zhang Z, Jia M (2010) Procedural fairness and cooperation in public-private partnerships in China. J Manage Psychol 25(5):513–538. https://doi.org/10.1108/02683941011048409
108. Colquitt JA, Wesson MJ, Porter COLH, Conlon DE, Ng KY (2001) Justice at the millennium: a meta-analytic review of 25 years of organizational justice research. J Appl Psychol 86(3):425–445. https://doi.org/10.1037/0021-9010.86.3.425
109. Luo Y (2007) The independent and interactive roles of procedural, distributive, and interactional justice in strategic alliances. Acad Manage J 50(3):644–664. https://doi.org/10.5465/AMJ.2007.25526452
110. Bies RJ, Shapiro DL (1987) Interactional fairness judgments: the influence of causal accounts. Soc Justice Res 1(2):199–218. https://doi.org/10.1007/BF01048016
111. Aibinu AA, Ling FYY, Ofori G (2011) Structural equation modelling of organizational justice and cooperative behaviour in the construction project claims process: contractors' perspectives. Constr Manage Econ 29:463–481. https://doi.org/10.1080/01446193.2011.564195
112. Fisher R (1983) Negotiating power: getting and using influence
113. Emerson RM (1962) Power-dependence relations 27(1):31–41
114. Blau P (2017) Exchange and power in social life. In: Exchange and power in social life. Routledge
115. Salancik GR, Pfeffer J (1977) Who gets power—and how they hold on to it: a strategic-contingency model of power. Organ Dyn (Winter):3–21
116. Wolfe RJ, McGinn KL (2005) Perceived relative power and its influence on negotiations. Group Decis Negot 14(1):3–20. https://doi.org/10.1007/s10726-005-3873-8
117. Lawler EJ (1986) Bilateral deterrence and conflict spiral: a theoretical analysis (iii). https://doi.org/10.1108/s0882-6145_2014_0000031014

118. Lawler EJ, Ford RS, Blegen MA (1988) Coercive capability in conflict: a test of bilateral deterrence versus conflict spiral theory author. Soc Psychol Q 51(2):93–107
119. Iklé (1967) How nations negotiate. Praeger
120. Chow PT, Cheung SO, Ka Wa Y (2015) Impact of trust and satisfaction on the commitment-withdrawal relationship. J Manage Eng 31(5):1–8. https://doi.org/10.1061/(ASCE)ME.1943-5479.0000331
121. Cohen A (2007) Commitment before and after: an evaluation and reconceptualization of organizational commitment. Hum Resour Manage Rev 17(3):336–354. https://doi.org/10.1016/j.hrmr.2007.05.001
122. Saridakis G, Muñoz Torres R, Johnstone S (2013) Do human resource practices enhance organizational commitment in SMEs with low employee satisfaction? Br J Manage 24(3):445–458. https://doi.org/10.1111/j.1467-8551.2012.00814.x
123. Pool S, Pool B (2007) A management development model: measuring organizational commitment and its impact on job satisfaction among executives in a learning organization. J Manage Dev 26(4):353–369. https://doi.org/10.1108/02621710710740101
124. Meyer JP, Allen NJ (1991) A three-component conceptualization of organizational commitment. Hum Resour Manage Rev 1(1):61–89. https://doi.org/10.1057/9780230501997_5
125. Chow PT, Kong F, Cheung SO (2012) Mediating and moderating effect of tension on withdrawal-commitment relationship in construction dispute negotiation. J Constr Eng Manage 138(10):1230–1238. https://doi.org/10.1061/(ASCE)CO.1943-7862.0000528
126. O'Reilly C, Chatman J (1986) Organizational commitment and psychological attachment: the effects of compliance, identification, and internalization on prosocial behavior. J Applied Psychol 71(3):492–499. https://doi.org/10.1037/0021-9010.71.3.492
127. Bandura A (1977) Self-efficacy: toward a unifying theory of behavioral change. Translating national policy to improve environmental conditions impacting public health through community planning, vol 84, no 2, pp 1–312. https://doi.org/10.1007/978-3-319-75361-4
128. Bandura A (2010) Self-efficacy. The Corsini encyclopedia of psychology. Retrieved from https://onlinelibrary.wiley.com/doi/epdf/10.1002/9780470479216.corpsy0836
129. Stajkovic AD, Luthans F (1998) Self-efficacy and work-related performance: a meta-analysis. Psychol Bull 124(2):240–261. https://doi.org/10.1037/0033-2909.124.2.240
130. Bandura A, Cervone D (1983) Self-evaluative and self-efficacy mechanisms governing the motivational effects of goal systems. J Pers Soc Psychol 45:1017–1028. https://doi.org/10.1037/0022-3514.45.5.1017
131. Sullivan BA, O'Connor KM, Burris ER (2006) Negotiator confidence: the impact of self-efficacy on tactics and outcomes. J Exp Soc Psychol 42(5):567–581. https://doi.org/10.1016/j.jesp.2005.09.006
132. Wood R, Bandura A (1989) Social cognitive theory of organizational management. Acad Manage Rev 14(3):361–384
133. Marilyn E, Kay C, Anna G (1991) Effects of self-efficacy and post-training intervention on the acquisition and maintenance of complex interpersonal skills. Pers Psychol 44(4):837–861
134. O'Connor KM, Arnold JA (2001) Distributive spirals: negotiation impasses and the moderating role of disputant self-efficacy. Organ Behav Hum Decis Process 84(1):148–176. https://doi.org/10.1006/obhd.2000.2923
135. Vroom VH (1964) Work and motivation
136. Pruitt DG (1983) Strategic choice in negotiation. Am Behav Sci 27:167–194
137. Zetik DC, Stuhlmacher AF (2002) Goal setting and negotiation performance: a meta-analysis. Group Process Intergroup Relat 5(1):35–52. https://doi.org/10.1177/1368430202005001537
138. Reif JAM, Brodbeck FC (2014) Initiation of negotiation and its role in negotiation research. Organ Psychol Rev 4(4):363–381. https://doi.org/10.1177/2041386614547248
139. Ren Z, Anumba CJ, Ugwu O (2005) Negotiation in a multi-agent system for construction claims negotiation. In: Agents and multi-agent systems in construction, vol 9514. https://doi.org/10.4324/9780203006979
140. Kahneman D, Tversky A (1979) Prospect theory: an analysis of decision under risk. Econometrica 47(2):263–292

141. Shalev J (2002) Loss aversion and bargaining. Theor Decis 52(3):201–232. https://doi.org/10.1023/A:1019674323804
142. Liberman N, Idson LC, Higgins ET (2005) Predicting the intensity of losses vs. non-gains and non-losses vs. gains in judging fairness and value: a test of the loss aversion explanation. J Exp Soc Psychol 41(5):527–534. https://doi.org/10.1016/j.jesp.2004.06.007
143. Tversky A, Kahneman D (1991) Loss aversion in riskless choice: a reference-dependent model. Q J Econ 106(4):1039–1061
144. De Dreu CKW, Carnevale PJD, Emans BJM, Van De Vliert E (1994) Effects of gain-loss frames in negotiation: loss aversion, mismatching, and frame adoption. Organ Behav Hum Decis Process 60(1):90–107. https://doi.org/10.1006/obhd.1994.1076
145. Ross L, Stillinger C (1991) Barriers to conflict resolution. Negot J 7(4):389–404. https://doi.org/10.1007/BF01000332
146. Mnookin R (1992) Why negotiations fail: an exploration of barriers to the resolution of conflict. Ohio State J Dispute Resolut 8(2):235. Retrieved from http://heinonline.org/hol-cgi-bin/get_pdf.cgi?handle=hein.journals/ohjdpr8&section=21
147. Bazerman MH (1983) Negotiator judgment: a critical look at the rationality assumption. Am Behav Sci 27:211–228
148. Neale MA, Bazerman MH (1985) The effects of framing and negotiator overconfidence on bargaining behaviors and outcomes. Acad Manage J 28(1):34–49. https://doi.org/10.5465/256060
149. Neale MA, Huber VL, Northcraft GB (1987) The framing of negotiations: contextual versus task frames. Organ Behav Hum Decis Process 39(2):228–241. https://doi.org/10.1016/0749-5978(87)90039-2
150. Bazerman MH, Magliozzi T, Neale MA (1985) Integrative bargaining in a competitive market. Organ Behav Hum Decis Process 35:294–313

The Paradox of Power Asymmetry and Voluntary Participation in Construction Dispute Mediation

Nan Cao and Sai On Cheung

Introduction

Mediation has been promoted for use in the construction industry as a form of alternative dispute resolution (ADR hereafter) so that fewer disputes require costly arbitration and litigation [1]. In Hong Kong, since the mid-1800s, the Hong Kong government has made serious attempts to make mediation the mainstream ADR mechanism for all forms of dispute. In particular, the mediation movement received a strong push under the 2009 Hong Kong Civil Justice Reform (CJR hereafter). Practice Direction 31 (PD 31 hereafter), requiring an attempt at mediation before trial for all civil disputes except in the area of construction, was released for civil disputes except for those relating to construction. Practice direction 6.1 (PD 6.1 hereafter) is specifically prepared for cases reaching the High Court Arbitration ad Construction List (HCCT hereafter). In general, HCCT cases involve cases involving construction. A Steering Committee on Mediation was established by the Hong Kong Judiciary with the aim of making recommendations on ways to promote a wider use of mediation in Hong Kong. In this regard, the Mediation Ordinance (Chap. 620 of the Laws of Hong Kong) was enacted in June 2012 and became effective on January 1, 2013. The Mediation Ordinance provides the regulatory framework over the use of mediation, especially for the confidentiality of the proceeding. Furthermore, the Hong Kong Apology Ordinance (Chap. 631 of the Laws of Hong Kong) was passed and came into effect on 1st December 2017. It is believed that by protecting an apology offer from legal responsibility, more progressive resolution attempts could be taken up by disputants such as those involving offering an apology. At the 2020

N. Cao (✉) · S. O. Cheung
Construction Dispute Resolution Research Unit, City University of Hong Kong, Hong Kong, China
e-mail: nancao7-c@my.cityu.edu.hk

S. O. Cheung
e-mail: Saion.cheung@cityu.edu.hk

S. O. Cheung (ed.), *Construction Dispute Research Expanded*, Springer Tracts in Civil Engineering, https://doi.org/10.1007/978-3-030-80256-1_9

December 2nd Guangdong-Hong Kong-Macao Greater Bay Area Legal Department Joint Meeting, the three governments agreed to establish a "Greater Bay Area Mediation Platform" with the aim of providing innovative and diversified legal services for enterprises in the Greater Bay Area. It can therefore be said that mediation has become the predominant ADR mechanism used in Hong Kong. One unique characteristic of mediation is voluntary participation. Moreover, with the contractual use of mediation and construction contracting parties not being on equal footing, genuine voluntary participation may not be possible. This study aims to examine the paradox of voluntariness and asymmetry in construction dispute mediation.

Use of Mediation to Resolve Construction Disputes

In view of the large number of disputes that occur, the Hong Kong Judiciary commissioned two pilot mediation schemes for property management and construction disputes in 2006 and 2008, respectively. Successful experiences were reported, and these pilot arrangements have now become standard practice. In addition, the Hong Kong Judiciary has established an Office of the Building Management Mediation Coordinator in the Lands Tribunal since January 2008 to encourage litigants to consider using mediation to resolve their building management disputes.

However, from the past ten years of experience, the adoption of mediation has not been particularly impressive. The number of cases and success rate of building management disputes have fluctuated in recent years (Fig. 7.1) [2]. From January 1st 2008 to December 31st 2012, a total of 589 cases were referred to mediators by the Building Management Mediation Co-ordinator's Office (BMMCO). As a result, 556 cases have undergone mediation, leading to 226 mediated settlements. Accordingly, the success rate is approximately 40%. Averaged data for 2008–2013 and annual data for 2013–2019 for the building management cases are shown in Fig. 1.

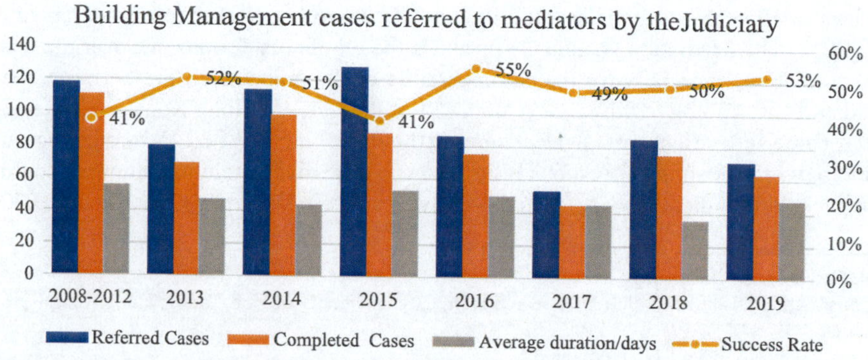

Fig. 1 Building management cases referred to mediators by the Hong Kong Judiciary

Another record also does not portray promising trend. According to the mediation reports filed with the Court of First Instance from 2011 to 2019, the settlement rate plateaued at approximately 50%, and the number of mediations conducted in 2019 underwent a sharp decline [3]. As shown in Table 1, the number of mediation certificates increased from 2011 to 2015. Since 2015, there has been no indication that mediation has gained popularity.

To examine the use of mediation in major construction disputes, the following summaries are collected. Table 2 presents the number of Construction and Arbitration Proceedings (HCCT)-related cases [4]. Tables 3 and 4 summarize the number of disputes handled by the Hong Kong International Arbitration Centre and the percentage of construction-related cases, respectively. Neither set of data indicates that there has been a broader use of mediation for construction disputes despite the aforementioned promotional efforts.

Table 1 Number of mediation related documents filed in the court of first instance[a]

	2011	2012	2013	2014	2015	2016	2017	2018	2019
Mediation certificate	2759	2977	2878	3271	3668	3623	3716	3590	2138
Mediation notice	1030	1146	1164	1223	1381	1380	1399	1248	958
Mediation response	949	1062	1031	1078	1258	1181	1249	1140	876
Mediation minutes	444	508	541	602	652	666	663	634	478
Settlement rate (%)	38	38	45	48	46	48	48	51	51

[a]It only includes cases commenced by the 5 CJR related case types in the Court of First Instance, i.e. Civil Action (HCA), Admiralty Action (HCAJ), Commercial Action (HCCL), Construction and Arbitration List (HCCT)

Table 2 Number of construction and arbitration proceedings (HCCT) related documents filed in the high court

	2011	2012	2013	2014	2015	2016	2017	2018	2019
HCCT	22	18	21	9	16	14	20	26	30

Table 3 Number of disputes involving HKIAC in recent 5 years

	2014	2015	2016	2017	2018	2019
Arbitration	252	271	262	297	265	308
Mediation	24	22	15	15	21	12
Adjudication	0	0	0	0	0	1

Table 4 Ratio of construction disputes involving HKIAC

	2014	2015	2016	2017	2018	2019
Construction dispute	–	22.2%	19.2%	19.2%	13.7%	14.8%

Voluntariness as the Necessary Condition for Successful Mediation

Mediation is a form of assisted negotiation [5]. Voluntariness is often considered its core feature. Disputing parties agree to engage in mediation when they prefer this process and have a genuine desire to resolve the problem at hand. The parties are also free to choose whether to use this method, when to us it and who will act as the mediator. The parties therefore have psychological ownership that is critical to ensuring that they will honour the settlement agreement reached. As a general rule, negotiating parties must have the will to negotiate; otherwise, there is a limited likelihood that the negotiation will be successful. In this regard, it has been quite notable that arrangements for mediation emphasize voluntary participation. Forced negotiation does not provide the necessary conducive platform for genuine attempts to settle. Moreover, there has been call for a mandatory use of mediation to accelerate its adoption. In Hong Kong, it is generally believed that any attempt to impose an involuntary process on a party may undermine the characterizing feature of mediation—voluntary participation. By analysing the current arrangements on construction mediation, an analysis of voluntary participation is illustrated.

Mediation Rules

The Hong Kong International Arbitration Centre (HKIAC) is the leading dispute resolution services provider in Hong Kong. Many dispute resolution activities have been organized by the HKIAC. In fact, the rules of the HKIAC are the most commonly used in Hong Kong. According to the HKIAC mediation rule, a failure by any party to reply within 14 days shall be treated as a refusal to mediate. Thus, mediation can only be conducted if all parties agree to mediate. The design of mediation is thus anchored in voluntary participation [6]. Many time delays could be avoided if disputants participated on their own accord. There would then be no issues related to compelling parties to mediate, and parties would be much more likely to make meaningful contributions, especially with good faith behaviours envisaged. In addition, in fully voluntary mediation, the parties are free to leave at any time.

Contractual Use of Mediation

It is now a standard dispute resolution process design to include mediation as an intermediate step between the contract administrator's decision and arbitration in construction contracts. In Hong Kong, most projects adopt standard forms of contracts with necessary modifications to suit their needs [5]. Normally, a three-tiered dispute resolution procedure is used. According to HKG General Conditions of Contract for Building Works/Civil Engineering Works/Design and Build Contracts Clause 86 and General Conditions of Contract for Term Contract for Building Works Clause 92/Civil Engineering Works Clause 89, when a dispute arises, it shall be reported to and settled by the designated contract administrator. If either party is dissatisfied with the decision made, they can refer the matter to mediation within 28 days of the decision. If the matter cannot or does not need to be resolved by mediation, any reference to arbitration shall be made in accordance with the Arbitration Ordinance within 90 days. A similar design is also adopted in the private building projects force of contract. More recently, the New Engineering Contract (NEC) has gained popular use for public works projects in Hong Kong. The 2017 NEC4 Dispute Resolution Service Contract (DRSC) offers three dispute resolution options (W1, W2, and W3), and Z-clauses that provide bespoke additional contract conditions can be added, allowing unique requirements for local dispute resolution practices. W1 and W2 under NEC4 use adjudication as the primary means of dispute resolution, W3 uses dispute avoidance, while mediation can be added to the Z-clauses as a construction dispute resolution tool in the NEC, such as adjudication and arbitration. This contractual use of mediation is quite different from its mandatory use because voluntary participation is retained under the contractual arrangements. It is now established that the mediation clause shall be specific enough so that objective criteria can be deduced to determine compliance or otherwise. As such, a mediation clause should specify the model and rules to be used. In addition, a clear time frame for its implementation, the nominating authority and the minimum amount of participation are essential items to be incorporated to develop an enforceable mediation clause for construction contracts.

Court Encouraged Mediation

According to section F of Hong Kong High Court Practice Direction 6.1, construction cases reaching the Hong Kong High Court are encouraged to attempt mediation as a possible cost-effective means of resolving disputes and to promote the use of mediation.

Upon receiving the Mediation Notice, the Respondent should respond to the Applicant in writing within 14 days, although he has the right to refuse to mediate. The principal way to encourage mediation attempts involves the imposition of cost sanctions where a party unreasonably refuses to attempt. However, if a party (1) has engaged in mediation to the minimum level of expected participation agreed upon by the parties beforehand or as determined by the Court or (2) has a reasonable explanation for nonparticipation, he should not suffer any adverse costs order. Thus, as long as the aforementioned procedures are completed, it can still be claimed that the parties have certain autotomy in deciding to mediate or not. As reported in Sect. 7.2, even though PD 6.1 came into effect in 2009, the number of construction disputes referring to mediation did not significantly increase. This may well be explained by parties having a final say to undertake mediation. The voluntary nature of mediation has thus been retained.

Court-Connected Mediation

There is an ongoing debate on whether courts should compel disputing parties to attempt mediation [5, 7–9]. Since the courts' association with mediation programs as in Canada, Australia, the United Kingdom and Singapore, a more balanced perspective on compulsory ADR has been developing. Court-annexed mediation is the most direct way to solicit attempts of mediation. However, an abuse of process can result. In fact, the Civil Justice Reform's (CJR's) Working Party has proposed court-annexed mediation; in its Interim in 2000, the proposal was finally rejected in the 2004 Final Report. Therefore, court-annexed mediation has not been implemented in Hong Kong [5, 7, 9]. Statutory use denotes that disputes will be automatically directed for mediation irrespective of the nature of the disputes. The negative effect is quite obvious, as parties may only take this condition perfunctorily [8]. Parties forced to mediate might not attend mediation in good faith, leading to extra costs and impeding mediation confidentiality when the case is finally heard in court [9]. Court-annexed mediation undermines the voluntary nature of mediation. The absence of voluntariness would worsen the relationship and make mediation less likely to succeed [5]. If parties are forced to mediate, settlement proposals may merely be formulated to satisfy mandatory requirements. Rules of law and justice may not even be on the agenda, which may address commercial issues in a way that lack clarity and certainty [7].

The Pillars of Successful Mediation

Despite an initial surge in the use of mediation when the Civil Justice Reform came into effect, the decline of this trend is quite disheartening. Examining dynamics of the willingness to mediate may unveil the underlying reasons for the slow uptake of construction mediation. Within this connection, the pillars of successful mediation are first considered. The removal of the pillars would likely hamper the chances of reaching settlement. More importantly, for this study, the role of voluntary participation in driving-mediated settlement is explored. Mediation is a voluntary, confidential and nonbinding dispute resolution process through which a neutral professional mediator helps parties reach amicable settlement. As mentioned in the previous section, voluntary participation is the first step in recognizing the applicability of the process and, most critically, the acceptance of the outcome. The freedom of exit at any time is definitely attractive to disputants who are not sure if the process is appropriate. Coercion runs against voluntariness and can have three facets: coercion to mediate, coercion to continue and coercion to settle. Cheung et al. [10] examined mediation from four aspects: nature, the neutral third party, settlement, and benefits. Those critical attributes shown in Fig. 2. The first aspect normally serves as the main reason for choosing mediation, and the second aspect is usually used to justify continuing mediation. The last two aspects are mostly related to the willingness to settle.

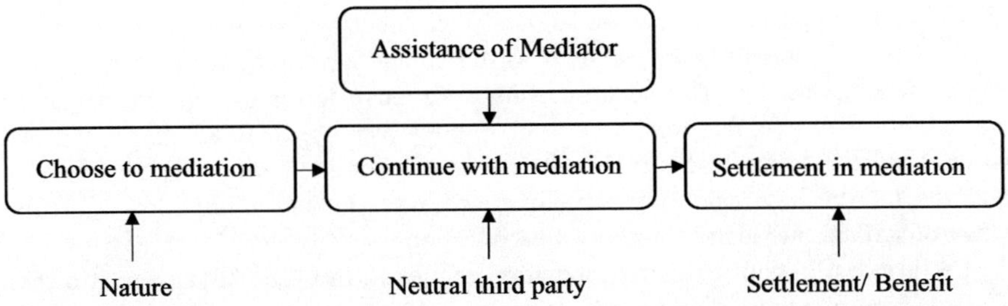

Fig. 2 Critical Attributes of the mediation process

Nature

Apart from voluntariness, confidentiality and enforceability are other major attractive attributes of the nature of mediation. The confidentiality ensured through mediation serves as an essential reason for disputing parties to use this method. Furthermore, discussions held during mediation are kept confidential. As a private proceeding, only the disputing parties know the happenings of a mediation. To many organizations, this is a very valuable characteristic. For this reason, parties feel much freer to express their views. The enforceability of mediation clauses can be a concern, as good faith provisions have been proven vulnerable in common law courts. Thus, far more detailed mediation arrangements are needed to ensure that parities will not find ways to avoid mediation. The associated downside is that parties will have fewer options. When the preferred choice of a party is not addressed, the level of voluntariness may be curtailed.

Neutral Third Party

One of the advantages of using mediation relates to the assistance of a professional experienced neutral mediator. If a dispute involves technical issues, the mediator should have at least some related technical knowledge. It is suggested that mediators of different backgrounds and with different specialties should be kept on the lists of mediators of nominating bodies. Mediators have a duty to act impartially and as such can be instrumental in assisting disputing parties in focusing on solving the problem at hand.

Settlement/Benefit

Parties' willingness to commit themselves to mediating should greatly improve the likelihood of reaching a settlement. Mediation is a non adversarial process through which disputing parties come to understand each other's needs and interests. Identifying common interests is the pathway to settlement. Mediation also offers a wider range of remedies than formal proceedings. Creative terms of settlement can be crafted. For example, the provision of an apology as a means to address emotional issues has been found to be extremely useful in communicating disputes. Lateral thinking is vital in considering settlement options. Mediation can allow a greater range of settlements than litigation and arbitration.

The greatest benefit of using mediation relates to the time advantage it offers. The duration of a mediation process is measured in hours or days instead of weeks or months, which is the case for arbitration and litigation. Fast resolution is one of the most notable advantages of mediation. Less time taken results in lower costs and resources being needed. Costs involved in settling include venue, mediator, documentation, and settlement costs. Mediation helps reduce the overall cost by ensuring that parties are working towards a settlement instead of focusing on winning. The disputants show substantial involvement in the resolution process, which should be articulated. The central idea of mediation is to manage the adversarial relationship between two parties and encourage them to work cooperatively. Many cases have demonstrated that the relationship between parties of an arbitration is often beyond repair. Mediation seeks to avoid this undesirable outcome.

Power Asymmetry Between Negotiating Parties

Another issue relating to using mediation concerns the potential diminishing focus on justice when unintended results are obtained due to a power imbalance. Equal footing is assumed in most negotiation theories, though this may not be the reality. For example, construction contracting parties show a substantial differential in power, limiting the applicability of renowned negotiation theories. In addition, voluntary participation in mediation will be affected by the existence of asymmetry between disputing parties. Is conflict further magnified when parties are 'forced' to enter mediation, as in the case of court-encouraged and court-annexed settings? The impact of asymmetry on commitment to mediation must therefore be investigated. Studies examining the effects of asymmetry of parties' perceptions of conflict on mediation outcome have been conducted [11, 12]. Some research has introduced specific styles of mediation that may be appropriate for disputing parties with uneven power [13]. Gewurz [13] also found that a skilful mediator could increase the probability of peaceful settlement by helping disputants overcome the problems associated with asymmetric information [14–16]. However, the above research focuses on nation to nation and neighbour to neighbour contexts. In construction, the function of incentivization in minimizing construction disputes by addressing asymmetry stems from disproportionate risk allocation, and power distribution between the contract and employer was reported by Zhu and Cheung [17]. Interdependence negates the aggregation of conflict. The more interdependent the parties are, the more likely compromising behaviour is to be practised to resolve the dispute [18].

Maintaining relationships is not confined to having disputes mediated. In fact, sustaining an enduring relationship among participants benefits the whole construction industry. Parties in dispute, however, must manage with conflicts arising from power asymmetry. This proposition is supported by research on organizational relationships ([19–22) and can be operationalized as a power balance being a prerequisite for joint effort.

Asymmetry in conflict situations is generally identified by the power differentials between the disputing parties. Although there are many sources or forms of asymmetry, many of which can be translated as power between parties [23, 24], in this study, three forms of asymmetry are believed to have an effect on the level of voluntary participation in construction dispute mediation: resources, information and expectations.

Resource Asymmetry (RA)

Resources can be measured by disposable capital and assets. In this regard, developers are often more resourced than contracting organizations. Resource asymmetry may be one of the reasons why parties involved in a construction dispute are not willing to attempt mediation simply because of the disadvantages derived from unpreparedness. A less-resourced party may not be treated equally as far as negotiation dynamics are concerned. Since mediation is private and its process is flexible, no procedural safeguard exists, such as an observance of natural justice principles. Abusive use of flexibility by a powerful party can be an issue. Thus, concerns over being bullied during the process may become a barrier open-hearted participation. The classification of RA is drawn from the literature. As indicated in Table 5, coercive resource asymmetry (CRA), reward resource asymmetry (RA-RA), expert resource asymmetry (ERA), referent resource asymmetry (RE-RA), and legitimate resource asymmetry (LRA) are the most reported forms of resource asymmetry.

Table 6 lists the characterizing behaviours that manifest in the respective forms of resource asymmetry.

Table 5 Key references of the five forms of resource asymmetry

CRA	RA-RA	ERPA	RE-RA	LRA	References
*	*	*	*	*	[25]
*	*	*	*	*	[26]
*			*	*	[27]
*	*	*	*	*	[28]
*	*		*	*	[29]
*	*	*	*	*	[30]
*	*	*	*	*	[13]

Information Asymmetry (IA)

Successful construction dispute mediation is based on effective communication and could be hampered by information asymmetry. Information asymmetry occurs when one party has more or better information than the other. Studies in law and economics have shown that information asymmetry creates an imbalance of power in an exchange whereby adverse selection, moral hazard and hold-up problems could result [33–35]. Studies of information asymmetry in construction projects have mainly focused on risk and contract management. For example, asymmetric information in construction projects can lead to inequitable risk allocation [36]. Information asymmetry creates communication risks according to principal-agent theory [37]. Xiang et al. [38] suggested that risk prevention can be effected by reducing asymmetric information among project stakeholders. In sum, asymmetric information generates adverse selection, moral hazard and hold-up problems among construction project participants. On the other hand, the components of these causes could be used to measure information asymmetry. The imbalance of information between disputants could be examined from three different perspectives: adverse selection, moral hazard and hold-up problems.

Adverse selection refers to a situation where a party with less information is concerned about an unfair settlement. In such circumstances, parties who have more information use it for their benefit at the expense of their counterparts. The fear of

Table 6 Lists the resource asymmetry types and their characterizing behaviours

PRA types	Resource asymmetry-based behaviours	References
Coercive resource asymmetry	We are more capable of delaying the payoff to the other side	[19, 31, 32]
	We are more capable of reducing the profit of the other side	
	We are more capable of withdrawing certain needed services from the other side	
	We are more capable of making the process more difficult for the other side	
Reward resource asymmetry	We are more capable of affording future work opportunities to the other side	
	We are more capable of providing economic incentives to the other side	
Expert resource asymmetry	We have more knowledge and expertise in designing or constructing new projects	
	We usually are given advice from the other side	
	We are more likely to recommend appropriate actions to the other side	

(continued)

Table 6 (continued)

PRA types	Resource asymmetry-based behaviours	References
Referent resource asymmetry	We adopt better exemplary project cost management methods	
	We adopt better exemplary project time management methods	
	We adopt better exemplary project quality management methods	
	We adopt better exemplary project safety management methods	
	We have better exemplary values	
Legitimate resource asymmetry	We more frequently use sections of contract agreements as a "tool"	
	We believe other side has a stronger right to request and expect that things be done according to its requirements, even when they are not referenced in the contract	
	We are more obliged to accept the other side's suggestions	

unfair settlement can prompt the concerned party to refuse mediation. Moral hazard occurs when parties act opportunistically and is characterized by maximizing self-interest seeking [39]. Moral hazard usually arises when a dispute situation is vague or when contracts are incomplete. The occurrence of moral hazard would prompt parties to withdraw from mediation. A hold-up problem arises when two parties are able to mediate most efficiently through joint effort but refrain from doing so due to concerns that they may have to concede too much to the more powerful party. Hold-up problems lengthen mediation time. Hesitation may cause parties to miss the opportunity to reach a settlement. Table 7 presents those three information asymmetry types and their characterizing behaviours.

Expectation Asymmetry (EA)

Expectation asymmetry could be interpreted as the difference between expectations and perceived performance. Conflict research often assumes that parties in dispute have the same view of the nature of the conflict at hand. Moreover, the parties may have different perceptions and expectations of the conflict. Some researchers reveal expectation asymmetry in the construction industry and inconsistencies in construction project practitioners' expectations and realities. Liu [41] noted the importance of disputing parties bridging the expectation-reality gap for proper risk assessment.

Table 7 Presents those information asymmetry types and their characterizing behaviours

PIA types	Information asymmetry-based behaviours	References
Adverse selection	We know more about the attributes of construction products, such as the number of construction projects, visual effects, etc.	[38, 40]
	We have a better understanding of the expected and actual progress of construction projects	
	We are more aware of construction costs, including labour costs, material costs, mechanical equipment costs, etc.	
	We are more capable of making things more difficult for the other side	
Moral hazard and hold-up problem	The contractor, in the construction of the project after signing the contract, exhibits cautiousness over the design of construction drawings, the quality of personnel involved in the construction process, the quality of building materials, construction methods and technologies, etc.	
	The owner supervises and regularly assesses the contractor's construction behaviour, which mainly involves establishing a problem responsibility system, a reward and punishment system, and an acceptance system to inspect the construction status and effort of the contractor	
	The owner's financial ability, including progress payments for construction projects being issued on time	

In terms of cooperation, a gap between reality and one's expectations about the likelihood of another's future actions decreases trust in the partnership eventually leads to inefficiency in construction management [42].

Mediation is less likely to have a successful outcome when disputing parties have asymmetric perceptions of the outcome [11, 43]. The above research covers only cases of effective mediation, and this may be an important limitation because conflict asymmetry may be directly related to the likelihood of both parties voluntarily participating. The parties involved in construction disputes are named "complainants" or "respondents". It has been found that complainants are more demanding and difficult to satisfy. In contrast, respondents are more yielding and less sure of themselves. The result implies that respondents usually end up with less than what they expected [44]. At the same time, it is less likely that both parties will accept the sharing of responsibility if their expectations differ. Additionally, if accepting one's responsibility is a necessary condition for parties to cooperate in a mediation session, it will also be an important condition for a party's willingness to participate in a mediation session in the first place. Moreover, asymmetry may inhibit the possibility of practising integrative conflict resolution, thus impairing the likelihood of successful mediation. Aversion attitudes demotivate participation in mediation. Most decisions have a status quo alternative—that is, doing nothing or maintaining one's current or previous decision. Status quo effects restrain decision-making in many settings [45]. During a negotiation, once an assessment of the dispute is made, changing one's position to accept a counterpart's proposal is often taken as a loss [46]. In an experimental setting found that asymmetrical conflicts are less likely to run a constructive course and are more likely to escalate into impasse or win-lose outcomes, showing that third parties should be aware of the structure of the conflict when they engage in mediation [47]. Overall, making both parties aware of the gap between the expectations and realities of a conflict and of taking responsibility is important not only for parties' cooperation in a mediation session but also to bring both parties to the mediation table [48].

Expectation asymmetry occurs at both the pre- and post contract stages. The inherent expectation asymmetry involved in risk aversion imposes an original restriction on the observed construction bid data, while experience influences the degree of risk aversion. Regarding the contractual form of procurement contracts under cost uncertainty, research provides that the contractor is more averse to ambiguity than the owner—the more ambiguity of belief there is, the lesser the power of the optimal incentive scheme becomes. A fixed-price contract is optimal if there is no ambiguity. If levels of ambiguity are high, a cost-plus contract is optimal. A cost-share scheme is sensible for conditions in between [49]. A previous study on international construction projects [50, 51] showed that (1) decision-makers are more risk-averse in opportunity situations than in threat situations, (2) decision-makers are extremely risk-averse when the chance of loss is too high, and (3) decision-makers are more risk-taking when decisions involve losses than when decisions involve gains. The effect of loss aversion is also explored, anchoring in commercial real estate prices; a study shows this effect to vary by type of market participant and cycle [52] As for problem fixing, Kahneman and Tversky [50] discuss how the psychophysics of value induce risk aversion in the domain of gains and risk-seeking in the domain of losses and point out that the distinction between decision and experience values is rarely explicit in decision theory because it is tacitly assumed that they coincide. This assumption suggests that disputants' voluntariness to participate can be improved by framing negative outcomes as having negative value [51]. Two experiments were conducted to verify that negatively framed bargainers generally show more voluntariness than their positively framed counterparts in most situations. A pair of positively framed bargainers reach more integrative settlements than a pair of negatively framed bargainers [53]. Neale and Bazerman [54] also found that a positive frame leads to more concessionary behaviours and successful performance than a negative frame. Risk aversion and empathy gaps belong to expectation asymmetry. Table 8 presents behaviours manifesting the effects of sexpectation asymmetry.

The Paradox of Asymmetry and Voluntary Participation in Construction Dispute Mediation

Voluntariness is considered to be a prerequisite to reaching settlement through mediation [61]. Mediation was introduced to Hong Kong as a voluntary resolution process. Even under Practice Direction 6.1, though the Court may apply adverse cost orders against parties who unreasonably refuse to mediate, the use of mediation may be deemed 'not quite voluntary'. Moreover, the take-up rate and settlement rate of mediation not be particularly impressive. Furthermore, the mediation programs operated in Australia, Canada and the UK hold that as long as the mediation outcome is self-determined, a positive effect can be detected regarding whether its use is mandatory. As such, the settlement rates of mandatory and voluntary mediation show no major differences [62, 63]. Nonetheless, some practitioners have suggested that forced mediation also has potential to erode access to justice, especially if the power to order compulsory mediation is exercised frequently [6, 64]. The HKSAR government appears to take a more moderate stance in encouraging the use of mediation while retaining its voluntary nature under PD 6.1.

The presence of power asymmetry in mediation has been identified by Gazal-Ayal and Perry [65] and Gewurz [13]. The more powerful party is likely to impose its will on the weaker party with a "take-it-or-leave-it" or "take-it-or-suffer" strategy. Moreover, the weaker party expects to mediate on equal footing [66]. The paradox between power asymmetry and voluntariness is thus formulated. There is no easy answer to this inherent dilemma between voluntariness and asymmetry. Some authors, however, argue that power symmetry is a favourable condition for effective negotiation [29, 67, 68]. This suggestion makes practical sense when the weaker party is desperate

Table 8 Presents expectation asymmetry types and their characterizing behaviours

PEA types	Expectation asymmetry-based behaviours	References
Benefit gap asymmetry	Perceived cost savings from participating in mediation are lower than expected	Aibinu and AI-Lawati [55], Karambayya et al. [56]
	Perceived time savings from participating in mediation are lower than expected	
	Perceived productivity improvements from participating in mediation are less significant than expected	
	Perceived business opportunity improvements from participating in mediation are less significant than expected	

(continued)

Table 8 (continued)

PEA types	Expectation asymmetry-based behaviours	References
Cost gap asymmetry	Perceived higher share proportion of the initial cost of mediation (e.g., the on-time charges of the mediator, the cost of renting a conference room, etc.) than expected	
	Perceived higher share proportion of the additional cost of mediation (e.g., relevant service charges to hire a lawyer or other professionals to provide professional advice during the mediation process, etc.) than expected	
Security gap asymmetry	More perceived possibilities of submission data or document leakage than expected	
	Perceived confidentiality of mediation process lower than expected	
	More perceived possibilities of minimum expectation resolution leakage than expected	
Fairness gap asymmetry	The procedure rules are more inclined to the other side than expected (e.g., the favourable sequence for asking questions and offering rebuttals, keeping track of new information changes and options, etc.)	
	The substantive outcomes are more inclined to the other side than expected	
	The third-party favours the other side over us more than expected (e.g., mediator considers another side's feelings and opinions more, etc.)	
Framing gap asymmetry	Our options are presented with more positive connotations than expected	Pinkley [57, 58], Kahneman and Tversky [59], Segal [60]
	Our options are presented with more negative connotations than expected	

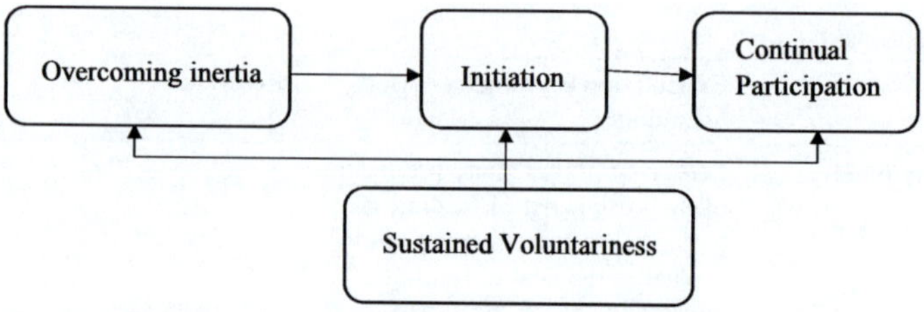

Fig. 3 The role of voluntariness in construction dispute mediation

for an early settlement whereby the more powerful party is likely to be in the upper hand and drive for a settlement in his favour even at the expense of the weaker party. Thus, a faster settlement may be result. Power asymmetry is taken as a reality check for the weaker party despite the unintended result of conflict being suppressed [28, 69, 70]. Inspired by these differing perspectives, this study calls to examine the effect of power asymmetry on the level of voluntary participation in construction dispute mediation (CDM hereafter). Figure 3 illustrates the role of voluntariness in underpinning successful mediation.

Notably, conceptualizing power asymmetry and voluntariness in CDM would help detect power asymmetry from different aspects and voluntary/involuntary behaviours present during the dispute mediation process. Balancing the power asymmetry relationship is an effective way to improve voluntary participation and the possibility of achieving successful dispute settlement. Mediation is a form of assisted negotiation; it is therefore vitally important to include the role of the mediator in the relationship framework between asymmetry and voluntary participation. In fact, previous research has illustrated the function of mediators in facilitating dispute settlement through the appropriate use of tactics to address bottlenecks such as dispute sources and disputants' attitudes [71–73]. This study also aims to explore mediators' interventions to address the power asymmetry relationship and thus make voluntary participation sustainable throughout mediation. Against this background, the following questions are addressed in this study: (i) Does power asymmetry affect construction disputing parties' voluntary participation in mediation? (ii) What underlying constructs of involuntary behaviours hinder amicable dispute mediation? (iii) In what ways can mediators help mediation address the paradox between asymmetry and voluntary participation?

The literature on voluntary negotiation is growing, but there has yet to be a definition of voluntariness that encapsulates the central idea of participation at one's own will in construction dispute mediation. A voluntary participation framework has been used extensively in construction dispute mediation. Voluntary participation embraces intentional action, the absence of controlling influences and no-role restriction. Voluntariness is a multidimensional concept that cuts across several domains. Based on the literature review conducted for this study, a summary of voluntary manifestations is presented in Table 9.

Table 9 Based on the literature review conducted for this study, classifications of voluntariness used in previous studies is summarised

Dimensions of voluntariness	Manifestations	References
Intentional action	The party in the performance of actions uses intentional action	[74]
The absence of persuasion	No side persuading another side believes something through the merit of reasons proposed	
The absence of coercion	No side intentionally forces another side or uses a credible and severe threats of harm to control another side	

(continued)

This study contributes to the study of construction dispute mediation in Hong Kong. The assumption of voluntary participation is seldom questioned. However, there is obvious and notable power asymmetry between disputing parties. Acknowledging this inherent paradox and devising appropriate intervention by the mediator would bring construction dispute mediation to higher adoption level.

Intervention of Mediator in Bridging Power Asymmetry

There are different schools of thought on whether asymmetry hampers voluntary participation. Marquardt and Wiedman [78] found that managerial participation in stock market offerings is negatively associated with information asymmetry. Wu and Babcock [79] found that under the green payment program, farmers may use information asymmetry to obtain favourable combinations of production and subsidies. The concepts of principal-agent theory were applied to design the green payment program. Considering the possible information asymmetries between the government and farmers, the voluntary and self-selecting participation rates of the program were projected. Marquardt and Wiedman [78] suggested that firms change their disclosure activity before offerings to reduce information asymmetry to hype the stock. Similarly, in analysing an international sample of 575, Martínez-Ferrero et al. [80] proposed a bidirectional relationship between voluntary disclosure and asymmetric information. It was found that more asymmetric information leads to more voluntary information disclosure practices, which can boost investor participation confidence. Similar to the ability to retain users, this issue is an important concern for social network sites. Shi et al. [81] found that expectation asymmetry negatively influences users' motivation to use Facebook. Analogously, customers normally enter into a service with certain expectations regarding the level of service they are likely to receive. Customers' expectations can thus be attributed to the causes of service failure [82]. In terms of partnerships, information and decision-making power imbalances between nurses and patients may inhibit partnerships in care and create more

Table 9 (continued)

Dimensions of voluntariness	Manifestations	References
The absence of information manipulation	There is no use of nonpersuasive means to alter a side's understanding of a situation	
The absence of reward manipulation	No side motivates another side to do what the agent of influence intends	
Inducement	No offers to provide incentives are made	[75]
Persuasion	No application of interpersonal pressure or by an exhortation to self-interest or community norms is applied	
Force	No enforcement by nonconsensual intervention or the issuance of threats is used	
Understanding of the proposed program	Potential participants have an adequate understanding of specific aspects of the proposed program or even of the program in general	[76]
Social norms	No side considers decision making by the other side as the social norm	
Social relations	Cross-cutting interpersonal and contextual domains does not make it difficult to say no	
Value	There is a willingness to mediate the dispute for shared value	
Diminished capacity	Supply or funding chains are disrupted	[77]
Goals	There is a willingness to mediate the dispute to achieve a mutual goal	
Manipulation	The choice of an action is free from constraints imposed by other persons or social institutions	
Inducements	The voluntariness of the disputants is undermined by "inducements" or "offers" designed to encourage the parties to enter mediation	

imbalance with less subsequent patient input, harming cooperation in patients' health care.

On the other hand, asymmetry in resources can develop trust between team members and solicit voluntary participation [83]. Building on social comparison theory, Wang et al. [84] found that a moderate level of reward resource asymmetry is likely to involve a higher percentage of participatory employees than firms with either very low or very high levels of inequity. The statistical analysis results of Benk and Budak [85] reveal that Turkish taxpayers support the effects of referent and legitimate power asymmetry and promote trust and that voluntary tax compliance thus results.

From the contingent perspective, Cowan et al. [86] proposed a power–benefit matrix for interfirm relationships and illustrated that resource asymmetry affects cooperation willingness from the relationship benefits perspective. If the earned benefit is below expected, higher power will lead to opportunism,if the benefit is at or exceeds expectations, the weaker party is likely to tolerate it. Fuchs and Lippi [87] report descriptive and econometric evidence that national divergence matters for policy decisions to voluntarily participate. It was found that an optimal policy responds to a country's incentive to remain in or leave the monetary union. A study of African small exporters also found [88] that relational resource asymmetry positively impacts the innovative participation of small suppliers. Aiello et al. [89] considered both intergroup and interpersonal approaches to power and examined how social dominance orientation influences coordination between supervisors and subordinates. The authors' results show that the stronger participants were in social dominance orientation (SDO), the greater their willingness to participate under harsh levels of resource asymmetry than under lower levels of resource asymmetry became.

In general, there is more research supporting the idea that information and expectation asymmetry create obstacles to rather than encouraging full participation in mediation. Furthermore, the effect of resource asymmetry between disputants on mediation willingness is more complex; both pros and cons opinion have been reported; and further investigation of resource asymmetry under more detailed classifications, such as coercive, reward and resource asymmetry, on voluntary participation, is needed. In contrast, power asymmetry stems from expertise, reference, and legitimacy and may encourage participation in mediation. Referring to the moderating factor, mediator tactics and techniques should help deal with different power asymmetries in construction mediation. A conceptual framework summarizing the paradox of asymmetry and voluntary participation in construction dispute mediation is shown in Fig. 4.

When a power imbalance exists between parties, this will inevitably influence their willingness to participate in mediation. In this case, the mediation outcome may not be ideal. In what ways could a mediator help alleviate such concerns? Ippolito and Pruitt [44] attempted to address this connection and found that mediator efforts to balance power discrepancies are not effective. In addition, it is controversial when a

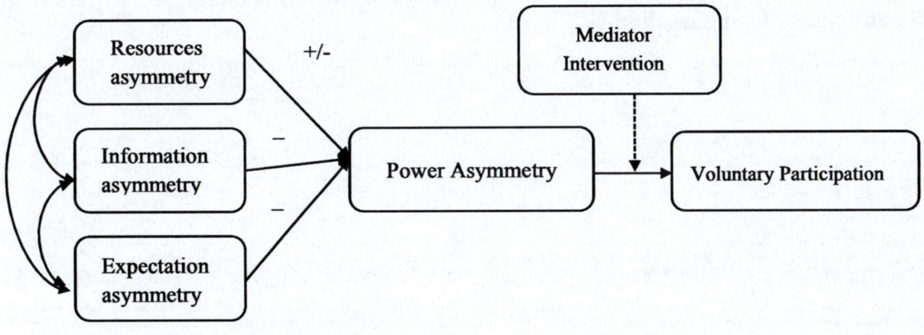

Fig. 4 Paradox and mediator intervention

mediator tries to rectify power asymmetry because this suggests that the mediator is departing from a neutral stance. Therefore, it is recommended that the focus should be on how mediation tactics can be used contingent on the circumstances of power relations and on the willingness to mediate. Table 10 lists mediator strategies for power asymmetry and voluntary participation.

To sustain voluntary participation, the mediator may intervene in an effort to balance the power asymmetry throughout mediation to ensure good mediation outcomes, and the weaker party will thus not be pushed into unfair agreement. Six categories of disputants' power asymmetry-related tactics and 13 categories of disputants' voluntary participation-related tactics used in 5 main stages of mediation were identified through content analysis. The subsequent discussion will provide insight into the relationship between mediator intervention and the measurement of disputants' voluntary characterizing behaviours.

Summary

The presence of asymmetry between parties of construction contracting is real. Complete absence of asymmetric power can't be expected due to the nature of the transaction. The relationship between owners and contractors is typified as one of principal-agent. When dispute between them arises, the asymmetry between them may affect their voluntariness in undergoing mediation. From the perspective of organizational economics, asymmetry in construction contracting would give rise to the use of one-side contracts. As such, ex post practice of opportunism seems likely. Empirical research has shown that it is important to have voluntary participation so that the mediated settlement will be honoured. This study offers the conceptual lens to analyse the paradox between asymmetry and voluntary participation. Contractual use of mediation is the prevalent approach with voluntary participation being the central design consideration. Power asymmetry may be a problem as a unwilling party would be pushed to attempt mediation. Likewise, court encouraged or court-annexed approach may have the same effect-undermining voluntary participation. This study raises the need to revisit the assumption of voluntary participation in construction dispute mediation.

Table 10 Lists the mediator potential Intervention on power asymmetry and voluntary participation

Mediation stages	Attributes	Mediator tactics
Introduction	Related to disputant's power asymmetry	Ensure that all documents deemed conducive to mediation have been disclosed to the mediator
Venue and arrival		Ensure the attendance of representatives of all parties in advance
Opening phase		Explore the genuine concerns of the parties
		Encourage parties to show respect, cooperate and create a productive atmosphere for later interactions
Individual session		Guide the parties to review their interests and concerns
		Challenge each party to identify a solution that will satisfy all of the parties
Introduction	Related to disputant's voluntary participation	Ensure that the case is appropriate for mediation and is appropriate to mediate at the given point in time
Venue and arrival		Properly manage the booking of required rooms and visual facilities and individual mediation session wait times
Opening phase		Introduce all attendees to create a more personalised atmosphere and to highlight the differences between mediation and litigation
		Allows parties to express their views and feelings in a controlled confrontation for negotiation and later compromise
		Establish authority and control from at the start of mediation by demonstrating confidence, judgement and familiarity with the process
Individual session		Show an understanding of the issues and empathy for the situation faced by the parties
		Prioritise issues and address less contentious issues first to create a sense of cooperation and progress

(continued)

Table 10 (continued)

Mediation stages	Attributes	Mediator tactics
		Help the parties develop a realistic understanding of different alternatives they have in resolving the disputes
		Remind the parties of the cost and time consequences of continued litigation or arbitration
Dealing with deadlock		Shift from deadlock on a substantive issue to discuss procedural ways of moving forward
		Advise the parties to carry out an early neutral evaluation (ENE) that effectively resolves technical issue deadlock
		Advise the parties to take legal advice from a jointly appointed lawyer about the legal issues
		Recall what can be agreed on between the parties within the existing authority and resources

References

1. McKenzie DM (2015) The role of mediation in resolving workplace relationship conflict. Int J Law Psychiatry 39:52–59. https://doi.org/10.1016/j.ijlp.2015.01.021
2. The Hong Kong Judiciary (2019a) Mediation statistics for building management cases. Retrieved from https://mediation.judiciary.hk/en/figuresandstatistics.html
3. The Hong Kong Judiciary (2019b) Mediation statistics for civil justice reform related cases. Retrieved from https://mediation.judiciary.hk/en/figures_and_statistics.html
4. Legal Reference System (2020) Cases reached construction and arbitration proceedings in the high court of Hong Kong special administrative region. Retrieved from https://legalref.judiciary.hk/lrs/common/ju/judgment.jsp?L1=HC&L2=CT&AR=14#A14
5. Cheung SO (2016) 4 Court-connected mediation in Hong Kong. In: Court-connected construction mediation practice: a comparative international review. Routledge, pp 55–77
6. Katz LV (1993) Compulsory alternative dispute resolution and voluntarism: Two-headed monster or two sides of the coin. J Disp Resol 1
7. Hilmer SE (2013) Mandatory mediation in Hong Kong: a workable solution based on Australian experiences. China-EU Law J 1(3–4):61–96. https://doi.org/10.1007/s12689-012-0016-y
8. Leung RHM (2014) Hong Kong mediation handbook. Sweet & Maxwell
9. Meggitt G (2018) The cases for (and agaist) complusory court-connected mediation in Hong Kong. In: 5th Asian law institute conference, pp 22–23
10. Cheung SO, Suen HC, & Lam TI (2002) Fundamentals of alternative dispute resolution processes in construction. J Constr Eng manage 128(5):409–417
11. Jehn KA, Rupert J, Nauta A (2006) The effects of conflict asymmetry on mediation outcomes: satisfaction, work motivation and absenteeism. Int J Confl Manag 17(2):96–109. https://doi.org/10.1108/10444060610736594

12. Ufkes EG, Giebels E, Otten S, van der Zee KI (2012) The effectiveness of a mediation program in symmetrical versus asymmetrical neighbor-to-neighbor conflicts. Int J Confl Manag 23(4):440–457. https://doi.org/10.1108/10444061211267290
13. Gewurz IG (2001) (Re)designing mediation to address the nuances of power imbalance. Confl Resolut Q 19(2):135–162. https://doi.org/10.1002/crq.3890190203
14. Caputo A (2013) A literature review of cognitive biases in negotiation processes. Int J Confl Manag 24(4):374–398. https://doi.org/10.1108/IJCMA-08-2012-0064
15. Downie BM (1991) When negotiations fail: causes of breakdown and tactics for breaking the stalemate. Negot J 7(2):175–186. https://doi.org/10.1111/j.1571-9979.1991.tb00612.x
16. Li K, Cheung SO (2020) Alleviating bias to enhance sustainable construction dispute management. J Cleaner Prod 249:119311. https://doi.org/10.1016/j.jclepro.2019.119311
17. Zhu L, Cheung SO (2020) Power of incentivization in construction dispute avoidance. J Leg Aff Disput Resolut Eng Constr 12(2):03720001. https://doi.org/10.1061/(ASCE)LA.1943-4170.0000368
18. Gao Y (2018) Effects of interdependence asymmetry of parties in construction conflicts. Doctor of Philosophy, The University of Hong Kong, p 991044058175703400. https://doi.org/10.5353/th_991044058175703414
19. Coleman PT, Kugler K, Mitchinson A, Chung C, Musallam N (2010) The view from above and below: the effects of power and interdependence asymmetries on conflict dynamics and outcomes in organizations: the view from above and below. Negot Confl Manage Res 3(4):283–311. https://doi.org/10.1111/j.1750-4716.2010.00062.x
20. Barakat A, Chernobai A, & Wahrenburg M (2014) Information asymmetry around operational risk announcements. J Bank & Finance 48:152–179
21. Campo S (2012) Risk aversion and asymmetry in procurement auctions: Identification, estimation and application to construction procurements. J Econ 168(1):96–107
22. Mnookin RH (2003) Strategic barriers to dispute resolution: A comparison of bilateral and multilateral negotiations. Harv Negot L Rev 8(1)
23. Golan G (2019) Asymmetry in cross-conflict collaboration: is there a gender factor? (2011). In: Golan G (ed) Galia golan: An academic pioneer on the soviet union, peace and conflict studies, and a peace and feminist activist: with a foreword by William Zartman and a Preface by George Breslauer. Springer International Publishing, pp 237–253. https://doi.org/10.1007/978-3-319-95213-0_12
24. Mitchell CR (1991) Classifying conflicts: asymmetry and resolution. Ann Am Acad Pol Soc Sci 518(1):23–38. https://doi.org/10.1177/0002716291518001003
25. French JR, Raven B, Cartwright D (1959) The bases of social power. Classics Organ Theory 7:311–320
26. Raven BH (1964) Social influence and power. California Univ Los Angeles
27. Cuevas JM, Julkunen S, Gabrielsson M (2015) Power symmetry and the development of trust in interdependent relationships: the mediating role of goal congruence. Ind Mark Manage 48:149–159. https://doi.org/10.1016/j.indmarman.2015.03.015
28. Yu J, Kilgour DM, Hipel KW, Zhao M (2015) Power asymmetry in conflict resolution with application to a water pollution dispute in C hina. Water Resour Res 51(10):8627–8645. https://doi.org/10.1002/2014WR016257
29. Wolfe RJ, Mcginn KL (2005) Perceived relative power and its influence on negotiations. Group Decis Negot 14(1):3–20. https://doi.org/10.1007/s10726-005-3873-8
30. Barnhizer DD (2005) Inequality of bargaining power. U Colo l Rev 76:139
31. Frost DE, Stahelski AJ (1988) The systematic measurement of French and Raven's bases of social power in workgroups. J Appl Soc Psychol 18(5):375–389. https://doi.org/10.1111/j.1559-1816.1988.tb00023.x
32. Handley SM, Benton WC (2012) Mediated power and outsourcing relationships. J Oper Manag 30(3):253–267. https://doi.org/10.1016/j.jom.2011.11.004
33. Balakrishnan S, Koza MP (1993) Information asymmetry, adverse selection and joint-ventures: theory and evidence. J Econ Behav Organ 20(1):99–117. https://doi.org/10.1016/0167-2681(93)90083-2

34. Bernardo AE, Cai H, Luo J (2001) Capital budgeting and compensation with asymmetric information and moral hazard. J Financ Econ 61(3):311–344. https://doi.org/10.1016/S0304-405X(01)00065-4
35. Schmitz PW (2006) Information gathering, transaction costs, and the property rights approach. Am Econ Rev 96(1):422–434. https://doi.org/10.1257/000282806776157722
36. Schieg M (2008) Strategies for avoiding asymmetric information in construction project management. J Bus Econ Manag 1:47–51
37. Ceric A (2014) Strategies for minimizing information asymmetries in construction projects: project managers' perceptions. J Bus Econ Manag 15(3):424–440. https://doi.org/10.3846/16111699.2012.720601
38. Xiang P, Zhou J, Zhou X, Ye K (2012) Construction project risk management based on the view of asymmetric information. J Constr Eng Manag 138(11):1303–1311. https://doi.org/10.1061/(ASCE)CO.1943-7862.0000548
39. Dembe AE, Boden LI (2000) Moral hazard: a question of morality? New Solutions J Environ Occup Health Policy 10(3):257–279
40. Succar B (2009) Building information modelling framework: A research and delivery foundation for industry stakeholders. Autom Constr 18(3):357–375. https://doi.org/10.1016/j.autcon.2008.10.003
41. Liu LA (2004) Shared mental models in negotiation. Vanderbilt University
42. Cheung S-O, Ng TST, Wong S-P, Suen HCH (2003) Behavioral aspects in construction partnering. Int J Project Manage 21(5):333–343. https://doi.org/10.1016/S0263-7863(02)00052-2
43. Jehn KA, Rispens S, Thatcher SMB (2010) The effects of conflict asymmetry on work group and individual outcomes. Acad Manag J 53(3):596–616. https://doi.org/10.5465/amj.2010.51468978
44. Ippolito CA, Pruitt DG (1990) Power balancing in mediation: outcomes and implications of mediatosr intervention. Int J Confl Manag 1(4):341–355. https://doi.org/10.1108/eb022688
45. Samuelson W, Zeckhauser R (1988) Status quo bias in decision making. J Risk Uncertain 1(1):7–59. https://doi.org/10.1007/BF00055564
46. Cheung SO, Li K, Chow OY (2020) Reactive devaluation as a psychological impediment to construction dispute negotiation. J Manag Eng 36(4):04020025. https://doi.org/10.1061/(ASCE)ME.1943-5479.0000787
47. De Dreu CK, Kluwer ES, Nauta A (2008) The structure and management of conflict: fighting or defending the status quo. Group Process Intergroup Relat 11(3):331–353
48. Poitras J (2005) A study of the emergence of cooperation in mediation. Negot J 21(2):281–300
49. Mukerji S (2003) Ambiguity aversion and cost-plus procurement contracts
50. Kahneman D, Tversky A (2013a) Choices, values, and frames. In: Handbook of the fundamentals of financial decision making: Part I. World Scientific, pp 269–278
51. Kahneman D, Tversky A (2013b) Prospect theory: an analysis of decision under risk. In: Handbook of the fundamentals of financial decision making: Part I. World Scientific, pp 99–127
52. Bokhari S, Geltner D (2011) Loss aversion and anchoring in commercial real estate pricing: empirical evidence and price index implications. Real Estate Econ 39(4):635–670. https://doi.org/10.1111/j.1540-6229.2011.00308.x
53. Bottom WP, Studt A (1993) Framing effects and the distributive aspect of integrative bargaining. Organ Behav Hum Decis Process 56(3):459–474
54. Neale MA, Bazerman MH (1985) The effects of framing and negotiator overconfidence on bargaining behaviors and outcomes. Acad Manag J 28(1):34–49
55. Aibinu AA, & Al-Lawati AM (2010). Using PLS-SEM technique to model construction organizations' willingness to participate in e-bidding. Automation in construction 19(6):714–724
56. Karambayya R, & Reilly AH (1992) Dual earner couples: Attitudes and actions in restructuring work for family. J Organ Behav, 13(6):585–601
57. Pinkley RL (1990) Dimensions of conflict frame: Disputant interpretations of conflict. J appl psychol 75(2):117

58. Pinkley RL (1992) Dimensions of conflict frame: Relation to disputant perceptions and expectations. Int J Confl Manage
59. Kahneman D, & Tversky A (2013) Choices, values, and frames. In Handbook of the fundamentals of financial decision making: Part I. 269–278
60. Segal U (1987) The ellsberg paradox and risk aversion: An anticipated utility approach. Int. Econ Rev 175–202
61. Salanke M (2015) Voluntariness of mediation and cost sanctions for parties refusal to consider mediation: an oxymoron? Academia. Edu) www.Academia.Edu/5169933. Accessed 10
62. Hanks M (2012) Perspectives on mandatory mediation. UNSWLJ 35:929
63. Quek D (2009) Mandatory mediation: an oxymoron-examining the feasibility of implementing a court-mandated mediation program. Cardozo J Confl Resol 11:479
64. Wissler RL (1997) The effects of mandatory mediation: empirical research on the experience of small claims and common pleas courts. Willamette L Rev 33:565
65. Gazal-Ayal O, Perry R (2014) Imbalances of power in ADR: the impact of representation and dispute resolution method on case outcomes. Law Soc Inquiry 39. https://doi.org/10.1111/lsi.12063
66. Landau, Pfetsch (2000) Symmetry and asymmetry in international negotiations. Int Negot 5(1):21–42. https://doi.org/10.1163/15718060020848631
67. Dwyer FR (1984) Are two better than one? Bargaining behavior and outcomes in an asymmetrical power relationship. J Consum Res 11(2):680–693. https://doi.org/10.1086/209004
68. Rindt J, Mouzas S (2015) Exercising power in asymmetric relationships: the use of private rules. Ind Mark Manage 48:202–213. https://doi.org/10.1016/j.indmarman.2015.03.018
69. Moul W (2003) Power parity, preponderance, and war between great powers, 1816–1989. J Conflict Resolut 47(4):468–489. https://doi.org/10.1177/0022002703252980
70. Shapira O, Introduction I (2008) Exploring the concept of power in mediation: mediators' sources of power and influence tactics 24:37
71. Carnevale PJ, Pegnetter R (1985) The selection of mediation tactics in public sector disputes: a contingency analysis. J Soc Issues 41(2):65–81. https://doi.org/10.1111/j.1540-4560.1985.tb00855.x
72. Cheung SO, Chow PT, Tse JKM (2015) Performance of mediator tactics in building management disputes. J Manag Eng 31(2):04014033. https://doi.org/10.1061/(ASCE)ME.1943-5479.0000230
73. Yiu KTW, Cheung SO (2007) A study of construction mediator tactics—Part II: the contingent use of tactics. Build Environ 42(2):762–769. https://doi.org/10.1016/j.buildenv.2005.09.005
74. Nelson RM, Beauchamp T, Miller VA, Reynolds W, Ittenbach RF, Luce MF (2011) The concept of voluntary consent. Am J Bioeth 11(8):6–16. https://doi.org/10.1080/15265161.2011.583318
75. Appelbaum PS, Lidz CW, Klitzman R (2009) Voluntariness of consent to research: a conceptual model. Hastings Cent Rep 39(1):30–39. https://doi.org/10.1353/hcr.0.0103
76. Kamuya D, Marsh V, Molyneux S (2011) What we learned about voluntariness and consent: Incorporating "background situations" and understanding into analyses. Am J Bioeth 11(8):31–33. https://doi.org/10.1080/15265161.2011.583328
77. Brunk CG (1979) The problem of voluntariness and coercion in the negotiated plea. Law Soc Rev 527–553. https://doi.org/10.2307/3053267
78. Marquardt CA, Wiedman CI (1998) Voluntary disclosure, information asymmetry, and insider selling through secondary equity offerings*. Contemp Account Res 15(4):505–537. https://doi.org/10.1111/j.1911-3846.1998.tb00569.x
79. Wu J, Babcock BA (1995) Optimal design of a voluntary green payment program under asymmetric information. J Agric Resour Econ 20(2):316–327
80. Martínez-Ferrero J, Ruiz-Cano D, García-Sánchez I-M (2016) The causal link between sustainable disclosure and information asymmetry: the moderating role of the stakeholder protection context. Corp Soc Responsib Environ Manag 23(5):319–332. https://doi.org/10.1002/csr.1379
81. Shi N, Lee MKO, Cheung CMK, Chen H (2010) The continuance of online social networks: how to keep people using facebook? In: 2010 43rd Hawaii international conference on system sciences, pp 1–10. https://doi.org/10.1109/HICSS.2010.369

82. Rebecca Yen H, Gwinner KP, Su W (2004) The impact of customer participation and service expectation on Locus attributions following service failure. Int J Serv Ind Manag 15(1):7–26. https://doi.org/10.1108/09564230410523312
83. de Jong SB, Van der Vegt GS, Molleman E (2007) The relationships among asymmetry in task dependence, perceived helping behavior, and trust. J Appl Psychol 92(6):1625–1637. https://doi.org/10.1037/0021-9010.92.6.1625
84. Wang T, Zhao B, Thornhill S (2015) Pay dispersion and organizational innovation: The mediation effects of employee participation and voluntary turnover. Hum Relat 68(7):1155–1181. https://doi.org/10.1177/0018726715575359
85. Benk S, Budak T (2012) Power and trust as determinants of voluntary versus enforced tax compliance: empirical evidence for the slippery slope framework from Turkey. Afr J Bus Manag 6. https://doi.org/10.5897/AJBM11.2157
86. Cowan K, Paswan AK, Van Steenburg E (2015) When inter-firm relationship benefits mitigate power asymmetry. Ind Mark Manage 48:140–148. https://doi.org/10.1016/j.indmarman.2015.03.013
87. Fuchs W, Lippi F (2006) Monetary union with voluntary participation1. Rev Econ Stud 73(2):437–457. https://doi.org/10.1111/j.1467-937X.2006.00382.x
88. Matanda MJ, Ndubisi NO, Jie F (2016) Effects of relational capabilities and power asymmetry on innovativeness and flexibility of sub-sahara Africa small exporting firms. J Small Bus Manage 54(1):118–138. https://doi.org/10.1111/jsbm.12134
89. Aiello A, Pratto F, Pierro A (2013) Framing social dominance orientation and power in organizational context. Basic Appl Soc Psychol 35(5):487–495. https://doi.org/10.1080/01973533.2013.823614

The Values of Apology in Incentivizing Construction Dispute Settlement

Sai On Cheung and Liuying Zhu

Introduction

An Apology Ordinance (AO hereafter) was enacted in July 2017. This is in response to push of the Hong Kong Government in developing Hong Kong into the dispute resolution hub in the South East Asia region. For this, mediation has been identified to be the primary alternative dispute resolution mechanism for use in Hong Kong. To promote the use of mediation, a steering committee on mediation was established in 2012 by the then secretary for justice. Having an apology legislation was one of the key recommendations of the committee. It was believed that making apology can be a valuable settlement option in mediation. Moreover, it is of equal importance to protect an apology offeror who has taken step to enhance chance of settlement. In this connection, apology ordinance has been enacted in the United States, Canada, Australia and England [1]. Hong Kong has taken similar step and enacted CAP 631 of the Laws of Hong Kon on 13th July 2017. The aim of the AO is to promote and encourage the making of apologies as a means to prevent the escalation of disputes and thereby facilitate their amicable resolution.

Negotiation has proven to be the most efficient means to resolve dispute because of the time and cost efficiency [2]. Combating settlement barriers [3] would enhance the chance of amicable negotiation [4–7]. Levi [8] suggested that offering an apology might reignite the desire for settlement by healing emotional injury. This study advocates that offering an apology can bring catalytic effect on reciprocating positive responses. Kelman [9] pioneered this concept and proposed the theory of Response Restriction (RR). In dispute negotiation, this would mean one's negotiation attitude

S. O. Cheung (✉) · L. Zhu
Construction Dispute Resolution Research Unit, City University of Hong Kong, Hong Kong, China
e-mail: Saion.cheung@cityu.edu.hk

L. Zhu
e-mail: liuyinzhu3-c@my.cityu.edu.hk

S. O. Cheung (ed.), *Construction Dispute Research Expanded*, Springer Tracts in Civil Engineering, https://doi.org/10.1007/978-3-030-80256-1_10

can be changed by the behaviours of the negotiating counterpart [9]. The first part of the study applies the theory of RR to construction dispute negotiation (CDN hereafter) and has four parts:

(a) To conceptualize apology in CDN;
(b) To propose a relationship framework between apology and response in CDN:
(c) To test the relationship framework developed in (b); and
(d) To analyze the implications of offering apology in CDN.

The second part of the study focuses on the use of apology in mediation. Section "Part One of the Study—Application of the Theory of RR in CDN" of the AO emphasize the AO wishes to encourage a wider use of mediation by offering legal protection on the use of apology. In Hong Kong, mediation has become an integral part of the construction contractual dispute resolution regime. The part of the study aims to identify the prerequisite conditions that lead to settlement of disputes in mediation; and explore the incentivizing effects of an apology in harvesting the prerequisites.

Part One of the Study—Application of the Theory of RR in CDN

The Conceptual Bases of the Study

Resolving construction dispute almost always starts with negotiation [10]. Most negotiation studies are premised that negotiators are having full control of the negotiation including the form and process. Moreover, if negotiators only focus on their own wills, settlement would be unlikely. Working together is an important ingredient of a negotiated settlement [11, 12]. The theory of response restriction (RR) [9] described RR as "*any action on the part of A (a person or group, e.g., the communicator) which limits B's (the recipient's) choice of behaviour and thus influences B towards a response that is favoured by A*". In CDN, disputing party's attitude and opinion would change respective to the actions taken from the counterpart. If the response to an apology is a positive one, the gap between the parties is narrowed by overcoming precedent barriers. Three constructs are involved: (1) Barriers against settlement; (2) Forms of apology and (3) Responses towards an apology. The relationships among these three constructs are presented in Figs. 1 and 2.

Barriers Against Settlement

Three forms of barrier against settlement had been identified by Korobkin [13]. These are institutional, organisational and psychological. For this study, psychological barriers are the most relevant. Ross and Ward [3] outline the following ways

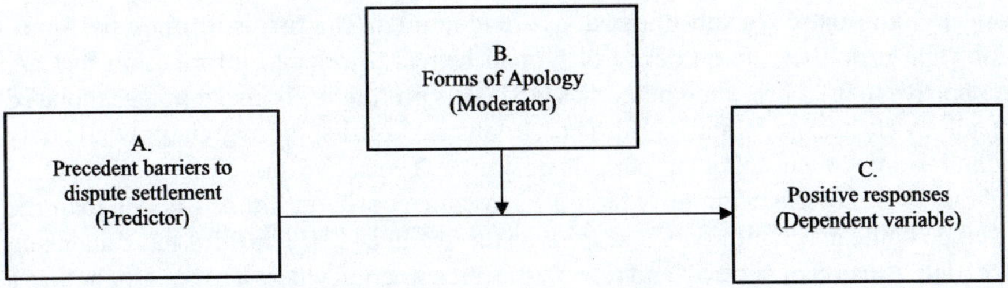

Fig. 1 An apology-response restriction relationship (apology as a moderator)

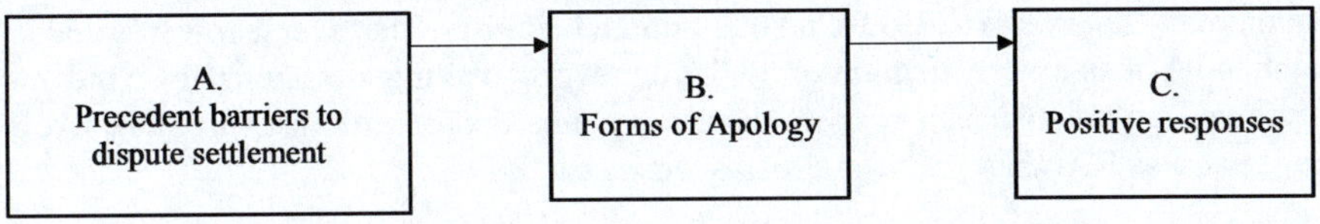

Fig. 2 An apology-response restriction relationship (apology as a mediator)

decisions are influenced by psychological barriers: (1) cognitive and motivational processes, (2) feelings of gain or loss; (3) risk evaluation; (4) information interpretation; and (5) priority setting. Based on these, the psychological barriers against settlement in CDN are categorized as: (1) dissonance arising from the past; (2) optimistic bias; (3) loss aversion; (4) subjective construal and (5) devaluation of adversary's concessions.

(a) Dissonance Arising from the Past

Festinger [14] used Cognitive dissonance theory to explain that humans are reluctant to take action that is contrary to their belief because one is inclined to maintain one's cognitive consistency. In short, one's behaviours are reflective of their beliefs, values, or feelings. This will ensure that they will feel comfortable with their behaviours. From another angle, it is not easy to change one's course of action as this would imply some forms of inconsistency when compared with the past. Applying this conception to dispute negotiation, Ross and Ward [3] found that a negotiator would refrain from retracting from his previous stances of the dispute. Settlement will thus be less likely if the negotiator had been maintaining a negative position.

(b) Optimistic Bias

When one is overly optimistic about their positions, they are having unwarranted confidence of their judgments [15]. In dispute negotiation, it is not surprising to find negotiators are having strong faith in their cases [16]. Unwarranted optimism may lead to irrational decisions [17]. Lichtenstein et al. [17] further found that more difficult tasks are likely to derive excessive confidence. Ma [15] explained that over-confidence has both internal (cognitive) and external (environmental) dimensions.

Under cognition, three sub-dimensions are identified. The first is 'differential attention' that describes the tendency of human beings to refer to information that are supportive to their preferred outcomes [13]. This preference is also termed as 'biased recall' by Hastorf and Cantril [18]. Thus, a construction dispute negotiator would only attend to information that support his position. The second sub-dimension is called 'above-average effect' under which a negotiator considers himself better than the average while his counterpart is worse than the average [13]. The third sub-dimension is called 'illusion of control' and refer to the phenomenon where a negotiator believes that he has the power to control the outcome of the negotiation [13].

(c) Loss Aversion

Kahneman and Tversky [19] found that human beings are risk averse towards uncertainties. A loss-averse negotiator will thus avoid making decisions to avoid his perceived potential loss. Prospect theory explains that a settlement is more likely be viewed as loss than gain for a loss-averse [3].

(d) Subjective Construal

Construal theory explains how one perceive, comprehend, and interpret external stimulations. Two subcategories of subjective construal effects are identified. These are consensus effect [20] and fundamental attribution error [21]. False consensus effect can be viewed as a kind of cognitive bias that explains why humans are inclined to believe that their positions are more reasonable and better than those of the counterpart. Ross [20] added that the bias is not necessary supported by facts. Fundamental attribution error is related to one's perception that may be instigated by speculations [21]. In CDN, subjective construal would crystallise one's position notwithstanding there may not have been sufficient objective analyses.

(e) Devaluation of Adversary's Concessions

Reactive devaluation [22] is the theoretical base that explains the habitual devaluation of others' opinions and suggestions. Is a kind of cognitive bias against proposed by his/her counterpart are always against his/her benefits [14, 23]. The effect of reactive devaluation is drawing premature boundary of the solution set [24]. Ross [25] added that attitude polarization would be aggregated by the attitude of reactive devaluation in downgrading of other proposals with one's proposal becomes the sole one that is considered to be worthwhile [26].

Assisted by the afore-mentioned literature review, the potential barriers against settlement have been identified. It is proposed that apology ca n be used as the stimulating action of the Kelman's RR framework. In offering an apology, the apologizer (A) is prompting an altitude change of the recipient of the apology (B). Under the theory of RR, making apology can channel the disputing parties to positive responses whereby narrowing the gap between them.

Forms of Apology

An apology can be seen as an expression of repentance or sorrow [27]. "I am sorry" is the most commonly used expression that can be taken as an admission of liability or simply a gesture of courtesy, good faith or sympathy. Orenstein [28] asserted that a genuine apology should entail the followings. First, the grievance should be acknowledged. Second, there has been violation respective to specific rule or norm. Third, the harm inflicted is understood. Fourth, certain admission of responsibility is expressed. Fifth, element of regret for the injury is presented. Sixth, there is a wish for maintaining relationship. Seventh, the wrongful act will not be repeated. Eighth, compensation will be offered to the injured.

The positive effect of an apology in dispute negotiation is to patch psychological harm [29]. It is therefore not suggested that apology can have effect in every negotiation. It is also important to catch the favourable time if making an apology. The following section discusses the forms of apology that may be made. Levi [8] suggested two generic forms of apology: tactical and win–win. Operationalising this categorisation in the context of construction dispute negotiation, four forms of apology of potential use are proposed. These are (1) Ice-breaking apology, (2) Conciliatory apology; (3) Reality-checking apology and (4) Congruence-driving apology.

(a) Ice-breaking Apology

Ice-breaking apology aims to unleash deadlock by conveying empathy to the counterpart. Pruitt [30] opined that ice-breaking apology has attitude-restructuring effect that may remove stereotyping of the apology offeror. Ideally, the attitude change is one on trust [31]. If this is materialised, ice-breaking apology would successfully improve the conditions for negotiation between the disputing parties [32].

(b) Conciliatory Apology

Conciliatory apology is attached with acknowledging conducting misbehaviour and seeks the understanding of the counterpart [32]. Conciliatory apology would also explicitly address the sufferer's feelings explicitly and invite acceptance of the apology. Such apology embraces conciliatory intents and aims for reduction in anger. This would create the environment for the suffer to render forgiveness [33].

(c) Reality-checking Apology

Reality checking is commonly used by third-party neutrals to help disputing parties to be pragmatic [34]. In mediation, reality checking tactics are used to avoid parties anchoring on unrealistic expectations. Through checking on the likely outcome if the current impasse is maintained, the parties are invited revisit their positions by stepping into the shoes of the counterparts [8]. Wen parties are taking a pragmatic approach, rational decisions are more likely.

(d) Congruence-driving Apology

By congruence-driving, the apology is made for a win–win settlement [35]. If the prospect of having settlement can be construed, settlement focused efforts can be engendered [36]. Offering a congruence-driving apology indicates the desire to end the dispute and the willingness to pursue common goals. Schweitzer et al. [37] suggested that congruence-driving apology must be sincere and accompanied with (1) candour, (2) remorse, and (3) commitment to change.

Responses Towards an Apology

Kelman [38] identified six forms of positive sentiment that underpin the reciprocation of positive responses to an apology: (i) sense of security; (ii) identity; (iii) recognition; (iv) autonomy; (v) sense of justice; and (vi) cooperative image.

(a) Sense of Security

Burton and Sandole [39] advocated that conflict is manifestation of deprivation of human needs [40]. Under Maslow's hierarchy of needs [41], security is the need next to the basic physiological needs. Security is sensed by being safe, orderly with stable, predictable and free from anxiety environment [41]. Offering an apology can be interpreted as the commitment of the offeror not endangering the apology offeree [37].

(b) Identity

Identity can be recognised as making sense of one's own value or worth as a person [41]. Attacking the identity of the counterpart can be an aggressive negotiating behaviour [42]. In the alternative, if identity is acknowledged, this show certain respect on the counterpart [43]. This face-saving act may reciprocate positive response [44].

(c) Recognition

Recognition may be the other side of the same coin of identity [45, 46]. Wolf [47] suggested that recognition is a form of respect. While disrespect breeds conflict because it would trigger a victim's anger and self-protective reactions. Burton [45] pointed out that recognition is useful in suppressing frustration. Showing respect thus will improve the relationship [47]. Murphy [48] found also that offering an apology could enhance relationship because of the recognition accorded.

(d) Autonomy

Deci and Ryan [49] claimed that autonomy is reflected by the freedom to make choices. This is also termed as self-determination theory [50]. In construction, autonomy is observed by the degree to which the job allows freedom, independence, and discretion to an individual has on the procedure and schedule [51]. A sincere apology would uphold one's autonomy [52].

(e) Sense of Justice

According to the Fairness theory [53], sense of justice is important to most disputants. Adams's [54] equity theory advocates that humans are intrinsically demanding an equitable balance between input and output. When one is unfairly treated and suffered loss, the resulting sense of injustice would prompt retaliation [55]. Thibaut and Walker [56] advocated that by restoring the victim to the prior position before the damage, sense of justice can be fostered. An apology can pacify a victim for the sense of justice posed [57]. By offering a sincere apology, the feeling of unfairness can be alleviated [58].

(f) Cooperative Image

Disputing parties are typically having "enemy image" [59]. This is non-conducive because of the confrontations inherent with being enemy. An apology is able to instil a sense of cooperation as explained by the Evolutionary Game theory [60, 61]. By directing efforts in engaging in cooperative relationship, it can help in avoiding opportunism [62]. Nonetheless, willingness to expose to exploitation is a bold step to canvas trust [63].

Relationship Framework Between Apology and Response in CDN

The roles of apology in conflict resolution have been investigated by Witvliet et al. [64] who found that an apology can alter the outcome expectations. Brown [32] added that such changes are effected by the messages directed towards the relationship between the disputants. Moreover, situational factors such as emotion and sentiment cannot be underestimated. In the experiment of Robbennolt [65] the reactions of 556 subjects in different pre-set scenarios of using apology were analysed. It was found that an apology can alter the perceptions and value of the apology recipients in two ways: (1) the injured parties became more amenable to discuss settlement and (2) the desire of the injured party to settle was raised. These findings supported that settlement can be promoted by apologising through overcoming certain barriers. Applying the theory of response restriction, apology can be a moderator for positive response by relishing formerly barriers of settlement". The first hypothesis of this study is as follows:

H1: An act of apology has positive moderating effect on construction dispute settlement.

However, another school of thought suggests that apology can give direct effect and hence can be a mediator of positive responses under certain circumstances. The positive effect of an apology lies in its ability to heal psychological harm and canvasses forgiveness [29]. Many psychologists found that an act of apology could relish barriers against settlement. For example, Darby and Schlenker [66] proposed that apologies could reduce the negative repercussions of the injured. This proposition is supported by the work of Bennett and Earwaker [67] and of Hodgins and Liebeskind [68]. The second hypothesis of apology in CDN is as follows:

H2: An act of apology has positive mediating effect on construction dispute settlement.

Figure 2 shows the hypothesised relationship.

There are four key components in this part of the study: (1) the Kelman's RR Theory; (2) the precedent barriers to dispute settlement; (3) the forms of apology and (4) the positive responses of CDN. Based on that, two hypotheses are developed as afore-stated.

Effects of Offering Apology in CDN

Research Approach

A quantitative deductive approach is applied. A data collection questionnaire was developed. To maximize the return rate, both on-line and paper-based questionnaire were used. Data from construction professionals were collected at seminars, conferences and public lectures held in Hong Kong.

Measures

The questionnaire has three parts and includes 42 questions. Part 1 is used to collect personal particulars of the respondents. In Part 2, the respondents were asked to indicate the likelihood of the 8 dispute behaviours using a seven-point Likert-scale (1 = strongly disagree; 2 = disagree; 3 = slightly disagree; 4 = neutral; 5 = slightly agree; 6 = agree; 7 = strongly agree). In Part 3, respondents were asked to score the degree of agreement on the extent of likelihood on how the four forms of apology would affect the outcome of disputes in negotiation.

Table 1 shows the matrix of questions set in Part 3.

Data Analysis

Multiple regression analysis (MRA) which is a statistical technique that allows one to evaluate the relationship between dependent variables and a number of independent variables [69] is applied In this study to analyse the relationship among precedent barriers, forms of apology and positive responsess. in this study. MRA has been widely applied in psychology, social science and behavioural science research to test the interactive effect on the independent variables [70–72].

(a) Moderated Regression Analysis

Table 1 The matrix of questions set in part 3

	I. S/He tried to embrace your sense of sympathy	II. S/He tried to abscond the responsibility	III. S/He was urged by his/her advisor to be realistic	IV. S/He would honour the terms of the settlement
i. According to your experience to construction dispute settlement, please indicate the frequency of happening as in right hand side by your counterpart	← 1 2 3 4 5 6 →	← 1 2 3 4 5 6 →	← 1 2 3 4 5 6 →	← 1 2 3 4 5 6 →
ii. According to your behaviour pattern, please indicate your degree of likelihood of the following statements *If my counterpart act as in right hand side, my feeling would be:*	Strongly Disagree, Disagree, Slightly Disagree, Neutral, Slightly Agree, Agree, Strongly Agree ← →	Strongly Disagree, Disagree, Slightly Disagree, Neutral, Slightly Agree, Agree, Strongly Agree ← →	Strongly Disagree, Disagree, Slightly Disagree, Neutral, Slightly Agree, Agree, Strongly Agree ← →	Strongly Disagree, Disagree, Slightly Disagree, Neutral, Slightly Agree, Agree, Strongly Agree ← →
1 "A sense of security is regained."	1 2 3 4 5 6 7	1 2 3 4 5 6 7	1 2 3 4 5 6 7	1 2 3 4 5 6 7
2 "My identity is a significance."	1 2 3 4 5 6 7	1 2 3 4 5 6 7	1 2 3 4 5 6 7	1 2 3 4 5 6 7
3 "Being understood."	1 2 3 4 5 6 7	1 2 3 4 5 6 7	1 2 3 4 5 6 7	1 2 3 4 5 6 7
4 "My will is respected."	1 2 3 4 5 6 7	1 2 3 4 5 6 7	1 2 3 4 5 6 7	1 2 3 4 5 6 7
5 "A sense of justice is raised."	1 2 3 4 5 6 7	1 2 3 4 5 6 7	1 2 3 4 5 6 7	1 2 3 4 5 6 7
6 "Cooperative relationship is rebuilt."	1 2 3 4 5 6 7	1 2 3 4 5 6 7	1 2 3 4 5 6 7	1 2 3 4 5 6 7

The outcome of moderated regression analysis predicts a dependent variable Pi by two independent variables Oj and Ak [73]. A total of 120 moderated regression models (devised from the combination of five dispute settlement behaviours, six apology expectations and four forms of apology) were resulted. The significance of the moderating effect is indicated by Fisher Z test (F test) [73–76]. Critical values of such test were obtained from F-distribution table with significant level at $\alpha = 0.05$. Referring to the F-distribution table, the moderating analyses result with an F value > 3.96 would be treated as statistically significance [73].

(b) Mediated Regression Analysis

Regression models were also conducted for mediated regression analysis [77]. Separate coefficients for each equation were further estimated and tested [78]. In this study, 120 mediated regression models were also examined to explored whether apology is a mediator of positive responses.

Findings and Discussions

The study investigates if and how apology can engender positive responses in construction dispute negotiation. The research findings are presented seriatim:

The Relationship Between Apology and Responses in CDN

Personal particulars

A total of 251 questionnaires were distributed and 103 effective responses were received representing a response rate of 42% that is considered to be reasonably good when compared with typical response rate of 25–30% in construction management studies [79]. Table 2 shows the personal particulars:

The Measurement of Part 2 and Part 3

Table 3 presents the descriptive statistics of the measurement of dispute resolution behaviours. With reference to Table 3, the mean scores of all the questions are higher than 4 of a Likert scale of 7. The standard deviations of the variables are all close to 1, meaning the degree of dispersion are relatively small suggesting that the respondents are in general do not have notable divergent views.

(a) Inter-group analysis

Kruskal–Wallis H test [82] was used to determine if there are statistically significant differences between two or more groups of an independent variable on a continuous or ordinal dependent variable. The null hypothesis of the K-W test reveals that no differences were detected within different groups of company type, project nature and respondents' working experiences.

(b) Correlation analysis

Pearson correlation analysis was used to test the initial consistency of the three major factors. With reference to Tables 4 and 5, the question structure in this study was subjected to revision with due regard to the correlation results. Scale inversion was applied to certain questions before conducting correlation analysis. "Q2.1: I would not initiate negotiation even settlement seemed possible." is a question in a negative expression. It is expected that the correlation between the variables would be positive after the inversion of scale. However, the result was contrary to the expectation. The negative expression of Q2.1 perhaps confused the understanding of respondents. Thus, the question Q2.1 was taken out from further analysis. Moreover, "Q2.6: The

Table 2 Personal particulars

Background	Type	%
Company type	Developers	30
	Consultants	37
	Contractors	33
Project nature	Building	62
	Civil	28
	Others, include maintenance, fitting-out, infrastructure and all related construction projects	10
Working experience	<5 years	33
	5–10 years	33
	>10 yrs	34

Table 3 Descriptive statistics of the measurement of dispute resolution behaviours

Part 2: Degree of behaviour likelihood if construction dispute arises—from 1 to 7	Min.	Max.	Mean.	Std.
Cognitive Dissonance [3, 14]			**4.14**	**1.21**
Q2.2 I would only negotiate when there was foreseeable gain	2.00	6.00	4.14	1.21
Overconfidence effect [15–18]			**4.29**	**1.08**
Q2.3 I would be very confident in my assessments	2.00	6.00	4.65	1.07
Q2.4 I would take the subjective perspective rather than the objective one	2.00	6.00	3.92	1.08
Prospect Theory [3, 19]			**4.55**	**1.02**
Q2.5 I would be mindful of being cheated	2.00	6.00	4.72	1.02
Q2.6.1 The gain would be more important	3.00	7.00	4.39	1.01
Construal Theories [20, 21]			**4.73**	**0.94**
Q2.7 My first impression of the counterpart would direct my judgement	2.00	6.00	4.40	0.98
Q2.8 Past experience would direct my judgement	2.00	6.00	5.07	0.90
Reactive Devaluation [3, 22, 25]			**4.32**	**0.91**
Q2.9 I think the settlement proposal proposed by the counterpart would be exaggerated	2.00	6.00	4.34	0.92
Q2.10 Dispute resolution was a forceful option	2.00	6.00	4.30	0.89
Part 3(i): Frequency of happening in construction dispute resolution experience—from 1 (hardly ever) to 7 (always)				
Q3.1.1 S/He tried to take advantage of your sense of sympathy (Ice-breaking apology)	2.00	6.00	4.22	1.03
Q3.1.2 S/He tried to abscond the responsibility. (Conciliatory apology)	1.00	7.00	4.33	1.20
Q3.1.3 S/He was swapped by her/his business consulting lawyer. (Reality-checking apology)	1.00	7.00	3.86	1.10
Q3.1.4 S/He honoured the terms of the settlement. (Congruence-driving apology)	1.00	7.00	4.21	1.26
Ice-breaking apology [8, 32]				
** S/he tried to embrace your sense of sympathy*			**4.20**	**1.19**
Q3.2.1.1 "A sense of security is regained." [40, 41]	2.00	7.00	4.10	1.10

(continued)

Table 3 (continued)

Q3.2.1.2 "My identity is a significance." [43]	2.00	7.00	4.16	1.18
Q3.2.1.3 "Being understood." [45–48]	1.00	7.00	4.17	1.22
Q3.2.1.4 "My will is respected." [50]	1.00	6.00	4.18	1.20
Q3.2.1.5 "A sense of justice is raised." [53, 54, 56, 80]	2.00	7.00	4.35	1.19
Q3.2.1.6 "Cooperative relationship is rebuilt." [59, 81]	1.00	7.00	4.26	1.27
Conciliatory apology [8, 32]				
*** S/he tried not to abscond the responsibility***			**3.75**	**1.19**
Q3.2.2.1 "A sense of security is regained."	1.00	7.00	3.65	1.22
Q3.2.2.2 "My identity is a significance."	1.00	7.00	3.73	1.23
Q3.2.2.3 "Being understood."	1.00	6.00	4.12	1.02
Q3.2.2.4 "My will is respected."	1.00	6.00	3.64	1.08
Q3.2.2.5 "A sense of justice is raised."	1.00	6.00	3.69	1.40
Q3.2.2.6 "Cooperative relationship is rebuilt."	1.00	7.00	3.65	1.21
Reality-checking apology [8]				
*** S/he was urged by his/her advisor to be realistic***			**3.98**	**1.03**
Q3.2.3.1 "A sense of security is regained."	2.00	7.00	4.17	1.22
Q3.2.3.2 "My identity is a significance."	2.00	6.00	3.98	0.90
Q3.2.3.3 "Being understood."	2.00	7.00	3.90	1.01
Q3.2.3.4 "My will is respected."	1.00	6.00	3.96	1.07
Q3.2.3.5 "A sense of justice is raised."	2.00	6.00	4.07	0.96
Q3.2.3.6 "Cooperative relationship is rebuilt."	2.00	6.00	3.77	1.03
Congruence-driving apology [8, 37]				
*** S/he would honour the terms of the settlement***			**4.72**	**0.99**
Q3.2.4.1 "A sense of security is regained."	3.00	7.00	4.72	0.97
Q3.2.4.2 "My identity is a significance."	2.00	7.00	4.59	1.04
Q3.2.4.3 "Being understood."	3.00	7.00	4.59	0.90
Q3.2.4.4 "My will is respected."	2.00	7.00	4.78	0.97
Q3.2.4.5 "A sense of justice is raised."	3.00	7.00	4.77	1.01
Q3.2.4.6 "Cooperative relationship is rebuilt."	2.00	7.00	4.85	1.06

Table 4 Descriptive statistics of frequency of happening of forms of apology

Questions	Revision	Reasons for revision
Q2.1 I would not initiate negotiation even settlement seemed possible	Deletion	Negative correlation shown; misunderstanding of respondents
Q2.6 The loss would be more important Q2.7 The gain would be more important	Combination to Q2.6.1 "The gain would be more significant than the loss."	Both questions were asking the same object in two directions

loss would be more important." and "Q2.7: The gain would be more important." aims to compare the sense of loss and gain through literature loss aversion [19]. With the checking on the mean score of Q2.7 and Q2.6 (i.e., Q2.7 > Q2.6), these two questions were combined to "Q2.6.1: The gain would be more significant than the loss." for further analysis.

(iii) Multiple regression analysis on moderating effect

The results of the Moderating analyses are summarised in Table 6. Congruence-driving apology offers the most significant moderating effects (i.e., F-value >3.96) and are highlighted with an asterisk (*). Form of apology that is having medium significant and least significant moderating effects is conciliatory and reality-checking respectively. With reference to Fig. 3, high congruence-driving apology solicits higher positive responses than that of low- congruence-driving apology.

(iv) Multiple regression analysis on mediating effect

120 mediated regression models devised from the combination of five dispute settlement behaviours, six apology expectations and four forms of apology were identified. The results of mediated regression analyses are summarised with change of coefficients between a22 and a32, and the significance are shown in Table 7.

Discussions

The moderating and mediating effect in bringing positive responses of the four forms of apology are discussed here-follow:

Moderating Effect

The level of moderating effect is considered by the number of moderating significances with regard to the precedent barriers. It is found that not all forms of apology are having significant moderating effect in soliciting positive responses. Table 8

Table 5 The correlation matrix of Part 2

	Q2.1	Q2.2	Q2.3	Q2.4	Q2.5	Q2.6	Q2.7	Q2.8	Q2.9	Q2.10	Q2.11
Q2.1	1	−0.351[a]	−0.120	−0.299[a]	−0.003	−0.062	0.003	−0.096	−0.012	−0.143	−0.164
Q2.2	−0.351[a]	1	0.241[a]	0.085	0.180	0.165	0.235[b]	0.231[b]	0.274[a]	0.349[a]	0.268[a]
Q2.3	−0.120	0.241[a]	1	0.008	0.299[a]	0.151	0.159	0.038	0.188[b]	0.116	0.030
Q2.4	−0.299[a]	0.085	0.008	1	0.087	0.223[b]	0.181	0.240[a]	0.086	0.193[b]	0.139
Q2.5	−0.003	0.180	0.299[a]	0.087	1	0.159	0.351[a]	0.184[b]	0.440[a]	0.354[a]	−0.069
Q2.6	−0.062	0.165	0.151	0.223[b]	0.159	1	0.393[a]	0.066	0.313[a]	0.200[b]	0.131
Q2.7	0.003	0.235[b]	0.159	0.181	0.351[a]	0.393[a]	1	0.169	0.430[a]	0.299[a]	0.143
Q2.8	−0.096	0.231[b]	0.038	0.240[a]	0.184[b]	0.066	0.169	1	0.227[b]	0.330[a]	0.176
Q2.9	−0.012	0.274[a]	0.188[b]	0.086	0.440[a]	0.313[a]	0.430[a]	0.227[b]	1	0.489[a]	0.269[a]
Q2.10	−0.143	0.349[a]	0.116	0.193[b]	0.354[a]	0.200[b]	0.299[a]	0.330[a]	0.489[a]	1	0.184[b]
Q2.11	−0.164	0.268[a]	0.030	0.139	−0.069	0.131	0.143	0.176	0.269[a]	0.184[b]	1

[a]Correlation is significant at the 0.01 level (2-tailed)
[b]Correlation is significant at the 0.05 level (2-tailed)

Table 6 Summary of the F value and significance level in moderating regression analyses results

Moderator variables—Forms of apology (A_k)	Dependent variables—Positive responses (P_i)					
	P1	P2	P3	P4	P5	P6
Predictor variables—Precedent barrier to settlement (O_j) Cognitive Dissonance						
Ice-breaking apology	0.31	0.17	−0.90	−0.93	−0.41	2.40
Conciliatory apology	2.38	**7.16***	0.20	−0.77	2.46	−0.82
Reality-checking apology	−0.63	1.02	−1.04	−0.81	−0.79	−0.99
Congruence-driving apology	**6.53***	**10.16***	**6.89***	**6.50***	**7.71***	**8.51***
Overconfidence effect						
Ice-breaking apology	−1.80	−0.89	0.20	−0.52	−0.88	−1.01
Conciliatory apology	−0.43	0.61	−0.39	−0.62	0.66	−0.72
Reality-checking apology	0.32	−1.00	−0.73	−0.93	−0.91	−0.20
Congruence-driving apology	**11.13***	**7.89***	**10.86***	**7.47***	**11.66***	**9.47***
Prospect theory						
Ice-breaking apology	−0.52	−0.41	−0.57	−0.31	−0.89	0.41
Conciliatory apology	−1.05	2.60	−0.40	0.98	−0.11	1.25
Reality-checking apology	**4.22***	−0.80	**5.69***	−0.91	2.56	0.61
Congruence-driving apology	**6.11***	**7.18**	**5.43***	**4.09***	**5.89***	**5.10***
Construal theory						
Ice-breaking apology	−0.81	−1.10	−0.45	−0.51	−0.50	0.20
Conciliatory apology	−2.83	−0.99	−0.78	−0.68	2.31	−0.59
Reality-checking apology	1.07	−1.00	−0.65	−0.30	−0.90	−0.59
Congruence-driving apology	**7.37***	**6.52***	**10.75***	**6.88***	**9.21***	**9.47***
Reactive devaluation						
Ice-breaking apology	−0.63	−1.03	−0.46	−0.95	−0.50	2.37
Conciliatory apology	**4.67***	**5.16***	2.65	2.06	−0.93	0.10
Reality-checking apology	1.15	−0.84	−0.21	−0.62	−0.93	−0.98
Congruence-driving apology	**6.07***	**6.11***	**7.89***	**7.55***	**8.06***	**8.61***

Noted $F > 3.92$ sig*; $p < 0.05$ sig* [83]; P1 represents "A sense of security is regained"; P2 represents "My identity is a significance"; P3 represents "Being understood"; P4 represents "My will is respected"; P5 represents "A sense of justice is raised" and P6 represents "Cooperative relationship is rebuilt"

summarises the significant level of apologies having moderating effect on positive responses

With reference to Table 8, congruence-driving apology recorded the most frequent significant moderating effect. Congruence-driving apology was found to have moderating effect on all six positive responses under the five precedent barriers against dispute settlement. Cheung and Yiu [84] advocated that construction disputes can have a human dimension whereby congruence-driving apology may bring about win–win solution for the disputing parties. Conciliatory apology was found to have

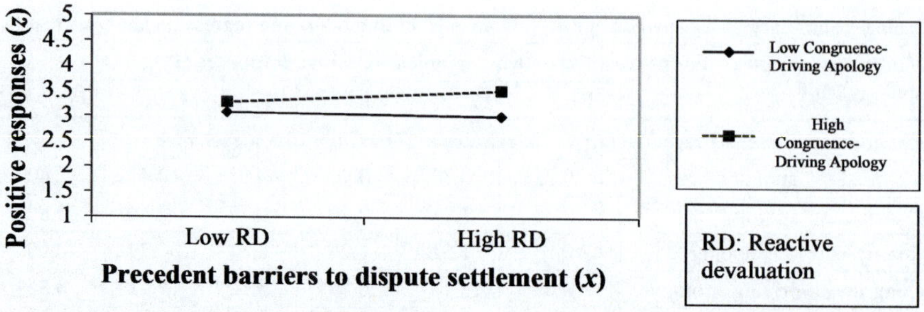

Fig. 3 Interaction effect of precedent barriers to dispute settlement (*x*) and forms of apology (*y*) on positive responses in construction dispute negotiation (*z*)

medium moderating effect and moderate the sense of identity when the settlement barrier is cognitive dissonance [14] and reactive devaluation [22]. When disputants' conflicts are deeply rooted, polarisation of position as a state of cognitive dissonance is very likely. A simple compromise may not be sufficient to break the deadlock. Nonetheless, conciliatory apology that addresses the concerns of the apology offeror [32] would reduce hostility through raising the sense of security. Reality-checking apology has recorded the least number of moderating effects. Reality-checking shows moderating effect with respect to the barriers of sense of security and recognition [19]. Moreover, reality-checking may be more effective if used by third-party neutrals because a disputant may feel safer to make apology as advised. This would be of particular impact for those risk-averse disputants.

For ease of comparison, the six positive responses that can be moderated respective to the forms of apology with respect to the five settlement barriers are summarised in Table 9.

From Table 9, autonomy, sense of justice and cooperative image are the positive responses that can be moderated by congruence-driving apology (5 out of 20). Recognition can be moderated by both reality-checking apology and congruence-driving apology (6 out of 20). Sense of security and identity can also be moderated by reality-checking apology and conciliatory apology under the settlement barriers explained by cognitive devaluation and Reactive devaluation respectively (7 out of 20), which appears to be the more likely response when apology is offered.

Mediating Effect

Ice-breaking apology is the most significant in mediating sense of identity by addressing most of the precedent barriers except risk avoidance [19]. Ice-breaking apology involves acknowledging the suffering of her/his counterpart. Brown [32] suggested that trust can thereby be gained from the counterpart. It is further believed that trust between the disputing parties can subdue the uncertainties and alleviate the worries that hinder disputants from settling their differences. The elevated level of

Table 7 Summary of the coefficients between a_{22} and a_{32} in mediating regression analyses

Mediator variables—Forms of apology (A_i)	Dependent variables—Positive responses (P_k)					
	P1	P2	P3	P4	P5	P6
Predictor variables—Precedent barriers to settlement (O_j) Cognitive Dissonance						
Ice-breaking apology	0.189→0.108	**0.555→0.384***	0.243→0.087	0.231→0.165	0.244→0.208	0.206→0.165
Conciliatory apology	0.006→0.032	0.064→0.037	0.05→0.04	0→0.009	0.008→0.025	0.226→0.213
Reality-checking apology	0.216→0.163	0.111→0.07	0.146→0.09	0.065→0.011	0.342→0.006	0.143→0.116
Congruence-driving apology	0.141→0.078	0.23→0.176	0.051→0.042	0.076→0.029	0.226→0.171	0.076→0.007
Overconfidence effect						
Ice-breaking apology	0.223→0.161	**0.348→0.167***	0.266→0.146	0.202→0.140	0.054→0.013	0.14→0.091
Conciliatory apology	0.294→0.266	0.177→0.153	0.086→0.077	0.26→0.26	0.308→0.286	0.25→0.237
Reality-checking apology	0.188→0.129	0.133→0.092	0.213→0.159	0.194→0.148	0.207→0.177	0.137→0.109
Congruence-driving apology	0.183→0.126	0.12→0.069	0.194→0.191	0.194→0.155	0.398→0.354	0.244→0.304
Prospect theory						
Ice-breaking apology	0.218→0.169	0.137→0.014	**0.191→0.095***	0.094→0.036	0.018→0.033	0.147→0.110
Conciliatory apology	0.25→0.22	0.167→0.143	0.193→0.188	0.043→0.035	0.292→0.268	0.224→0.210
Reality-checking apology	0→0.049	0.046→0.13	0.099→0.055	0.071→0.032	0.071→0.043	0.152→0.131
Congruence-driving apology	0.13→0.079	0.06→0.011	0.004→0.005	0.168→0.133	0.293→0.249	0.167→0.113
Construal theory						
Ice-breaking apology	0.129→0.044	**0.189→0.045***	0.201→0.055	0.111→0.024	0.007→0.079	0.133→0.078
Conciliatory apology	0.162→0.101	0.022→0.042	0.119→0.111	0.118→0.116	0.212→0.172	0.092→0.057
Reality-checking apology	0.194→0.142	0.147→0.112	0.265→0.220	0.105→0.058	0.143→0.112	0.04→0.008
Congruence-driving apology	0.235→0.185	0.181→0.134	0.034→0.025	0.244→0.210	0.328→0.285	0.268→0.216

(continued)

Table 7 (continued)

Mediator variables—Forms of apology (A_i)	Dependent variables—Positive responses (P_k)					
	P1	P2	P3	P4	P5	P6
Reactive devaluation						
Ice-breaking apology	0.259→0.195	**0.486→0.262***	0.334→0.181	0.288→0.239	0.123→0.037	0.172→0.117
Conciliatory apology	0.269→0.227	0.13→0.086	0.244→0.252	0.189→0.197	0.255→0.222	0.21→0.194
Reality-checking apology	0.36→0.31	0.285→0.255	0.253→0.193	0.217→0.163	0.253→0.226	0.041→0.005
Congruence-driving apology	0.203→0.128	0.166→0.095	0.014→0.001	0.143→0.088	0.207→0.138	0.161→0.078

Noted $p < 0.1$ sig* [83]; P1 represents "A sense of security is regained"; P2 represents "My identity is a significance"; P3 represents "Being understood"; P4 represents "My will is respected"; P5 represents "A sense of justice is raised" and P6 represents "Cooperative relationship is rebuilt"

Table 8 The significant level of apologies having moderating effect on positive responses

Level of moderating effect	Forms of apology	Significant moderating effect on soliciting positive responses						
		L1	L2	L3	L3	L4	L5	L6
Most	Congruence-Driving apology	****	****	****	****	****	****	****
Medium	Conciliatory apology	**	***					
Least	Reality-checking apology	*		*				

***Significant under all five precedent barriers to dispute settlement
***Significant under Reactive Devaluation [22] and Cognitive Dissonance [14]
**Significant under Reactive Devaluation [22]
*Significant under Prospect theory [19]

Table 9 Number of apologies with significant moderating effect on positive responses (Developed from Table 8 by counting the number of significant effect)

Positive responses					
Sense of security	Identity	Recognition	Autonomy	Sense of justice	Cooperative image
Number of apologies with significant moderating effect under 5 precedent barriers					
7/20	7/20	6/20	5/20	5/20	5/20

sense of security would help to promote dispute settlement. The other three forms of apology do not record any significant mediating effect in soliciting positive responses. It can be concluded that offering an apology is not effective in directly bringing about positive responses from the apology recipient. In this connection, Yiu and Cheung [76] found that apology is among the range of negotiating tactics that can be deployed. In general, tactics that address the tangible issues are more likely to derive direct and spontaneous positive responses.

Implications on Dispute Management

In general, apology would have mediating and/or moderating effect in soliciting positive responses if given in the right context like emotion is at stake. For discussion purposes, "versatile" is used to describe an apology that records the highest number of moderating effects on the positive responses [76]. Accordingly, congruence-driving is the most "versatile" form of apology among the four forms of apology studied in this study. To make a congruence-driving apology effective, it is important to express the apology appropriately by the right people, at the right time and the right place [8]. It is therefore suggested that, when dispute arise, the disputant should.

Offer an Apology Based Specifically Targeting the Bottleneck

When emotion is the bottleneck against settlement, Schweitzer et al. [37] pointed out that a sincere apology may provide the breakthrough. In such situation, acknowledging fault and regret would be welcome by the counterpart. Levi [8] suggested that investment in self-image, interpersonal orientation and gender will enhance the usefulness of an apology.

Offer an Apology at the Right Time

Maintain the momentum of a negotiation is the gateway to settlement. One key difference between congruence-driving apology and ice-breaking apology is when to make the offer [8]. The findings of this study suggest that these two forms of apology can have quite different outcomes. Usually, a sincere apology before a complaint may suppress the conflict.

Offer an Apology in an Appropriate Manner

The content of an apologising statement and the delivering tone are pivotal in creating positive effects [32]. It must be focused in addressing the emotion of the injured and avoid legalistic analysis. Sincere acknowledgment of fault and regret for the damaging conduct must be an integral part of the apology [37]. The apologetic wordings to be used must suit the context and the extent of damages caused. This study found that congruence driving apology is an added device that couple apology with the desire to drive win–win solution [85]. Moreover, because of the delicacy involved, this form of apology must be undertaken with great skill with the counterpart being an attentive listener. Sincerity is certainly a critical component of the apology.

The Second Part of the Study

In construction, mediation is one of the commonly used alternative dispute resolution mechanisms that encompasses a softer way for the disputing parties to reach negotiated settlement through the help of a mediator. The part of the study is reported in four parts: (1) prerequisite conditions towards dispute settlement in mediation; (2) principle of Apology; (3) data collection and analysis; and (4) discussion and recommendations.

Prerequisite Conditions Towards Dispute Settlement in Mediation

In Sect. "Discussions", it has been suggested that apology offer initiated by mediator may be more receptive to a prospective apology offeror. In his part of the study, the incentivising effects of an apology in dispute settlement through mediation is explored. First, the prerequisite conditions for such settlement are examined. Chau [86] advocated the followings: (i) willingness to settle, (ii) desirability to continue amicable business relationships, (iii) intention to save cost and time, (iv) reliability of a negotiated accommodation, and (v) cultural tradition. These conditions can further be grouped psychological and perception of the dispute.

Psychological Prerequisites

(a) Willingness to Settle

Willingness to negotiate, communicate and bargain are the determining ingredients for a successful mediation [87]. When the disputing parties are having the desire to find a solution that is acceptable to all, the job of the mediator will become much easier. On the other hand, if the parties do not want to settle the dispute, a mediation even performed, will just perfunctory. Cohen [88] believed that willingness is indeed the necessary condition for a workable mediation [89]. Parties' willingness to deal with the dispute and accept the facilitation by the mediator will direct how the parties to prepare for the negotiation [89]. Such willingness is also vital in determining whether the parties will accept the consequence/result of mediation [89].

(b) Release of Emotional Distress

The bounded rationality theory [90] explained that individuals are "Passionate economists". Their rationality is somehow limited before of the limitation of knowledge they possess. Thus, it is not uncommon for them to settle with "satisfying" instead of "optimal" outcomes. Humans naturally prefer having emotional comfort.

This preference would take a disputant to decide on options that would enable them to be relieved from emotional and psychological effects of the wrong act.

Perception of the Dispute

Different view of the issue is the crunch of a dispute. Naturally, finding common ground is the key to drive a settlement. Without common ground, it is difficult to move forward with the negotiation. In mediation, the third-party neutral is uniquely positioned to identify common grounds through soliciting wish list from the disputing parties. Once common ground is uncovered, there will be rooms for moving the negotiation forward [91]. Experienced mediators will also be skilful in noting issues arising from cultural differences [92] and hidden barriers due to mis-communication [93].

Levi [8] found also the prospect of a mediation will also depend on the perceptive view of the complexity of the issue in dispute. To this ends, Felstiner [94] suggested that the escalation of disputes typically goes through "naming" to "blaming" to "claiming". Escalation would be attached with greater demand [86]. Lewicki et al. [95] distinguishes elements of mediation into tangible and intangible. Tangible elements are those quantitative issues like damages, while intangible elements are soft issues such as reputation, relation [96], sense of guilt or shame [97]. Intangible elements are thus more subtle and not conspicuous. Accordingly, the prerequisite conditions for dispute settlement in mediation are listed in Table 10.

Settlement Incentivising Power of Apology

Lazare [109] explained how exchange of apology works: "What makes an apology work is the exchange of shame and power between the offender and the offended." The function of an apology is to establish the moral accountability of the wrongdoer. An apology is not simply No saying sorry. It should also recognize the consequence of saying so. By offering an apology, the apologizer runs the risk of admitting the wrong and the potential liability [110]. Beyond the injured, an apology may be seen by the community at large of admission of errors/mistakes [103].

Moreover, offering an apology can be useful in dispute negotiation. Equity theory [111] projects that inequity is a source of conflict. Inequity may arise when one's output/input ratio is higher than others. This argument is based on the belief that output/input ratio among individuals should be compatible. When a wrongdoer harms the other party, the latter is inflicted with pain and distress [112]. Restoration of these negative effects should be addressed [113]. It is proposed that incentivising behaviours can be planned for this purpose would assist the correction [114]. In this regard, apology may be an incentivizing agent to restore the equity [115]. For example, disputants are expected to be better able to recover from emotional and

Table 10 Prerequisite conditions for dispute settlement in mediation

No.	Description	Key references
1	Psychological prerequisites	
1.1	*Willingness to settle through mediation*	
V1	Both parties are willing to settle	[88]
V2	Sincere intent to resolve dispute	[8]
V3	Willingness to enter to mediation	[8]
V4	Intention to save money	[8]
V5	Willingness to solve the problem in an amicable way	[89, 98]
V6	Willingness to have more communication with disputants	[99]
V7	Willingness to maintain relationship	[100]
1.2	*Release of emotional distress*	
V8	Demand for emotional comfort	[101]
V9	Demand to focus on the issue rationally	[102]
V10	Demand to achieve equal footing	[103]
V11	Motivation to repair relationship	[8]
2	Perceptions of the dispute	
V12	Disputants have similar organizational culture	[104]
V13	Disputants have similar communication style	[105]
V14	The dispute is easy to solve	[94]
V15	Solving the dispute is costless	[106]
V16	The dispute contains not many intangible issues to solve	[107]
V17	The dispute is very complex	[108]
V18	The dispute involves a large number of parties	
V19	The overall likelihood of settlement is high	[8]

psychological effects of the wrong act, while offenders may reconcile and repair damaged relationships.

In addition, the more sincere an apology, the more effective it will be in reducing the tension between the disputants. Lowering tension would pave the path for accepting proposed compensation [116]. For protracted disputes, the resources consuming situation demands a sensible solution [117]. How to effect such a change, Folger [53] suggested something that breaks the pattern may work—enhancing the communication between the disputing parties [106]. A mediator can help in developing a new communication channel [8]. Hiltrop [107] suggested that tangible elements are more negotiable than intangible elements. Thus, intangible elements would need extra effort to sooth the emotional issues such as tension and distress. A sincere apology if accepted would transform an accuser-accused relationship into disputing parties of equal footing.

In summary, apology would be regarded as a plausible act to fulfil prerequisite conditions and to address perceptions on the disputes. Its usefulness is anchored

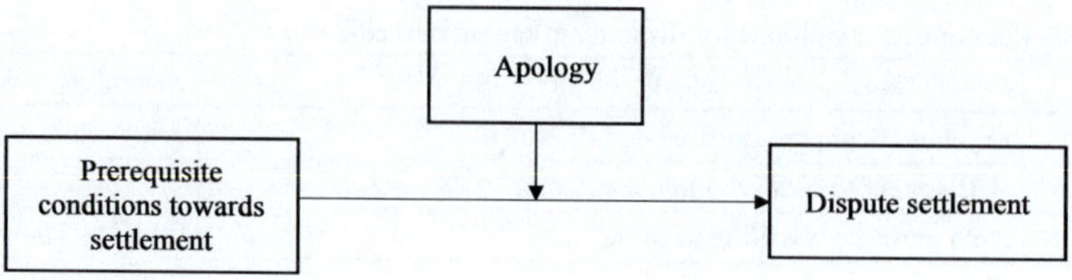

Fig. 4 Conceptual framework of incentivization effect of apology on dispute settlement

on the ability to address the emotional distress of the injured. If legal protection is accorded to the apology offeror, it can be envisaged that wider use of apology can be expected. Figure 4 shows the conceptual framework of the use of apology in mediation.

Data Collection and Analysis

To verify the framework, data was collected through a specially designed questionnaire that has three parts. Part A collects the particulars of the respondents. Part B has 15 questions (v1 to v19) listed in Table 11. The respondents were asked to indicate their degree of agreement on the alone statements in a Likert Scale of 1 (not agree) to 7 (totally agree). Part C repeats the questions in Part B, but the respondents were asked to respond with reference to the following scenario within which the use of apology is incorporated:

> You were the client and commissioned the construction of a 2-storey building. The contractor was given 8 months to complete the works. The contractor hired a sub-contractor to do certain works. After 3 months, the sub-contractor was in trouble in other project and the progress of the works was slow. The project could not be completed on time. There were different opinions about whether extension of time should be granted. The contractor wanted extension of time as this was out of his anticipation, but you thought it is the responsibility of the contractor to supervise the subcontractor. You were upset and did not want further delay. You and the contractor agreed to mediate the dispute on extension of time. At the meeting, the contractor softened a bit and took the following actions. He expressed his regret to you, showed his remorse for problem caused by the sub-contractor. Under the Hong Kong Apology Ordinance (CAP. 631), a regret would not be considered as an admission of responsibility.

By comparing the respondents' responses to Part B and Part C, the effects of the apology on settlement was investigated. A total of 78 responses from construction practitioners in Hong Kong were received. Table 11 gives also the descriptive statistics.

In Table 12, x and y represent the two categories of prerequisite condition towards dispute settlement in mediation; x represents the average means of the psychological prerequisites while y represents the perception of the dispute. It is suggested that both high score of x and y would imply dispute settlement is more likely.

Table 11 Descriptive statistics of the responses

No.	Description	N	Min.	Max.	Mean	Std.
Part B: Behaviour in resolving construction dispute						
V1	I have the willingness to settle the construction dispute	78	3	7	5.53	0.89
V2	I have sincere intent to resolve the construction dispute	78	3	7	5.33	0.85
V3	I desire and appreciate mediation than litigation	78	4	7	5.38	0.87
V4	I have the intention to save money	78	4	7	6.19	0.81
V5	I want to settle the dispute in an amicable way	78	4	7	5.53	0.75
V6	I have proper communication	78	2	7	4.92	1.00
V7	I concern developing relationship	78	3	7	5.36	0.95
V8	I want to offer emotional comfort	78	2	7	4.83	1.16
V9	I want to focus on the issue rationally	78	3	7	5.37	0.88
V10	I want to have equal footing with the counterpart	78	2	7	4.92	1.08
V11	I have motivation to repair relation ship	78	3	7	5.15	0.99
V12	I find similar culture with the counterpart (e.g., company goal, political view)	78	1	7	4.32	1.31
V13	I find similar communication style with the counterpart	78	1	7	4.18	1.26
V14	The dispute leads to severe effect/consequence	78	3	7	4.95	0.85
V15	The dispute can consume less time/labour etc. for settlement	78	2	7	4.45	1.22

(continued)

Table 11 (continued)

No.	Description	N	Min.	Max.	Mean	Std.
V16	The dispute includes intangible issues to be solved	78	1	7	4.94	1.17
V17	The dispute is complex	78	2	7	5.32	1.13
V18	The dispute involves a large number of parties	78	1	7	4.91	1.22
V19	The settlement can fulfil the desired outcome	78	2	6	4.76	0.82
Part C: Scenario case						
V1	I have the willingness to settle the construction dispute	78	2	7	5.41	0.93
V2	I have sincere intent to resolve the construction dispute	78	3	7	5.40	0.98
V3	I desire and appreciate mediation than litigation	78	3	7	5.27	0.91
V4	I have the intention to save money	78	3	7	5.58	0.88
V5	I want to settle the dispute in an amicable way	78	2	7	5.53	1.28
V6	I have proper communication	78	3	7	5.55	0.83
V7	I concern developing relationship	78	1	7	4.38	1.14
V8	I want to offer emotional comfort	78	1	7	4.28	1.19
V9	I want to focus on the issue rationally	78	4	7	5.63	0.79
V10	I want to have equal footing with the counterpart	78	2	7	5.41	1.01
V11	I have motivation to repair relationship	78	4	7	5.88	0.76
V12	I find similar culture with the counterpart (e.g., company goal, political view)	78	2	7	6.05	1.03
V13	I find similar communication style with the counterpart	78	4	7	5.54	0.82

(continued)

Table 11 (continued)

No.	Description	N	Min.	Max.	Mean	Std.
V14	The dispute leads to severe effect/consequence	78	3	7	5.10	0.77
V15	The dispute can consume less time/labour etc. for settlement	78	2	7	5.00	0.98
V16	The dispute includes intangible issues to be solved	78	1	6	4.59	1.09
V17	The dispute is complex	78	1	7	5.60	1.17
V18	The dispute involves a large number of parties	78	1	7	4.71	1.02
V19	The settlement can fulfil the desired outcome	78	2	7	5.56	0.88

Table 12 Score clusters of x and y

Scope	Part B			Part C		
	$x < 5$	$5 \leq x < 6$	$x \geq 6$	$x < 5$	$5 \leq x < 6$	$x \geq 6$
$y > 5$	2	14	5	0	10	6↑
$4.5 < y \leq 5$	5	14	3	5	28↑	4
$4 \leq y \leq 4.5$	8	14	1	2	13	4
$y < 4$	5	6	1	1	4	0

With reference to Table 12, it can be observed that the trends of the prerequisite conditions are mostly improved in Part C. In Part C, rising trends for both psychological prerequisites and perception of the dispute with the responses mostly located in these two clusters: $4.5 < y \leq 5$ and $5 \leq x < 6$.

For some more details, the number of responses with $x < 5$ are lower for Part C when compared to Part B. This suggests the level of having the psychological prerequisites is generally higher when apology was used. Also, the numbers of data in the ranges of $5 \leq x < 6$ and $x \geq 6$ also increases. Furthermore, for Part C, the respondents in cluster $5 \leq x < 6$ showed also better perception of the dispute when apology was involved in the dispute negotiation. To further investigate the improvements in Part C, the comparison of the mean scores is shown in Fig. 5. It can be noted that oonly four questions in Part B are having higher mean score than that of Part C. Three of them (v4, v12 and v18) in fact show only very slight decrease in Part C. For v17, the question setting is "The dispute is very complex", which is considered as a reverse question. The percentage (−12%) means that within apology behaviour, the respondents recognized that the complexity of the dispute is reduced.

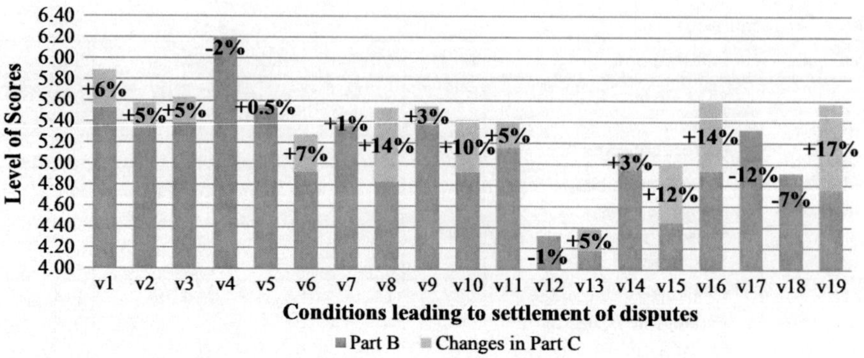

Fig. 5 Comparison of the mean scores

Accordingly, the general pattern is that with the use of apology, the conditions for a settlement are better embraced.

Discussions and Recommendations

The settlement incentivizing power of an apology was generated by its ability to embrace the two categories of prerequisite conditions for settlement. The use of apology was incorporated in the Part C, scenario case, of the data collection questionnaire. The contributions of the apology is the most notable in releasing emotional distress (question v8~11) and improving the cognitions of the likelihood towards dispute settlement (question v 14~19). The apology appears to have enhanced the perception of the apology offeror by the recipient, the overall improvement in the negotiation environment are conducive to derive greater intention to settle. The following recommendations are made based on the findings:

There is No Loss to Apologize in Mediation with the Protection Rendered by the AO

Mediation is a voluntary process through which disputing parties, assisted by a neutral third-party, to iron out their differences. A mediator facilitates and enhances the communication between the parties. Most importantly, an experienced mediator would help the parties to identify common grounds to build the foundation for a settlement. Settlement is only possible if parties share the will to put an end to the dispute. Humans are in general sympathetic, and an apology can soften the psychological barriers against settlement. Protected by the Apology Ordinance, making apology in Hong Kong will not expose the parties to legal responsibility. Thus, there is no loss to apologize and the potential gain is an improved negotiation environment.

It is likely that both sides are having some wrongs and exchanging frank apology would remove skepticism and mistrust.

Offer Apology When Emotional Distress is at Stake

To develop an amicable environment for dispute settlement, the emotion of the parties should be controlled. An apology from a wrong doer would be a more effective option than financial concessions when emotion is the barrier against settlement. Emotional comfort is somehow intangible and cannot be easily alleviated through monetary compensation. Offering an apology is one of the empathetic means to sooth distress and tension. It would be instrumental when people need an apology to heal their broken heart.

Disputants Should Be on Equal Footing

Equity theory [54] advocates that dealing in equal footing is expected in human activities. This human nature is deep-rooted and should also be observed in dispute negotiation. Disparity in resources is inevitable in construction contracting. The protract and resource lade legal proceedings are intimidating to most small and medium sized contractors. In this connection, this group of organizations are quite disadvantaged in formal dispute resolution processes. The less resourced party may turn to more aggressive and uncompromising approach should opportunity arises. Therefore, avoiding unequal footing is useful. The use of apology can balance the positions of the disputing parties. A recipient of apology would definitely feel better and have better impression on the apology offeror. The skill of mediators can be relied on to improve the effectiveness.

Summary

The enactment of the first-ever apology ordinance (AO) in 2017 marked the start of a new wave of mediation movement in Hong Kong. The AO aims to alleviate the concern of disputing parties in making apology. This study pioneers use of apology in construction dispute resolution through (i) examining the mechanism through which apology can enhance settlement; and (ii) the power of apology in incentivizing settlement in construction dispute mediation. The first part of the study has been developed based on Kalman's Response Restriction Theory [9]. It is proposed that offering an apology would solicit positive responses of the counterpart by overcoming certain barriers against settlement of the dispute. Two hypotheses are developed: (i) an act of apology has positive moderating effect on construction dispute settlement, and (ii) an act of apology has positive mediating effect on construction dispute settlement. Four forms of apology were summarized from literature- ice-breaking, conciliatory, reality-checking and congruence-driving. It is found that congruence-driving apology is the most versatile in moderating positive responses. For mediating effect, only ice-breaking apology offers significant mediating ability in addressing most of the settlement barriers. The second part of the study examines if enacting an Apology Ordinance in Hong Kong will be useful in promoting the use of construction dispute mediation. The objective of the study was achieved by (1) identifying the prerequisite conditions that lead to settlement of disputes in mediation and (2) exploring the significant effects of apology on these components. Two main influencers to settle dispute are (i) psychological prerequisites that include the willingness to settle and the demand of the disputants to obtain emotional comfort; and (ii) perception of the dispute that can be the difficulty, efficiency and complexity. Apology is found to be instrumental in soothing human distress and mediators are ideal facilitators in formulating and delivering of apology proposals.

Acknowledgements The empirical work of this chapter has been reported in a paper entitled "The value of apology in construction dispute negotiation" of the International Journal of Construction Management and a paper "How apology incentivizes construction dispute settlement" of the Journal of Legal Affair and Dispute Resolution in Engineering and Construction. Special thanks to Miss Ka In Yu and Miss Sui Yan Tong in helping with the empirical studies.

References

1. Cheung SO, Cheung HC (2016) Apology ordinance: a push for mediation in Hong Kong. In: RICS COBRA conference, Toronto, Canada, 20–22 Sept 2016
2. Cheung SO, Pang HY (2013) Anatomy of construction disputes. J Constr Eng Manage ASCE 139(1):15–23
3. Ross L, Ward A (1995) Psychological barriers to dispute resolution. Adv Exp Soc Psychol 21:255–304
4. Allred KG, Mallozzi JS, Matsui F, Raia CP (1997) The influence of anger and compassion on negotiation performance. Organ Behav Hum Deci Process 70(3):175–187
5. Barry B, Friedman RA (1998) Bargainer characteristics in distributive and integrative negotiation. J Personal Soc Psychol 74(2):345–359
6. Carnevale PJ, Isen A (1986) The influence of positive affect and visual access on discovery of integrative solution. Organ Behav Hum Decis Process 37(1):1–13
7. De Dreu CKW, Van Kleef GA (2004) The influence of power on the information search, impression formation, and demands in negotiation. J Exp Soc Psychol 40(3):303–319
8. Levi DL (1997) The role of apology in mediation. N Y Univ Law Rev 72(5):1165–1209
9. Kelman HC (1953) Attitude change as a function of response restriction. Hum Relat 6(3):185–214
10. Illankoon IMCS et al (2019) Causes of disputes, factors affecting dispute resolution and effective alternative dispute resolution for Sri Lankan construction industry. Int J Constr Manage 0(0): 1–11, Taylor & Francis. https://doi.org/10.1080/15623599.2019.1616415
11. Cheung SO (1999) Critical factors affecting the use of alternative dispute resolution processes in construction. Int J Proj Manage 17(3):189–194
12. Enshassi A, Choudhry RM, El-ghandour S (2009) Contractors' perception towards causes of claims in construction projects. Int J Constr Manag 9(1):79–92. https://doi.org/10.1080/15623599.2009.10773123

13. Korobkin R (2006) Psychological impediments to mediation success: theory and practice. Ohio State J Dispute Resolut 21(2):281–290
14. Festinger L (1957) A theory of cognitive dissonance, 1st edn. Stanford University Press, Stanford
15. Ma N (2008) Overconfidence—its theory and implications in dispute resolution. ADR Bull 10(4):1–6
16. Dunning D, Griffin DW, Milojkovic JD, Ross L (1990) The overconfidence effect in social prediction. J Pers Soc Psychol 58(4):568–581
17. Lichtenstein S, Fischhoff B, Phillips LD (1982) Calibration of probabilities: the state of the art to 1980. In: Kahneman D, Slovic P, Tversky, A (eds) Judgment under uncertainty: Heuristics and biases, 1st edn. Cambridge University Press, New York
18. Hastorf AH, Cantril H (1954) They saw a game; a case study. Psychol Sci Public Interest 49(1):129–134
19. Kahneman D, Tversky A (1979) Prospect theory: an analysis of decision under risk. Econometrica 47(2):263–292
20. Ross L (1977) The intuitive psychologist and his shortcomings: distortions in the attribution process. Adv Exp Soc Psychol 10:173–220
21. Griffin DW, Dunning D, Ross L (1990) The role of construal processes in overconfident predictions about the self and others. J Personal Soc Psychol 59(6):1128–1139
22. Ross L, Stillinger C (1991) Barriers to conflict resolution. Negot J 8(4):389–404
23. Osgood CE, Tannenbaum PH (1955) The principle of congruity in the prediction of attitude change. Psychol Rev 62(1):42–55
24. Curhan J, Neale M, Ross L (2004) Dynamic valuation: preference changes in the context of face-to-face negotiation. J Exp Soc Psychol 40(2):142–151
25. Ross L (1995) Reactive devaluation in negotiation and conflict resolution. In: Arrow K, Mnookin R, Ross L, Tversky A, Wilson RB (eds) Barriers to conflict resolution, 1st edn. New York, pp 27–42
26. Lord CG, Ross L, Lepper MR (1979) Biased Assimilation and attitude polarization: the effects of prior theories on subsequently considered evidence. J Pers Soc Psychol 37(11):2098–2109
27. Taft L (2000) Apology subverted: the commodification of apology. Yale Law J 109(5):1135–1160
28. Orenstein A (1999) Apology excepted: incorporating a feminist analysis into evidence policy where you would least expect it. Sw U L Rev 221–239
29. Department of Justice (2016) Consultation paper enactment of apology legislation in Hong Kong: report and 2nd round consultation, Department of Justice, Steering Committee on Mediation, Hong Kong
30. Pruitt DG (1981) Negotiation behaviour, 1st edn. Academic Press, New York
31. Li Q, Yin Z, Chong H, Shi Q (2018) Nexus of inter-organizational trust, principled negotiation, and joint action for improved cost performance: survey of chinese megaprojects. J Manag Eng. https://doi.org/10.1061/(ASCE)ME.1943-5479.0000634
32. Brown JG (2004) Apology: The Role of Apology in Negotiation. Marquette Law Rev 87(4):665–673
33. Tabak BA, McCullough ME, Luna LR, Bono G, Berry JW (2012) Conciliatory gestures facilitate forgiveness and feelings of friendship by making transgressors appear more agreeable. J Personal 80(2)
34. Frosch J (1966) A note on reality constancy. In: Psychoanalysis: a general psychology: essays in honor of Heinz Hartmann. International Universities Press, New York
35. Cheung SO, Chow PT, Yiu TW (2015) Performance of mediator tactics in building management disputes. J Constr Eng Manag. https://doi.org/10.1061/(ASCE)ME.1943-5479.0000230
36. Ring PS, van de Ven AH (1994) Developmental processes of cooperative interorganizational relationships. Acad Manag Rev 19(1):90–118
37. Schweitzer ME, Brooks AW, Galinsky AD (2015) The organizational apology. Harvard Bus Rev 44–52

38. Kelman HC (1996) Negotiation as interactive problem solving. Int Negot 725 1(1):99–123
39. Burton JW, Sandole JD (1987) Expanding the debate on generic theory of conflict resolution: a response to a critique. Negot J 3(1):97–99
40. Demirdöğen ÜD (2011) A social-psychological approach to conflict resolution: interactive problem solving. Int J Soc Inq 4(1):215–228
41. Maslow AH (1943) A theory of human motivation. Psychol Rev 50(4):370–396
42. Ting-Toomey S, Kurogi A (1998) Facework competence in intercultural conflict: an updated face-negotiation theory. Int J Intercultural Relat 22(2):187–225. https://doi.org/10.1016/S0147-1767(98)00004-2
43. Deutsch M (2003) Cooperation and conflict: a personal perspective on the history of the social psychological study of conflict resolution. In: West MA, Tjosvold DJ, Smith KG (eds) International handbook of organizational teamwork and cooperative working, 1st edn. Wiley, San Francisco, pp 9–43
44. Holtgraves T (1997) Styles of language use: Individual and cultural variability in conversational indirectness. J Personal Soc Psychol Personal Process Individ Differ 73(3):624–637
45. Burton JW (1998) Conflict resolution: the human dimension. Int J Peace Stud 3(1):1–5
46. Fraser N (2001) Recognition without Ethics? Theory Cult Soc 18(2–3):21–42
47. Wolf R (2011) Respect and disrespect in international politics: the significance of status recognition. Int Theory 3(1):105–142
48. Murphy M (2011) Apology, recognition, and reconciliation. Hum Rights Rev 12(1):47–69
49. Deci EL, Ryan RM (1985) Intrinsic motivation and self-determination in human behaviour, 1st edn. Plenum Press, New York
50. Deci EL, Vansteenkiste M (2004) Self-determination theory and basic need satisfaction: Understanding human development in positive psychology. Recherché di Psicologia 27(1):17–34
51. Hackman JR, Oldham GR (1976) Motivation through the design of work: Test of a theory. Organ Behav Hum Perform 16:250–279
52. Nobles M (2008) The politics of official apologies, 1st edn. Cambridge University Press, New York
53. Folger R, Cropanzano R (2001) Fairness theory: Justice as accountability. In: Greenberg J, Cropanzano R (eds) Advances in organizational justice, 1st edn. Stanford University Press, Stanford, CA, pp 1–55
54. Adams JS (1965) Inequity in social exchange. Adv Exp Soc Psychol 2:267–299
55. Folger R, Skarlicki DP (1999) Unfairness and resistance to change: hardship as mistreatment. J Organ Change Manage 12(1):35–50
56. Thibaut JW, Walker L (1975) Procedural justice: a psychological perspective, 1st edn. Erlbaum, Hillsdale
57. Wright M (2004) The British journal of criminology. Br J Criminol 44(2):290–292
58. Schweiger DM, DeNisi AS (1991) Communication with employees following a merger: a longitudinal field experiment. Acad Manag J 34(1):110–135
59. Bronfenbrenner U (1961) The mirror image in Soviet- American relations: a social psychologist's report. J Soc Issues 17(3):45–56
60. Hofbauer J, Sigmund K (1998) Evolutionary games and population dynamics, 1st edn. Cambridge University Press, Cambridge
61. Sigmund K (2010) The calculus of selfishness, 1st edn. Princeton University Press, Princeton
62. Han TA, Pereira LM, Santos FC, Lenaerts T (2013) Good agreements make good friends. Sci Rep 3(2695):1–7
63. Nesse RM (2001) Evolution and the capacity for commitment. Russell Sage Foundation series on trust, 1st edition, Russell Sage, New York
64. Witvliet CVO, Worthington EL, Wade NG (2002) Victims' heart rate and facial EMG responses to receiving an apology and restitution. Psychophysiol Suppl 39:88
65. Robbennolt JK (2006) Apologies and settlement levers. J Empir Leg Stud 3(2):333–373
66. Darby BW, Schlenker BR (1982) Children's reactions to apologies. J Pers Soc Psychol 43(4):742–753

67. Bennett M, Earwaker D (1994) Victim's response to apologies: the effects of offender responsibility and offense severity. J Soc Psychol 134(4):457–464
68. Hodgins HS, Liebeskind E (2003) Apology versus defence: antecedents and consequences. J Exp Soc Psychol 39:297–316
69. James LR, Brett JM (1984) Mediators, moderators, and tests for mediation. J Appl Psychol 69(2):307–321
70. Luo Y, Wang Z, Zhang H, Chen A (2016) The influence of family socio-economic status on learning burnout in adolescents: mediating and moderating effects. J Child Fam Stud 25(7):2111–2119
71. Mezquita L, Camacho L, Ibanez MI, Villa H, Moya-Higueras J, Ortet G (2015) Five-factor model and alcohol outcomes: Mediating and moderating role of alcohol expectancies. Personal Individ Differ 74:29–34
72. Vanhalst J, Luyckx K, Raes F, Goossens L (2012) Loneliness and depressive symptoms: the mediating and moderating role of uncontrollable ruminative thoughts. J Psychol 146(1–2):259–276
73. Jaccard J, Turrisi R, Choi KW (1990) Interaction effects in multiple regression. Sage, Thousand Oaks, California, U.S
74. Fisher RA (1924) On a distribution yielding the error functions of several well known statistics. In: Proceedings of the international congress of mathematics, vol. 2. University of Toronto Press, pp 805–813
75. Wong PSP, Cheung SO, Leung KY (2008) Moderating effect of organizational learning type on performance improvement. J Manag Eng 24(3):162–172
76. Yiu TW, Cheung SO (2007) A study of construction mediator tactics—Part II: The contingent use of tactics. Build Environ 42(2):762–769
77. Judd CM, Kenny DA (1981) Estimating the effects of social interventions, 1st edn. Cambridge University Press, New York, N.Y., U.S
78. Baron RM, Kenny DA (1986) The moderator–mediator variable distinction in social psychological research: conceptual, strategic, and statistical considerations. J Personal Soc Psychol Interpersonal Relat Group Process 51(6):1173–1182
79. Easterby-Smith M, Thorpe R, Lowe A (1991) Management research: An introduction. Sage, London
80. Greenberg J, Cropanzano R (2002) Advances in organizational justice, 1st edn. Stanford University Press, Stanford
81. Axelrod R (1984) The evolution of cooperation, 1st edn. Basic Books, New York
82. Kruskal WH, Wallis WA (1952) Use of ranks in one-criterio variance analysis. J Amer Statist Assoc 47(260):583–621
83. Hair JF, Black WC, Babin BJ, Anderson RE (2010) Multivariate data analysis. Seventh, Pearson Education. Seventh. Pearson Education. https://doi.org/10.1016/j.ijpharm.2011.02.019
84. Cheung SO, Yiu TW (2006) Construction disputes inevitable? IEEE Trans Eng Manage 53(3):456–470
85. Phillips-Alonge O (2019) The influence of partnering on the occurrence of construction requirement conflicts and disputes. Int J Constr Manag 19(4):291–306
86. Chau KW (2007) Insight into resolving construction disputes by mediation/adjudication in Hong Kong. J Prof Issues Eng Educ Pract 133(2):143–147
87. Brogan M, Spencer D (2006) Mediation law and practice. Cambridge University Press, New York
88. Cohen C (1992) Mediation: giving law a human face. De Rebus 1992:126–128
89. Rubin JZ (1981) Introduction. In: Dynamics of third party intervention: Kissinger in the Middle East. Praeger,. New York
90. Simon H (1972) Theories of bounded rationality, decision and organization. North-Holland Publishing Company
91. Ladd PD (2000) Mediation, conciliation and emotions. University Press of America Inc., Oxford

92. Huntington SP (1996) The clash of civilizations and the remaking of world order. Simon & Schuster, New York
93. Fischer G, Ostwald J (2003) Knowledge communication in design communities. Kluwer Academic Publishers, Netherlands
94. Felstiner WLF (1981) The emergence and transformation of disputes: naming, blaming, claiming. Law Soc Rev 15
95. Lewicki RJ, Barry B, Saunders DM (2004) Essentials of negotiation. McGraw-Hill Education, New York
96. Aubert V (1963) Competition and dissensus: two styles of conflict and conflict resolution. J Conflict Resolut 7:26–42
97. Schneider CD (2000) What it means to be sorry: the power of apology in mediation. Mediat Q 17(3):265–280
98. Philips BA (1994) Finding common ground: a field guide to mediation. Hells Canyon Publishing, Austin, TX
99. Lodge AJ (2001) Legislation protecting confidentiality in mediation: armor of steel or eggshells? Santa Clara Law Rev 41(4):1094–1121
100. Lewicki RJ, Bunker B (1995) Trust relationships: a model of trust development and decline. Jossey-Bass, San Francisco
101. Lee T (2000) Apology subverted: commodification of apology. Yale Law J 109(5):1135
102. Callahan R (2005) Facework in mediation: The need for "Face" time. Bepress Legal Series. Working Paper 837
103. Robbennolt JK (2003) Apologies and legal settlement: an empirical examination. Mich Law Rev 102(3):460–516
104. Lucke K, Rigaut A (2002) Cultural issues in international mediation. https://www.nottingham.ac.uk/research/groups/ctccs/projects/translating-cultures/documents/journals/cultural-issues-mediation.pdf
105. Horton WS, Keysar B (1996) When do speakers take into account common ground? Cognition 59:91–117
106. Kagel S, Kelly K (1989) The anatomy of mediation: What makes it work. Bureau of national affairs. Bureau of National Affairs, Washington
107. Hiltrop J (1989) Factors associated with successful labour mediation. Jossey, San Francisco
108. Henderson DA (1996) Mediation success: an empirical analysis. Ohio State J Dispute Resolut 11:105–147
109. Lazare A (1995) Go ahead, say you're sorry. Psychol Today 28(1):40–42
110. Minow M (1998) Between vengeance and forgiveness: Facing history after genocide and mass violence. Beacon Press, Boston
111. Adams JS, Freedman S (1976) Equity theory revisited: comments and annotated bibliography. Adv Exp Soc Psychol 9(C):43–90. https://doi.org/10.1016/S0065-2601(08)60058-1
112. Bornstein BH, Schopp RF, Wiener RL, Willborn SL (2008) Civil juries and civil justice: psychological and legal perspectives. Springer, New York
113. Walster E, Berscheid E, Walster GW (1973) New directions in equity research. J Pers Soc Psychol 25:151–176
114. Ho B (2005) A theory of apologies. Business (March):1–37. https://doi.org/10.1093/acprof:oso/9780198238515.001.0001
115. O'Hara EA (2004) Apology and thick trust: what spouse abusers and negligent doctors might have in common. Heinonline 1(1):8–23. https://doi.org/10.3868/s050-004-015-0003-8
116. Wagatsuma H, Rosett A (1986) The implications of apology: law and culture in Japan and the United States. 20 Law Soc
117. Plous S (1993) The psychology of judgement and decision making. McGraw-Hill, New York

Reality Check of Construction Dispute Negotiation Conditions

Market Competition and Dispute Occurrence

Sai On Cheung and Lu Shen

Introduction

The making of suicidal bids arises from extremely competitive construction markets characterized by low market entry thresholds [1]. Rooke et al. [1] and Cannon and Hillebrandt [2] observed that the return on capital that can be obtained by contractors can be larger than profit markups by increasing the turnover rate and through extensive subcontracting. The combination of fierce bidding competitions and insufficiently developed design at the bidding stage has led to tenders with negative profits [3]. The culture of exploiting claims and opportunism is often then legitimized, since it is believed that contractors have no other option in such a competitive environment [1]. The commercial reality is that securing a project outweighs all other considerations at the time of tender. Market competition may force contractors to become claim conscious ex post and to practice opportunism as circumstances arise. Unresolved claims develop into disputes [4].

Identifying Construction Disputes

Many studies have attempted to define disputes. Mururu [5] defined disputes as the formation of a position to maintain conflict, while Brown and Marriot [6] proposed that disputes should be regarded as conflicts that require resolution. In comparison, construction disputes are often more complex in nature and involve larger numbers of stakeholders, which increases the difficulty of developing a universally accepted definition. Spittler and Jentzen [7] argued that construction disputes are associated

S. O. Cheung (✉) · L. Shen
Construction Dispute Resolution Research Unit, City University of Hong Kong, Hong Kong, China
e-mail: Saion.cheung@cityu.edu.hk

S. O. Cheung (ed.), *Construction Dispute Research Expanded*, Springer Tracts in Civil Engineering, https://doi.org/10.1007/978-3-030-80256-1_11

with differences in the perspectives, interests and agendas of different people. The large number of stakeholders involved can worsen the situation. As suggested by Tillett and French [8], construction disputes are caused by the incompatibility of two or more groups' interests, needs or goals. As a result, the resolution of disputes in construction must deal appropriately with different causes [9].

The high levels of uncertainty involved in construction projects and the bounded rationality of human beings render construction contracts incomplete. Williamson [10] convincingly explained that incomplete contracts are inadequate to exhaustively deal with the eventualities that may arise from complex projects. Unfortunately, project happenings without preplanned solutions very easily result in disputes [11]. Although construction disputes are seldom defined, they are often identified with reference to the focused dimension. The most common identification is made by subject matter. In this regard, Diekman et al. [12] classified the causes of construction disputes into people, processes and products. Rhys-Jones [13] identified ten types of dispute causes: (1) management, (2) culture, (3) communication, (4) design, (5) economics, (6) tendering pressures, (7) law, (8) unrealistic expectations, (9) contracts, and (10) workmanship. In a similar fashion, Hewitt [14] classified construction disputes into six groups: (1) change of scope, (2) change of conditions, (3) delays, (4) disruptions, (5) acceleration, and (6) terminations. These categories can be further broken down to 59 different types, which can be summarized into six generic types: (1) determination of contracts, (2) payment related, (3) the site and execution of work, (4) time related, (5) final certificate and final payment, and (6) tort related. Heath et al. [15] identified seven major types of construction disputes: (1) contract terms, (2) payments, (3) variations, (4) extensions of time, (5) nominations, (6) renominations, and (7) availability of information. Conlin et al. [16] identified six causes of construction disputes: (1) payment, (2) performance, (3) delay, (4) negligence, (5) quality and (6) administration. Among the previous studies identifying major causes, Kumaraswamy [17] further ranked the causes based on the frequency and magnitude of disputes each can lead to. The ranked causes are listed by order: (1) variations due to site conditions, (2) variations due to client changes; (3) variations due to design errors; (4) unforeseen ground conditions; (5) ambiguities in contract documents; (6) variations due to external events; (7) interferences with utility lines; (8) exceptional inclement weather; (9) delayed design information; and (10) delayed site possession.

In terms of contractual disputes in the construction industry, Semple et al. [18] identified that site overhead, loss in productivity or revenue and financing costs are the major dispute types arising from construction contracts. On the other hand, Yates [11] proposed seven causes of contractual disputes: (1) variations, (2) ambiguities in contract documents, (3) inclement weather, (4) late issue of design information/drawings, (5) delayed possession of a site, (6) delays by other contractors employed by the developer, and (7) postponement of part of the project.

According to data collected by the UK Adjudication Reporting Center, the three most common types of construction disputes that reach adjudication include those involving the "valuation of variations", the "valuation of final account" and a "failure to comply with payment provisions" [19]. Regarding mediation in the UK, 72% of the mediation cases are related to payments, delays, quality issues and professional negligence [20]. In Hong Kong, the most common dispute subjects settled by mediation include variation, progress delays, the expectations of involved parties and problems arising between parties [21]. Spittler and Jentzen [7], on the other hand, argued that ambiguity in contract documents, adversarial attitudes and perceptions of unfairness among parties cause most of the disputes. Furthermore, it has been found that contractual provision itself can lead to disputes [18]. Similarly, Semple et al. [18] proposed that the two major sources of construction disputes are contracts and unpredictable events. To prevent disputes caused by uncertainties, contradictory contractual provisions must be avoided. Applying the transaction cost framework proposed by Mitropoulos and Howell [22], Williamson [10] proposed that the fundamental factors causing disputes include (1) project uncertainty, (2) contractual problems; and (3) opportunistic behaviours.

In a competitive market involving numerous bidders, the project owner will face a significant risk of awarding contracts to the bidder setting an unrealistic price [23, 24]. Pegged with the below cost bid, the contractor reverts to claims during construction to recoup the cost shortfall [24]. Several studies have pointed out that the setting of a bidding price depends on both actual costs and market competition [25, 26]. Rational contractors who intend to maximize their expected value will adjust profit markups to reflect market competition levels [27]. Since competitors are expected undercut the markup as well, Carr [27] suggests that the more competitors there are, the lower the markup will be. Carr [27] further argued that a contractor with lower actual costs is able to bid with a higher markup but still at a lower price than other competitors can. Without taking into consideration the possible gains from claims and disputes, it seems irrational and unlikely for any contractor to bid a price lower than its costs. However, more recent reports have found that competition can induce unrealistically low bids. Where cost factors are constant, an excessive number of contractors may lead to abnormally low bidding prices and associated excessive disputes [28]. Contractors adjust their markups in response to market competition levels, and [25, 29] argued that both actual and potential competitions can reduce bidder returns. Furthermore, it has been found that in an overtly competitive market, successful bidders in multiple-bidder contests earn significantly negative returns and fare worse than in single-bidder contests [29]. With sufficient competition in the market, the equilibrium market price decreases as the lowest bidding price drops, and the contractor that cannot efficiently reduce its costs will lose its competitive edge in the long term. Lo et al. [28] reported that contractors tend to compensate for the low bidding price by cutting corners [30] or bringing more disputes against the project owner [31]. Instead of suffering from the “Winner’s Curse” proposed by Capen et al. [32], contractors bid lower prices knowing that they are able to claim back the loss in bidding prices [31]. This arrangement results in contractors being able to remain competitive, secure more bid awards and achieve more dominant positions in the market by submitting tender prices lower than the actual estimated prices while generating more claims or cutting corners after the award is made [31]. Although fierce competition is an intrinsically appealing concept, it might lead to significant dispute burdens [31]. To enhance competitiveness, although risks should theoretically be priced into the bidding price, in practice, risks are mostly priced based on the contractual mechanism instead [33]. Claims can be well planned as early as in the tender stage by contractors lowering their bids with the shortfall to be recouped in ex post claims [1, 34]. Claims can thus be preplanned and allowed in the bids to legitimatize negative profit bidding [34]. However, such a strategy is believed to harm competition in the construction market, as only large contractors are able to afford the staff and expertise required to plan and estimate the claims to be made and the associated bidding prices [1]. Small contractors can be eliminated at the tender stage by repeated aggressive bidding offered by more resourceful competitors.

Applying game theory, Ho and Liu [35] argued that contractors tend to lower bidding prices when they expect profits from claims to be made during construction. From computer simulations, Lo et al. [28] found that when the market competition level reaches a certain point, the bidding price is inevitably lowered to even below the costs, and fierce competition is able to force contractors to make claims to restore the loss in profits. A number of studies have found that bidding strategies vary according to the conditions of the market. Carr and Sandahl [36] proposed that bidding strategies should be formed under the influence of job characteristics, the economic environment and competition conditions. Drew and Skitmore [37] proposed that factors at play include (1) the behaviours of contractors as a group (market conditions, the number of competitors, etc.), (2) the characteristics of individual contractors (contractor size, availability of staff, tenders in hand, etc.); and (3) the characteristics of contracts (size and type of project and client, project location, etc.). Flanagan [38] listed the factors affecting bidding behaviours, including (1) the size and value of the project, (2) the technical and managerial complexity of project completion; (3) regional market conditions; (4) the current and projected workload; (5) the type of client; and (6) the type of project. Drew et al. [26] found that in Hong Kong's construction industry, due to fierce competition, the bidding strategies of contractors are not clearly formed. Instead, contractors are encouraged to bid for various projects.

An experienced contractor will be able to identify mistakes in bills of quantities and exploit the same ex post. Rooke et al. [1] identified that a contractor will charge more for work items for which the quantities are expected to be increased and less for those with quantities that will be reduced during construction. This loaded pricing effort will embed potential profit without sacrificing bid competitiveness. Another way to do so is to make use of the possible delays the client may be responsible for and to maximize the cost of such delays. Contractors may deliberately bid for projects that are more likely to yield delays, claims and disputes. Rooke et al. [1] quoted a contractor estimator who indicated that "in construction, the only way to make a profit is when a contract goes wrong". Both proactive and reactive claims can be made during the course of construction. Rooke et al. [3] argued that subscription to market competition leads to an adaptive interdependent system and a system of personal relationships that achieves collective benefits that might be regarded as opportunist but that enables a tolerable level of performance. In summary, defining disputes in construction is not a straightforward task due to the wide range of possible causes involved. Proactive project management may help minimize their occurrence. For example, Cheung and Yiu [4] proposed several ways to address this issue. The authors suggested the use of prudent staffing policies, vigorous quality assurance, more realistic tender preparations and the separation of the roles of designs from contract administrations. Drew and Skitmore [37] regarded prequalification as a useful method with which clients can differentiate willing contractors. All of these methods cannot address the contracting behaviour issues arising from excessive market competition.

Competition in General

It has been commonly argued that the likelihood of anticompetition behaviours is closely related to the market concentration level [39–46]. Although the construction industry has always been characterized by a highly competitive market, empirical research supporting this proposition is rather rare [47].

Competition has been regarded as "a force that leads to an ideal solution of the economic performance problems" [48]. Construction contractors provide labour and skills that are highly similar and thus substitutable [47]. Contractors can choose to provide focused and specialized skills and expertise or a broad scope of services exploiting opportunities in various segments and accumulate reputations to rapidly enter and compete in new segments [49–51], especially given that in the construction contracting market, the fixed costs to enter one segment of the market are extremely low. Cheung and Shen [47] collected all procurement and tendering contract information of ten mega infrastructure projects in Hong Kong awarded to contractors as indicators of market shares. During the tendering stages of mega construction projects, centralized procurement was adopted in consideration of public accountability and cost efficiency. The adoption of centralized procurement raised concerns about an increase in contract size and in entry barriers for small- to medium-sized contractors. The two most commonly used methods to assess market concentration level are adopted. However, different results from using these two approaches are reported. The results for four-firm concentration ratios (CR4) show that the market is moderately concentrated, and the structure fits the oligopolistic market, while Herfindahl–Hirschman Index (HHI) results show that the market is fully competitive. However, as argued by Bikker and Haaf [52] and Rhoades [53], the inequality in market share distributions of contractors cannot be sufficiently reflected by HHI results. Significant inequality implies the power of larger firms to dominate the market price. Meanwhile, HHI results are too sensitive to the entry of small firms, while such new entrants hardly pose threats to existing large firms [54]. In combining the significant inequality of market share distributions and concentration index results from the CR4 and HHI, Cheung and Shen [47] argued that there exist two tiers of competition in the construction contracting market in Hong Kong. Ball [55] reported similar observations that a handful of large firms capture the majority of the project value, while a vast number of fringe firms compete for the rest. De Valence [56] argued that the existence of two levels of market is partially attributed to the additional barriers established through listings and prequalification. The contracting market in the construction industry thus includes two sectors. One sector includes a few large oligopolistic firms, and the other includes numerous small firms acting as price takers [47].

Cheung and Shen [47] found that the CR4 calculated in the mega project market to have reached 46, surpassing the threshold of 40, while the HHI value at 0.07167 was found to be much lower than the threshold of 0.15. The HHI value can also be interpreted using the concept of number equivalence, which is the inverse of the HHI value, as argued by Bishop [57] and Adelman [58]. The inverse-H identified in the study indicates that 14 firms of equal sizes exist in the same market. To examine the specificities of the market, the tests are repeated with the small sector distinguished by trades and employers. The results are consistent in that although the CR4 leads to the conclusion that the market is loosely concentrated, the HHI shows that the market is fully competitive. Listing and prequalifications are often believed to be prevent new entrants from accessing the market [59]. Bikker and Haaf [52] argued that the relative impacts of large or small firms on competition levels should determine the selection of concentration measures, especially as the HHI is vulnerable to market share inequality. Rhoades [53] suggested that the inequality of market shares implies the ability of leading firms to set market prices. The greatest criticism of the HHI is that it reacts too sensitively to the entry of any small firm, which usually does not have any impact on market competition or market structure [54]. In combining the results of the CR4 and HHI, Cheung and Shen [47] found it unlikely to classify the construction marketing market into the existing market structure types as a pure oligopoly market or fully competitive market. The divergence in the results indicates significant market share inequality [47]. Male [60] reported that although the entry barriers for lower-end firms are relatively low, merely requiring labour and a few pieces of equipment, entering the ranks of top-end companies is highly demanding, requiring previous work records and adequate financial support. In addition, it has been found that the size of incumbent firms has also become a barrier for smaller firms [61]. Specifically, two layers of barriers to entry are identified [56]. Cheung and Shen [47] extended the argument based on HHI and CR4 results by concluding that there exist two layers of entry barriers with the first layer being extremely weak, allowing numerous small firms to participate. Meanwhile, most small- to medium-sized contractors cannot surpass the second layer of entry barriers, leaving only a few larger firms qualified to compete for contracts of greater value. For such projects, competition is extremely limited [47].

Joint-Venture and Market Competition

Shen and Cheung [62] further reported a procurement strategy of forming joint ventures in the tendering and procurement stages in the construction contracting market. Previous studies report that forming joint ventures contributes to solving the anticompetitive effects of using centralized procurements and contract size expansions. Furthermore, forming joint ventures assists international contractors in easing their way into the local construction industry and accessing relevant knowledge and expertise [62] in addition to offering greater convenience to project owners as a form of single point responsibility [63]. Not only can the chances of winning bids be increased by combining the resources of joint venture partners, but also by reducing the capabilities of competitors, joint venture formations can contribute to success at the tendering stage, as joint ventures can be formed vertically with suppliers or subcontractors, which may prohibit these firms from providing services to competitors for the same project [64]. While forming joint ventures leads to reductions in the number of competitors [63, 65–67], some studies have reported that joint venture formation can lower entry barriers and allow new entrants into the market by rendering the participants more competitive. Such an advantage induces the formation of ad hoc joint ventures among contractors [64].

Many previous studies have found that by allowing joint ventures, bid rigging and price fixing are more likely to occur [64, 65, 67]. However, Rondeau et al. [63] found that prohibitions on joint venture formations increase the number of collusive bids due to a complex interaction between joint ventures and market competition. Infrastructure projects in the construction industry tend to be of large scale and great complexity, raising entry barriers to placing bids. Although one of the main motives for forming joint ventures is to prevent the entry of competitors into the market [68], joint ventures enable the entry of small- to medium-sized contractors through the accumulation of capital and the assembly of resources [42, 44, 69, 70]. The procompetition advantage of reductions in information costs has also been identified [71, 72].

Forming joint ventures is one of the preferred bidding strategies in the construction industry, as it also provides quicker and easier access to a new segment of the market [73]. Meanwhile, the value of reputations [74] and technology improvements can be shared among joint venture parents, while risks can be optimally distributed [73, 75]. However, forming joint ventures changes the competitive relationships and incentives among contractors. It has been found that joint venture partners or previous partners compete much less rigorously than otherwise [40, 42, 44, 76]. Especially where the joint venture parents are all fully capable contractors, the anticompetition effects become blatant [42]. Meanwhile, allowing joint ventures may raise entry barriers in terms of financial and technical qualifications [44, 77].

Concentration measures are used to capture the competition level of certain markets [48]. Commonly adopted methodologies in accessing market concentration levels include the four-firm concentration ratio (CR4) and Herfindahl–Hirschman index (HHI). The CR4 is commonly used by the Government Accountability Office (GAO) in the US, while the HHI is adopted by the Department of Justice (DOJ) and the Federal Trade Commission (FTC) in the US. The CR4 is defined as the sum of the market shares of the largest four firms in the market [78] and can be expressed as follows:

$$CR_4 = S_1 + S_2 + S_3 + S_4 \quad (1)$$

When the index equals 1, the market shares of the largest four firms make up the whole market, and when it approaches zero, the market is made up of numerous small fringe firms [52]. The HHI is the sum of the square market shares of all firms in the market [78], and the function can be expressed as follows.

$$HHI = \sum_{i=1}^{n} S_i^2 \tag{2}$$

where Si is the market share of the ith firm. The HHI value ranges from 0 to 1 and increases as the market concentration level rises. The HHI equals 1 when the market structure is a monopoly [48, 79].

The results obtained from conducting both tests of the CR4 and HHI are compared with the standards adopted by the DOJ, FTC (Table 1) and GAO (Table 2) in the studies done by Shen and Cheung [62] and Cheung and Shen [47]. Such thresholds are also widely adopted in scientific studies [80–82].

When the CR_4 is below 40% or the HHI value is lower than 0.15, the market is considered to be unconcentrated. When the CR_4 value is between 40 and 60% or the HHI value is between 0.15 and 0.25, the market is considered to be moderately concentrated or a loose oligopoly. When the CR_4 is larger than 60% or the HHI is larger than 0.25, the market is considered to be highly concentrated or a tight oligopoly (Tables 3, 4, 5, 6, 7, 8 and 9).

Shen and Cheung [62] conducted two tests. In the first test, joint ventures are regarded as new entities independent of the parent companies, while in the second test, the market shares of the joint venture firms are split into the parent firms. Furthermore, the contractors are divided into active and occasional contractors. Active contractors are defined as the top 10% of contractors based on the number of contracts obtained from Ten Mega Infrastructure Projects, while the remaining contractors are regarded as occasional. Shen and Cheung [62] found that the allowance of joint ventures lowers market concentration levels for active contractors, while it increases

Table 1 U.S. department of justice threshold

Market types	Thresholds	Remarks
Unconcentrated market	$HHI < 0.15$	Unlikely to have adverse competitive effects
Moderately concentrated market	$0.15 \leq HHI \leq 0.25$	Delta > 100 can raise significant competition concerns

Table 2 U.S. government accountability office threshold

Market types	Thresholds (%)
Unconcentrated market	$CR_4 < 40$
Loose oligopoly	$40\% \leq CR_4 \leq 60$
Tight oligopoly	$CR_4 > 60$

Table 3 Summary of ten mega projects

Project type	Project no	Project name	Project size (US Dollars)	Begin/Completion	Employer	Procurement methods
Transportation infrastructure	1	Project A	1.59 billion	2011–2015 (Expected)	Employer K	Selective
	2	Project B	10.3 billion	2012–2020 (Expected)	Employer K	Selective
	3	Project C	6.03 billion	2011–2018 (Expected)	Employer L	Selective (Listing)
Cross-border Infrastructure Projects	4	Project D	8.05 billion	2010–2017 (Expected)	Employer K	Selective
	5	Project E	2.74 billion	2009–2016 (Expected)	Employer L	Selective (Listing)
	6	Project F	Unknown	Not yet commence	–	–
	7	Project G	Unknown	Still in planning stage	Employers M & N	–
New urban development areas	8	Project H	Exceeding 6 billion	Still in planning	Employer O	–
	9	Project I	Exceeding 16.7 billion	2010–2021	Employers M & N	Selective (Listing)
	10	Project J	Unknown	Still in planning	Employers M & N	–

Table 4 Summary of sales revenues obtained by the firms

Contract value	Number of the firms
Above 10 billion	6
1–10 billion	17
0.1–1 billion	28
10 million—0.1 billion	17
1–10 million	2
Below 1 million	1

Table 5 Summary of firm size distribution

Rank	Firm name	Market share (%)	Rank	Firm name	Market share (%)
1	Firm A	12.29	25	Firm W	0.59
2	Firm B	12.16	26	Firm Z	0.58
3	Firm C	11.10	27	Firm A'	0.57
4	Firm D	10.71	28	Firm B'	0.56
5	Firm E	7.82	29	Firm C'	0.42
6	Firm F	7.56	30	Firm D'	0.38
7	Firm G	2.99	31	Firm E'	0.38
8	Firm H	2.89	32	Firm F'	0.38
9	Firm I	2.74	33	Firm G'	0.37
10	Firm J	2.40	34	Firm H'	0.36
11	Firm K	1.90	35	Firm I'	0.33
12	Firm L	1.83	36	Firm J'	0.31
13	Firm M	1.81	37	Firm K'	0.29
14	Firm N	1.79	38	Firm L'	0.26
15	Firm O	1.65	39	Firm M'	0.26
16	Firm P	1.59	40	Firm N'	0.22
17	Firm Q	1.54	41	Firm O'	0.22
18	Firm R	1.39	42	Firm P'	0.19
19	Firm S	1.21	43	Firm Q'	0.18
20	Firm T	1.17	44	Firm R'	0.17
21	Firm U	0.86	45	Firm S'	0.12
22	Firm V	0.81	46	Firm T'	0.11
23	Firm W	0.72	47	Firm U'	0.11
24	Firm X	0.65	48	Firm V'	0.10

Table 6 Results of concentration measures

	CR4 (%)	HHI
By company		
1. Overall	46	0.07167
2. Employer K	41	0.0730
3. Employer L, M & N	75	0.1747
Projects of employer K		
1. Overall	41	0.0730
2. Main Works	49	0.1026
3. E&M Works	55	0.125
Projects of employer L, M & N		
1. Overall	75	0.1747
2. Main Works	74	0.1787
By work trades—Main works		
1. Overall	51	0.08476
2. Employer K	49	0.1026
3. Employer L, M & N	74	0.1787
By work trades—E&M works		
1. Overall	55	0.103
2. Employer K	55	0.125

Table 7 Concentration level change for most active contractors

	Test I (7 firm)	Test II (7 firm)	Test I (9 firm)	Test II (9 firm)
CR_4 (%)	55.75	80.54	55.70	75.72
HHI	0.105689381	0.2004854312	0.1116355113	0.1616725451

Table 8 Concentration level change for 50 most occasional contractors

	Test I	Test II
CR_4 (%)	42.45	42.45
HHI	0.07406297	0.06844039

Table 9 Summary of comparisons

	Concentration Level	No. of firms	Average contract value	Technical Requirements
Active (Test II)	0.2004	7	1.6 billion	More demanding
Occasional (Test II)	0.684	50	451 million	Less demanding

the concentration for occasional contractors. From the concentration levels allowing joint ventures, it is found that for occasional contractors, forming joint ventures is a good way to penetrate the market and increase its competitiveness. Although allowing joint ventures increases the market concentration level for occasional contractors, the concentration level still remains extremely low since occasional contractors are mostly from fringe firms in the market with extremely small market shares. For active contractors, allowing joint ventures has limited impacts on the concentration or competition level since most joint ventures are made ad hoc for certain contracts instead of for long-term cooperative relationships. Shen and Cheung [62] argued that forming joint ventures in the construction contracting market is more often pursued as a temporary agreement to synergize resources for better capability and fewer risks than to actually introduce new entrants to the market. Meanwhile, sizing down and segmenting contracts to smaller ones to lower the entry barriers of smaller contractors are suggested to improve competition in the market [62].

Discussion

In the construction contracting market, the products and services offered are often homogeneous with similar cost functions. Akintoye and Skitmore [83] argued that the profitability of construction firms remains extremely low due to excessive market competition, but larger firms were found to be able to persistently enjoy a higher rate of return. It is suggested that unrestrained competition is likely to harm economic efficiency and innovation motives [83]. Ball et al. [84] found that in the UK, the construction market is made of a number of publicly listed middle-sized construction firms based on data ranging for 1990–1994 and rebut the argument that competition in the construction industry is excessive.

Only large firms can raise the financial resources needed to satisfy project financial requirements [84]. Clients can be easily tempted by a lower bidding price to trust projects with new entrants or firms that move resources to a new sector above the optimal level. The innovation cycle of the construction industry is extremely short, and therefore, instead of earning economic rents from initiating innovations at their own costs, most contractors just follow and adopt the newest techniques, materials or organization and management methods [84]. In the meantime, it has been found that although contracting firms value reputations as assets, the brand name has almost no effect on clients [84].

Cheung and Shen [47] found that the Hong Kong mega project contracting market can be separated into two sectors. One sector includes a small number of large contractors forming an oligopolistic market. The other sector includes numerous small firms that compete vigorously for contracts. Most of the reported studies on construction bidding describe competition associated with a winner's curse and an ex-post claim culture. The message seems to support the notion that excessive competition has led to cut-throat pricing and calculated claiming strategies. In this regard, firms with more resources are more capable of planning their bids with future claims anticipated. Therefore, high competition would need more rock bottom bids, and while some contracting organizations are able to execute postcontract claims, many less resourceful contractors may run into difficulties. Either way, claims and disputes become the only outlet as the outcome of planned action (oligopoly market of several large firms) or desperate attempts (free market of numerous small firms). Furthermore, due to the oligopolistic market structure, they might have greater incentives to engage in bid rigging instead of lowering bid values strategically. Whether forming joint ventures is allowed does not seem to directly affect dispute occurrence. However, if forming joint ventures changes market competitiveness, it might indirectly change dispute occurrence.

Summary

Construction contracting markets have been suffering from a culture of submitting suicidal bids and making unjustified postcontract claims. These conditions form a breeding ground for major disputes. Opportunistic behaviours such as raising unfounded claims have contributed to the occurrence of construction disputes. Previous studies have found that more competition may induce lower bidding prices. Experienced contractors are able to submit below lower-than-cost bids to secure projects with the aim of obtaining a dominant position in the market. Recovering losses through making claims and cutting corners postcontract are noted as obvious consequences. However, it is also found that only resourceful contractors are capable of planning and calculating the potential claims. Smaller contractors are unable to submit low bidding prices, as their ability to claim back such losses is quite limited. The unhealthy request is that lowering competition in the medium and long run. It is of interest to review construction market concentration in Hong Kong. This study adopted two concentration assessment tools: the CR and HHI. The CR4 results indicate that the market is moderately concentrated, while the HHI results suggest that the market is fully competitive. With a more detailed analysis, significant market share inequality is found among contractors and the mega project contracting market is found to be segmented by two layers of entry barriers. It is likely that larger contractors can try to make use of the oligopolistic market and to adopt certain anticompetitive behaviours. One unfortunate observation concerns the rise of sustentive disputes. Another common practice is the formation of joint ventures to bid mega projects. Would forming joint-ventures lower levels of competition?

With joint ventures allowed, occasional contractors are not only able to enter the second-tier market but can also access more resources and might be able to make use of the strategy of bidding lower than expected costs, introducing more disputes postcontract to compensate for the losses on contracts. In this case, the oligopolistic market structure of the second-tier market might be disrupted, and competition may intensify. On the one hand, the chances of anticompetitive conduct requiring an oligopoly, such as bid-rigging or price-fixing conduct, are lowered.

Acknowledgements The concentration ratios and HHI mentioned in this paper have been quoted from the papers "Concentration Analysis to measure competition in Megaprojects" in the ASCE Journal of Management in Engineering and "How forming Joint ventures may affect market concentration in construction industry" in the international Journal of Construction Management.

References

1. Rooke J, Seymour D, Fellows R (2004) Planning for claims: an ethnography of industry culture. Constr Manag Econ 22(6):655–662
2. Cannon J, Hillebrandt PM (1990) The modern construction firm. Palgrave Macmillan UK, London
3. Rooke J, Seymour D, Fellows R (2003) The claims culture: a taxonomy of attitudes in the industry. Constr Manag Econ 21(2):167–174
4. Cheung SO, Yiu TW (2006) Are construction disputes inevitable? IEEE Trans Eng Manage 53(3):456–470
5. Mururu N (1991) Anatomy of a construction dispute. Arbitration 5:262–264
6. Brown HJ, Marriot AL (1999) ADR: principles and practices, 2nd edn. Sweet & Maxwell, London
7. Spittler JR, Jentzen GH (1992) Dispute resolution: managing construction conflict with step negotiations. AACE Int Trans 1:D. 9.1
8. Tillett G, French B (1991) Resolving conflict: a practical approach. Sydney University Press, Sydney
9. Fenn P, Lowe D, Speck C (1997) Conflict and dispute in construction. Constr Manag Econ 15(6):513–518
10. Williamson OE (1979) Transaction-cost economics: the governance of contractual relations. J Law Econ 22(2):233–261
11. Yates DJ (1998) Conflict and dispute in development process: a transaction cost economic perspective. Working paper, Department of Real Estate and Construction, University of Hong Kong Retrieved from http://www.prres.net/proceedings/proceedings1998/Papers/Yates3Ai.PDF
12. Diekmann JE, Girard MJ, Abdul-Hadi N, Girard, Matthew J, Abdul-Hadi N (1994) University of Texas at Austin. Construction Industry Institute. (1994). In: DPI--disputes potential index : A study into the predictability of contract disputes. Construction Industry Institute, Austin
13. Rhys-Jones S (1996) How constructive is construction law? Constr Law J 10(1):28–38
14. Hewitt R (1986) Winning contract disputes: strategic planning for major litigation. Ernst & Young, London
15. Heath B, Hills B, Berry M (1994) The origin of conflict within the construction process. CIB Publication, vol. 171, Presented at First Plenary Meeting of TG 15, The Netherlands, pp 35–35
16. Conlin J, Langford D, Kennedy P (1996) The relationship between construction procurement strategies and construction contract disputes. Managing Constr Project Managing Risk 2:360–371
17. Kumaraswamy M (1997) Common categories and causes of construction claims. Constr Law J
18. Semple C, Hartman FT, Jergeas G (1994) Construction claims and disputes: causes and cost/time overruns. J Constr Eng Manag 120(4):785–795
19. Sheridan P (2003) Claims and disputes in construction. Constr Law J 12(1):3–13
20. Brooker P (2002) Construction lawyers' attitude and experience with ADR. Constr Law J 18(2):97–116
21. Yiu, K, Cheung, S (2004) Significant dispute sources of construction mediation. In: Paper presented at the proceedings of the 1st international conference on the world of construction project management, 27th–28th May, Toronto, Canada (CD-ROM Proceedings)

22. Mitropoulos P, Howell G (2001) Model for understanding, preventing, and resolving project disputes. J Constr Eng Manag 127(3):223–231
23. Grogan T (1992) Low bids raise hidden costs. Eng News-Rec 228(13):30–31
24. Ioannou PG, Leu S-S (1993) Average-bid method—competitive bidding strategy. J Constr Eng Manag 119(1):131–147
25. Carr RI (1982) General bidding model. J Constr Div 108(4):639–650
26. Drew D, Skitmore M, Lo HP (2001) The effect of client and type and size of construction work on a contractor's bidding strategy. Build Environ 36(3):393–406
27. Carr RI (1983) Impact of number of bidders on competition. J Constr Eng Manag 109(1):61–73
28. Lo W, Lin C, Yan M (2007) Contractor's opportunistic bidding behavior and equilibrium price level in the construction market. J Constr Eng Manag 133(6):409–416
29. De S, Fedenia M, Triantis AJ (1996) Effects of competition on bidder returns. J Corp Finan 2(3):261–282
30. Winch GM (2000) Institutional reform in British construction: partnering and private finance. Build Res Inf 28(2):141–155
31. Crowley LG, Hancher DE (1995) Evaluation of competitive bids. J Constr Eng Manag 121(2):238–245
32. Capen EC, Clapp RV, Campbell WM (1971) Competitive bidding in high-risk situations. J Petrol Technol 23(06):641–653
33. Laryea S, Hughes W (2011) Risk and price in the bidding process of contractors. J Constr Eng Manag 137(4):248–258
34. Mohamed KA, Khoury SS, Hafez SM (2011) Contractor's decision for bid profit reduction within opportunistic bidding behavior of claims recovery. Int J Project Manage 29(1):93–107
35. Ho SP, Liu LY (2004) Analytical model for analyzing construction claims and opportunistic bidding. J Constr Eng Manag 130(1):94–104
36. Carr RI, Sandahl JW (1978) Bidding strategy using multiple regression. J Constr Div 104(1):15–26
37. Drew D, Skitmore R (1993) Prequalification and C-competitiveness. Omega 21(3):363–375
38. Flanagan R (1982) Making good use of low bids. Chartered Quant Surveyor 14:226–227
39. Berg SV, Friedman P (1981) Impacts of domestic joint ventures on industrial rates of return: a pooled cross-section analysis, 1964–1975. Rev Econ Stat 63(2):293–298
40. Bresnahan TF, Salop SC (1986) Quantifying the competitive effects of production joint ventures. Int J Ind Organ 4(2):155–175
41. Green EJ, Porter RH (1984. Noncooperative collusion under imperfect price information. Econometrica J Econ Soc 52(1):87–100
42. Mead WJ (1967) The competitive significance of joint ventures. Antitrust Bull 12:819
43. Orr D, MacAvoy PW (1965) Price strategies to promote cartel stability. Economica 32(126):186–197
44. Pfeffer J, Nowak P (1976) Patterns of joint venture activity: implications for antitrust policy. Antitrust Bull 21:315
45. Stigler GJ (1964) A theory of oligopoly. J Polit Econ 72(1):44–61
46. Tong TW, Reuer JJ (2010) Competitive consequences of interfirm collaboration: how joint ventures shape industry profitability. J Int Bus Stud 41(6):1056–1073
47. Cheung SO, Shen L (2017) Concentration analysis to measure competition in megaprojects. J Manag Eng 33(1):04016023
48. Scherer FM, Ross D (1990) Industrial market structure and economic performance. Houghton Mifflin Company, Boston
49. Kale S, Arditi D (2002) Competitive positioning in United States construction industry. J Constr Eng Manag 128(3):238–247
50. Porter M (1980) Competitive strategy: Techniques for analyzing industries and competitors. Free Press, New York
51. Porter M (1985) Competitive advantage : Creating and sustaining superior performance. Free Press; Collier Macmillan, New York

52. Bikker JA, Haaf K (2002) Measures of competition and concentration in the banking industry: a review of the literature. Econ Fin Model 9(2):53–98
53. Rhoades SA (1995) Market share inequality, the HHI, and other measures of the firm-composition of a market. Rev Ind Organ 10(6):657–674
54. Hart PE (1975) Moment distributions in economics: an exposition. J R Statistical Soc Ser (Gen) 138(3):423–434
55. Ball M (1988) Rebuilding construction: economic change in the British. Routledge, London
56. De Valence G (2003) Market structure, barriers to entry and competition in construction markets. In: Paper presented at the international council for research and innovation in building and construction (CIB) Meetings, Workshops, Symposia, Conferences.
57. Bishop RL (1952) Elasticities, cross-elasticities, and market relationships. Am Econ Rev 42(5):780–803
58. Adelman MA (1969) Comment on the" H" concentration measure as a numbers-equivalent. Rev Econ Stat 51(1):99–101
59. McMillan J (1991) Dango: JAPAN'S price-fixing conspiracies. Econ Politics 3(3):201–218
60. Male S (1991) Strategic management in construction: conceptual foundations. In: Male SP, Stocks R (eds) Competitive advantage in construction. Butterworth-Heinemann, Oxford, pp 1–4
61. Mulhearn C, Vane HR (2001) Economics for business. Red Globe Press, London
62. Shen L, Cheung SO (2018) How forming joint ventures may affect market concentration in construction industry? Int J Constr Manag 18(2):151–162
63. Rondeau D, Courty P, Doyon M (2016) Simultaneous allocation of bundled goods through auctions: assessing the case for joint bidding. Am J Agr Econ 98(3):838–859
64. Branzoli N, Decarolis F (2015) Entry and subcontracting in public procurement auctions. Manage Sci 61(12):2945–2962
65. Albano GL, Spagnolo G, Zanza M (2009) Regulating joint bidding in public procurement. J Compet Law Econ 5(2):335–360
66. Calel R (2010) Auctioning conservation contracts in the presence of externalities, (Grantham Research Institute on Climate Change and the Environment Working Papers No.22). Retrieved from: http://eprints.lse.ac.uk/37395/1/Auctioning_conservation_contracts_in_the_presence_of_externalities(author).pdf
67. Tóth B, Fazekas M, Czibik Á, Tóth IJ (2014) Toolkit for detecting collusive bidding in public procurement, With examples from Hungary (Corruption Research Center Budapest Working Paper Series: CRC-WP/2014:02). Retrieved from: http://www.crcb.eu/wp-content/uploads/2015/04/Toth-et-al_CRCB_WP_v2_150413.pdf
68. Ustaömer TC, Durmaz V, Lei Z (2015) The effect of joint ventures on airline competition: the case of American airlines, British airways and Iberia joint business. Procedia, Soc Behav Sci 210:430–439
69. Kitch EW (1985) The antitrust economics of joint ventures. Antitrust LJ 54:957
70. Pate JL (1969) Joint venture activity, 1960–1968. Econ Rev 54:16–23
71. Shapiro C, Willig RD (1990) On the antitrust treatment of production joint ventures. J Econ Perspect 4(3):113–130
72. Werden GJ (1997) Antitrust analysis of joint ventures-an overview. Antitrust LJ 66:701
73. Norwood SR, Mansfield NR (1999) Joint venture issues concerning European and Asian construction markets of the 1990's. Int J Project Manage 17(2):89–93
74. Oliver C (1990) Determinants of interorganizational relationships: Integration and future directions. Acad Manag Rev 15(2):241–265
75. Ohmae K (1989) Managing in a borderless world. Harv Bus Rev 67(3):152–161
76. Reynolds RJ, Snapp BR (1986) The competitive effects of partial equity interests and joint ventures. Int J Ind Organ 4(2):141–153
77. Pitofsky R (1969) Joint Ventures under the Antitrust Laws: Some Reflections on the Significance of" Penn-Olin" . Harv Law Rev 82(5):1007–1063
78. Perloff JM, Karp LS, Golan A (2007) Estimating market power and strategies. Cambridge University Press, Cambridge

79. Hirschman A (1964) The paternity of an index. Am Econ Rev 54(5):761
80. Kwoka JE Jr (1985) The Herfindahl index in theory and practice. Antitrust Bull 30:915
81. Oinonen M (2010) Does EU merger control discriminate against small market companies?: diagnosing the argument with conclusions (International Competition Law Series). Kluwer Law International BV, The Netherlands
82. Shepherd WG (1982) Causes of increased competition in the US economy, 1939–1980. Rev Econ Stat 64(4):613–626
83. Akintoye A, Skitmore M (1991) Profitability of UK construction contractors. Constr Manag Econ 9(4):311–325
84. Ball M, Farshchi M, Grilli M (2000) Competition and the persistence of profits in the UK construction industry. Constr Manag Econ 18(7):733–745

Recognising the Importance of Interdependence

Liuying Zhu and Sai On Cheung

Introduction

Although project team members are used to identify the participants involved in construction projects, their relationships are often far from those expected from a team. Instead it is quite common to find them being defensive and noncooperative. When dispute arises, their working attitude would turn to confrontational [1]. Many industry reviews have recommended the use of cooperative working. This seems quite legitimate and in fact common sense because construction works are high interdependent. Latham [2] recommended the inclusion of 'duties of fairness, mutual co-operation, teamwork and shared financial motivation' in construction contracts. Partnering and alliancing have then been used extensively to provide the framework of cooperative contracting [3]. In Hong Kong, the first-ever Construction Industry Review was conducted in 2000, similarly partnering contracting was recommended [4]. This recommendation is based on the belief that only joint effort would achieve the best project performance. Thus, it is an undisputed fact that confrontation is counterproductive to project output. Moreover, successful partnering can only be possible with a major cultural change. As such, many partnering studies have focused on: (i) the development of conceptual frameworks [5, 6], (ii) the identification of critical success factors [7, 8], (iii) implementation issues [9, 10] and (iv) case studies [11, 12]. This study points to dig into the importance of interdependence in facilitating cooperative work. Ways to enhance interdependence are also discussed.

L. Zhu (✉) · S. O. Cheung
Construction Dispute Resolution Research Unit, City University of Hong Kong, Hong Kong, China
e-mail: liuyinzhu3-c@my.cityu.edu.hk

S. O. Cheung
e-mail: Saion.cheung@cityu.edu.hk

S. O. Cheung (ed.), *Construction Dispute Research Expanded*, Springer Tracts in Civil Engineering, https://doi.org/10.1007/978-3-030-80256-1_12

To achieve this aim, the study involves the following:

(1) Identifying interdependence and its drivers;
(2) Investigating the roles of interdependence in promoting cooperative work;
(3) Testing the relationship developed in (3); and
(4) Putting forward practical implications for interdependence enhancement.

Interdependence and Its Drivers in Cooperative Work

Williamson [13] advocated that most transaction can be characterised by three dimensions: uncertainty, asset specificity and frequency. Asset specificity describes a condition where investments in a transaction are specific and its value would be greatly diminished if the transaction was terminated. The extent of investment thus would be highly dependent of the transaction governance [14]. When asset-specific investments are made by a supplier, a 'lock-in' situation between the supplier and the buyer is likely to happen [14]. The devaluation of asset-specific investment is determined by the ability of the buyer to turn to other suppliers. Both parties thus are interdependent. High asset-specific investments can occur in construction contracting [15]. This situation often happens ex post, because the contracting parties are highly interdependent at the construction stage of the project [16]. For example, terminating construction contracts or switching contractors in the middle of the construction period is detrimental to both the employer and the contractor [17]. The contractor's asset-specific investments made have to be written off. The additional cost of seeking a completion contractor can be substantial and irrecoverable. The employer and contractor are therefore highly interdependent, particularly when the cost associated with ending the relationship is substantial for both of them [18].

Factors Leading to Interdependence

In this section, factors contributing to the establishment of interdependence are explored. It is advocated that the degree of interdependence among contracting parties can be enhanced when the following are in place. With due knowledge of the contributing factors, management tools can be devised to instil and foster interdependence between the contracting parties.

Aligned Objectives and Trust

Incentives are well known for its value in engendering extra efforts of contracting parties [10]. The effectiveness of incentives, however, depends on the parties as well as the attainability and value of the reward [19, 20]. Bresnen and Marshall [10] also pointed out that the evaluation of attainability may well only be perceptive.

Assuming the reward is commensurate to the required effort, the incentive would then have the function of aligning the goals of the parties in light of the project objectives. Lui et al. [21] provided valuable empirical support that demonstrates that a partnering relationship is a prerequisite condition to reaching common goals. Goal alignment would then have its maximum effect on construction project teams due to interdependence. The objectives of the individual members may not always be compatible [22]. In reality, sometimes these objectives can even be opposing as in many principal-agent relationships. Incentivisation can serve as the vehicle from which to set up a common objective that can only be met when the goals of the members are aligned [23]. Mayer et al. [24] further added that it is trusting relationship derived from interdependence that drives collaboration.

Relationship-Specific Investments and Previous Work Experience

Interdependence and relationship-specific investments go hand in hand and often have reciprocating effects [16]. When the parties are highly interdependent, there are little opportunities to replace an existing partner due to the excessive transaction cost. The loss of sunk investments is too great to separate with the current working partner. In terms of work experience, social exchange theory suggests that it is quite natural for organisations having previous fond relationship are most likely work together again [25]. Mutually dependent parties therefore would value their relationship more. Under Emerson's power-dependence theory (1962), the extent of mutual dependence is determined by the level of investment made in the relationship and the potential headache in changing the current working partner. Investing in a relationship represents long-term commitment to the partnership [26]. Investment in relationships recognises their interdependence. A positive experience of previous dealings will promote a desire to work together further, the more frequent dealings are, the higher the level of interdependence becomes [27]. Axelrod [28] described this process as a 'shadow of the future' that promotes collaborative efforts in anticipation of long-term relationships. Relationship-specific investments would suppress the desire for short-term returns and the practice of opportunism. Axelrod [28] further added that adopting cooperative behaviour in fact is an invitation to establish long-term relationship. When dealings are anticipated, contracting parties would take steps to take care of the interests of their parties [29].

Operationalizing Interdependence in Construction Contracting

In Section "Interdependence and Its Drivers in Cooperative Work" the casual linkage between interdependence and cooperation is articulated. This section discusses the realisation of interdependence in construction contracting.

Rationalism in Construction Contracts

One of the characterising features of construction contracting is the exposure to uncertainty. The resolution of the complex problems as raised must be addressed by project participants jointly [30]. Classical contract theory assumes the use of complete contracts to detail the contingent arrangements [23, 31]. This perspective approach is not feasible as adjustments during construction are almost inevitable and pre-planned actions may not be suitable. Instead, it would be more effective if adjustments can be figured out by the contracting parties cooperatively and collaboratively [23, 32]. In this regard, relational contracts have been used to promote cooperation between project team members [33, 34, 35]. A relational contract is a 'complex' or 'intertwined' contract [36] that gives the recognition that complex asset-specific investments that are difficult to list at the contact planning stage. Relational contracts allow for adjustments of the obligations of contracting parties in response to the changes as the project unfolds [37]. The adoption of cooperative attitudes to preserve contractual relationships characterises contractual relationships [36]. Relational contracts are therefore particularly useful for economic exchanges of intermediate and long durations [33].

Definition of Cooperation

The definitions of cooperation listed in Table 1 mainly focus on the relationships among team members at both the individual and organizational levels [30]. Setting mutual goals and benefits have been mentioned in most of the definitions. Cooperative behaviour can be obligatory or voluntary [38, 20]. Contractual obligations, such as job descriptions, sanctioning mechanisms and reward systems, are extrinsic motivational factors oriented towards cooperation. Hence, obligatory cooperation is the outcome of the contractual rewards and punishments. Effective contractual management thus can bring about obligatory cooperation. Heide and Miner [27] suggested that flexibility, information exchange, shared problem solving and restraint in the use of power are useful indicators of cooperation. Having mutual goals that require achievement through joint effort provides the necessary platform to establish interdependence. Moreover, interorganizational cooperation in construction is typically affected through the practice of partnering [39, 40]. There is no contractual obligation, instead, project participants work together to achieve common goals [3]. Mutual benefits harvested from accomplishment of common goals are the result pf cooperation. According to Deakin et al. [29], cooperation can exist in most contractual relationships because contracting parties have to work together to complete the transaction. Thus, contracts finalise common goals [41, 42].

Table 1 lists the definitions of inter-organisational cooperation summarized by Anvuur and Kumaraswamy [30].

Table 1 Definitions of inter-organizational cooperation (adopted and developed from Anvuur and Kumaraswamy [30])

No	Definitions	References
1	The presence of deliberate relations between otherwise autonomous organizations for the joint accomplishment of individual operating goals	[43]
2	Socially contrived mechanisms for collective action, which are continually shaped and restructured by actions and symbolic interpretations of the parties involved	[44]
3	The process by which individuals, groups and organizations come together to interact and form psychological relationships for mutual gain or benefit	[38]
4	Whether people act to promote the goals of the group	[20]
5	Seamless communication between parties as they work jointly towards a common goal	[3]
6	Behaviour that promotes the goals of the work group to which one belongs	[30]

Nonetheless, in construction, asset-specific investments make cooperative bonding valuable [45, 13]. Geyskens et al. [46] found that this feature should be the extra dimension that solidifies cooperative relationship. Katz [47] raised concern over the 'lock-in' condition without genuine cooperation. To this end, incorporating a state of interdependence would provide a secondary safeguard [48]. Interdependence arises when contracting parties realize that one's performance is dependent on the effort of the counterpart [49]. Thus, interdependent parties will be willing to cooperate [27, 50]. Interdependence between contracting parties could lead to inter-firm cooperation [43]. Cooperative work among construction organizations would minimize transaction costs [30]. Interdependent parties are then found to be more willing to refrain from opportunism [29, 51, 52].

Research Framework

Based on the theoretical deliberation reported in the Sections "Interdependence and Its Drivers in Cooperative Work" and "Operationalizing Interdependence in Construction Contracting", A conceptual interdependence-cooperation relationship framework (the Framework hereafter) is proposed (Fig. 1).

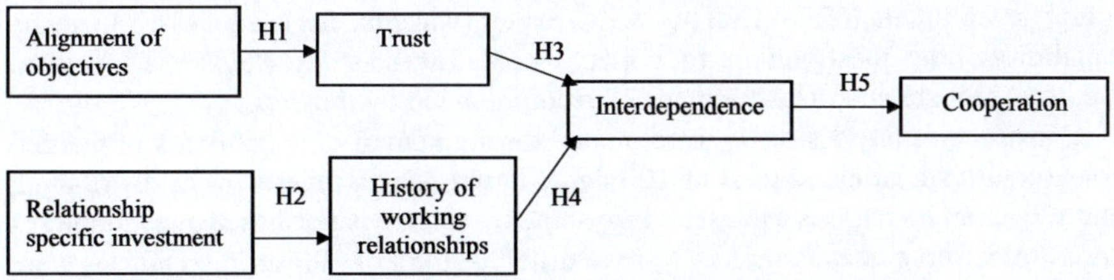

Fig. 1 Conceptual interdependence-cooperation relationship framework

The framework is supported by the following hypotheses:

H1: The alignment of objectives can help enhance the level of trust between contracting parties.

H2: The increase in relationship-specific investment promotes the history of working relationships between contracting parties.

H3: Trust developed between contracting parties is positively correlated to the interdependence of these two parties.

H4: The history of working relationships between contracting parties strengthen the extent of interdependence between them.

H5: Interdependence between contracting parties is positively related to their cooperative behaviour.

An empirical study was carried out to test the framework and the hypotheses. Research design is first introduced.

Testing of the Framework

Measurement statements used for data collection were developed based on the literature highlighted in the previous review (Table 2). Table 2 provides summary of the variables and measurement statements.

Structural equation modelling (SEM) was used to test the hypothesized relationships [60]. PLS-SEM was used in this study [60]. SPSS and SmartPLS 3 [61] were used to evaluate the measurement and structural models.

Data Collection

The data collection questionnaire has four parts. The first part covers particulars of the respondents, including information on company types, job positions and work experience. The third and fourth parts include questions developed from the Framework. Respondents were asked to indicate the accuracy of the measurement statements in describing happenings during the project. All measurement items were developed from Table 2 and measured on a 7-point Likert scale except for W1. W1 was measured using a categorical scale by year. A 7-point Likert scale is frequently used to assess recognition of a description [62] and allows for a wider but not excessive range for rubric development [62]. A trial run with 5 experts was first carried out to evaluate the suitability of the questionnaire for collecting data for the study. Revisions to several measurement statements were made as recommended by the experts.

Construction professionals were invited through email or at seminars of learned construction societies. A total of 100 paper-based questionnaires were distributed, and 130 email invitations were sent. Fifty-nine completed paper-based questionnaires and 81 electronic questionnaires were returned. A total of 100 valid responses were

Table 2 Summary of variables and measurement statements

Item	Descriptions	References
History of working relationship		
W1	Duration of previous working relationships	[53]
W2	Satisfaction with previous dealings	
W3	Overall project outcome of previous dealings	
Alignment of objectives		
A1	Proactively developing mutual incentives at the beginning of project	[54, 55]
A2	Proactively working towards mutual objectives during the project	
A3	Proactively achieving mutual beneficial outcomes at the end of project	
Trust		
T1	Trusting information provided by other project participants	[56, 57]
T2	Considering the interests of other project participants when making decisions	
T3	Keeping promises	
T4	Completing tasks competently	
T5	Receiving rewards commensurate with input efforts	
T6	Having a strong reputation in the industry	
Relationship-specific investment		
R1	Spending more time working from a common office	[58, 59]
R2	Making good use of resources and support provided by management staff	
R3	Spending more time building relationships with others	
Interdependence		
IN1	Time and financial costs would be incurred if delays occurred during construction	[56]
IN2	Time and financial costs would be incurred if project participants were to be replaced	
IN3	Project information and data would be lost if project participants were to be replaced	
Cooperative behaviour		

(continued)

Table 2 (continued)

Item	Descriptions	References
CB1	Modifying agreements with the occurrence of unexpected events	[27]
CB2	Flexibly responding to requests for changes	
CB3	Frequently exchanging information	
CB4	Continually informing other project participants of any changes	
CB5	Treating problems as joint rather than individual responsibilities	
CB6	Completing the project in joint manner	

finally obtained. The valid response rate of 42% is closed to the median rate (35.7%) of a survey conducted in the USA of 1607 organizational academic studies [63]. For factor analysis, Hatcher [64] recommends that the number of should at least be 5 times the number of variables or more than 100. The 100 responses obtained are therefore suitable for factor analysis. Table 3 provides respondents particulars.

Table 4 provides data description. The means of all variables (except W1) are all above the median (4) of the 7-point scale. All the standard deviation are close to 1, meaning that the degree of dispersion for the variables is similar, suggesting that the respondents generally agree with the measurement statements.

Table 3 Respondents particulars

		Num	%
Position	Engineer	30	30
	Quantity surveyor	43	43
	Project manager, architect, contract administrator, site supervisor, etc.	26	26
Organisation role	Developer	18	18
	Contractor	23	23
	Consultant	42	42
	Project manager	17	17
Project location	Hong Kong SAR	57	57
	Mainland China	26	26
	Singapore	10	10
	Australia, the US, the UK and Nigeria	7	7
Working experience (years)	1–4	39	39
	5–10	31	31
	≥10	30	30

Table 4 Data description

Item	Measuring statements	Mean	Std.
W1	Duration of previous dealings	2.12	1.31
W2	Degree of satisfaction of previous dealings	4.77	1.09
W3	Overall project outcome of previous dealings	4.81	1.05
A1	Proactively develop mutual incentives at the beginning of project	4.86	1.31
A2	Proactively work towards mutual objectives during the project	4.93	1.18
A3	Proactively achieve mutual beneficial outcomes at the end of project	4.91	1.04
R1	Spend more time in working in common office	4.50	1.24
R2	Make good use of resources and support provided by management staff	4.86	1.15
R3	Spend more time to build relationships with one another	4.70	1.22
T1	Believe in information provided by other project participants	4.64	1.10
T2	Consider interest of other project participants in making decision	4.76	1.09
T3	Keep their promises	4.70	1.11
T4	Fulfil their jobs competently	4.63	0.99
T5	Receive rewards that match their input efforts	4.70	1.05
T6	Have high reputation in the industry	4.74	1.11
CB1	Modify their agreements in occurrence of unexpected events	4.61	1.12
CB2	Respond to request of changes flexibly	4.70	1.09
CB3	Exchange information frequently	4.83	1.18
CB4	Keep informing other project participants of any changes	4.81	1.06
CB5	Treat problems as joint responsibilities, rather than individual responsibilities	4.58	1.31
CB6	Complete the project in joint manner	4.94	1.13
IN1	Time and money cost would be incurred if stoppage happens during construction	5.11	1.29
IN2	Time and money cost would be incurred if project participants were to be replaced	4.85	1.23
IN3	Project Information and data would be lost if project participants were to be replaced	4.91	1.23

IN1 has the highest mean among the measurement items (5.11), which can be explained by delays at construction sites during projects, commonly with severe consequences such as delays to completion and losses of potential work and profit. IN1 also presents a relatively high standard deviation. The respondents might have different views regarding the consequences of construction delays. Two other measurement items for interdependence, IN2 and IN3, also present relatively high mean scores, which suggests that the respondents generally perceive high costs to be involved in the replacement of project participants. The items also have the same standard deviation.

The alignment of objectives was measured by A1, A2 and A3 and all were highly rated. This result suggests that using incentive schemes to motivate contracting parties to work towards mutual objectives throughout a project period is plausible. The standard deviation of A1 is relatively high, suggesting that aligning objectives could be achieved at the early stage of construction.

W1 was measured by a categorized scale (1 = no previous dealings, 2 = 1–2 years, 3 = more than 2 years). The mean score of 2.13 (standard deviation = 1.31) suggests that on average, the respondents had approximately 1–2 years of experience working with other project participants. R1 and CB5 have mean scores of less than 4.60. R1 (mean = 4.50), which denotes more time spent working from a common office and are marked the lowest among the items. This ranking may reflect the nature of construction work, which is completed in stages involving different work trades. The relatively high standard deviation for CB5 (mean = 4.58, standard deviation = 1.31) suggests that the respondents may have quite differing views of how problems are perceived: as joint or individual.

Data Analysis

The proposed Framework is arranged as a structural model to be tested by structural equation modelling (SEM). SEM is a statistical multivariate technique commonly used to evaluate relationships among multiple variables [60]. SEM is a form of confirmatory factor analysis (CFA) as the structural model must be developed from strong theoretical bases. Partial least squares (PLS) SEM was used mainly due to our sample size of 100 [65, 66]. Normal distribution of the dataset is not a necessary condition for the application of PLS-SEM.

Preliminary analysis of data

As all measurements were collected from a single questionnaire, a common method variance (CMV) problem may exist, which could affect the hypothesised relations in the structural model. Two mitigating arrangements were adopted. First, the measurement items were randomized rather than sequenced without aligning with the respective constructs. This would minimize biased responses due to the positioning of the questions [67]. Second, Hartman's single-factor test was conducted to detect CMV problems in the dataset. The factor analysis results indicate that the first factor explains only 38.34% of the variance. If a significant factor is found to explain more than half of the variance in the factor analysis [68], CMV may exist. These results suggest that CMV problems are unlikely in this case.

Pearson's correlation test was then conducted to detect collinearity issues between the measurement items. Collinearity arises when two indicators are highly correlated [69]. Collinearity can affect the estimation of weights and statistical significance during PLS-SEM analysis [69]. Hence, redundant indicators should be considered for removal. Two pairs of indicators, W2 and W3 and CB5 and CB6, were found to have Pearson's coefficients of higher than 0.8. IN3 has low or negative correlations with most of the indicators. After a thorough consideration of the possible impacts of removal on the meaning of the constructs, W3, CB5 and IN3 were removed from the list of indicators.

The *Kruskal–Wallis H test* (K-W test) was performed to investigate whether different groups of respondents exhibited significant differences in their perceptions of cooperative behaviour. The *K-W test* was chosen over *the independent-samples t test and one-way ANOVA test,* as it is a nonparametric test that is not based on a normal distribution of the data. When $p < 0.05$, the null hypothesis H_0 of individual groups with no difference in perceptions of an indicator is to be rejected. The results show that different perceptions were found for W1 and R3:

W1: Duration of previous dealings.

Respondents with at least 10 years of experience reported a longer duration of previous dealings than groups with 1–4 years and 5–9 years of work experience. It is understandable that the likelihood of repeated dealings with the same contracting parties increases with more work experience.

R3: Spending more time building relationships with others.

The Hong Kong and Chinese samples show differences in perceptions of this statement. Maintaining strong 'guanxi' or relationships with project clients, especially government organizations, is important for the success of construction projects in China [70]. Confucians think that the reciprocity of interactions can derive sustained benefits through ongoing cooperation [53]. Moreover, Hong Kong has a long history of adopting Western practices. Such perceptive differences can be explained by differing cultural orientations.

Tables 5 and 6 present the results of the K-S tests.

Individual indicator reliability is evaluated by the indicator loadings. Hair et al. [62] used cross loading to assess discriminant validity. A threshold value of 0.70 is

Table 5 K-S test results for W1

Working experience	Years	Mean	SD
	1–4	2.12	1.09
	5–9	3.40	1.19
	≥10	3.20	1.20
K-S test	Sig	0.006	

Table 6 K-S test results for R3

Project location	Years	Mean	SD
	Hong Kong	5	1.15
	Mainland China	5.18	1.18
K-S test	Sig	0.0013	

applied. In addition, indicators with loadings of less than 0.40 are excluded. Accordingly, T5 was removed. Indicators A3, CB1, CB3, W3, T2 and T6 generated loadings of between 0.60 and 0.70. These indicators were individually evaluated and were excluded if deletion would have increased composite reliability and not affected the meaning of the construct. T2, T6 and CB3 were eventually excluded from further data analysis. Table 7 provides adjusted cross loadings of each indicator and the constructs.

Table 8 presents the results of Cronbach's alpha (α) and AVE tests.

The composite reliability of all constructs meets the threshold of 0.70 [71] and is the appropriate measure of internal consistency when using PLS-SEM [72]. A threshold of 0.7 is commonly used for Cronbach's α [71]. The Cronbach's α values for trust and interdependence range from 0.60 to 0.70. Chin [73] claims that a Cronbach's α value of 0.60 is acceptable. Accordingly, internal consistency reliability is therefore confirmed by these two indexes.

The convergent validity of the constructs was measured by the average variance extracted (AVE). The AVEs of the constructs are higher than the threshold value of 0.50 [74]. Thus, all constructs have adequate convergent validity.

Discriminant validity tests whether a construct is truly distinct from others [75]. Using cross loadings of indicators or applying Fornell and Larcker's criterion (1981) can establish discriminant validity. As summarized in Table 8, individual indicators have higher loadings (bolded values) than all cross loadings of other indicators, which demonstrates discriminant validity among the constructs [69]. Fornell and Larcker's criterion [74] is a more conservative criterion for discriminant validity. To satisfy the criterion, the square root of the AVE value of a construct has to be greater than its highest correlation with all other constructs. Comparisons between the square root of AVE values (bolded values) and Pearson's coefficient are shown in Table 8. No correlation coefficient of constructs was found to be greater than the square root of AVE values. All of the constructs are therefore different from each other.

Overall, the measurement data achieve satisfactory indicator reliability, internal consistency reliability, convergent validity and discriminant validity. Common method variance is thus unlikely to be at play. The measurement data are therefore sufficiently robust for testing the interrelationships between constructs.

PLS-SEM Analysis

SmartPLS 3 [61] was applied to evaluate the measurements and the structural model. The evaluation and reporting of results followed the recent guidelines developed by Henseler et al. [76] and Hair et al. [69]. Standardized path coefficients were obtained after running the PLS algorithm. The standardized path coefficient has a value of between −1 and +1. A higher path coefficient indicates more positive relationships between two constructs and vice versa.

The study adopted a 10% significance level in assessing the path coefficients. 10% is usually used in exploratory studies [69]. The significance of the path coefficient

Table 7 Cross loadings of the items

Item	A1	A2	A3	CB1	CB2	CB4	CB6	IN1	IN2	R1	R2	R3	T1	T3	T4	W1	W2
Alignment of objectives	**0.8**	**0.88**	**0.76**	0.44	0.6	0.5	0.67	0.3	0.23	0.54	0.72	0.67	0.62	0.56	0.44	0.3	0.41
Cooperative behaviour	0.54	0.6	0.57	**0.74**	**0.84**	**0.77**	**0.82**	0.32	0.27	0.47	0.54	0.64	0.53	0.62	0.41	0.15	0.27
Interdependence	0.25	0.28	0.26	0.25	0.32	0.24	0.27	**0.9**	**0.8**	0.21	0.19	0.24	0.36	0.14	0.24	0.16	0.23
Relationship-specific investments	0.65	0.74	0.56	0.42	0.63	0.44	0.65	0.25	0.18	**0.76**	**0.84**	**0.82**	0.58	0.57	0.45	0.3	0.33
Trust	0.51	0.73	0.44	0.41	0.59	0.6	0.55	0.36	0.17	0.49	0.59	0.6	**0.81**	**0.78**	**0.71**	0.24	0.19
History of working relationships	0.51	0.32	0.34	0.12	0.25	0.16	0.33	0.22	0.22	0.29	0.39	0.29	0.26	0.21	0.18	**0.72**	**0.81**

Table 8 The test results of reliability, AVE and interconstruct correlations

Construct	**Composite reliability (≥0.7)**	**Cronbach's Alpha (≥0.7)**	**AVE (≥0.5)**	Alignment of objectives	Cooperative behaviour	History of working relationships	Interdependence	Relationship-specific investment	Trust
Alignment of objectives	0.85	0.75	0.66	**0.81**					
Cooperative behaviour	0.87	0.80	0.63	0.70	**0.79**				
History of working relationships	0.74	0.65	0.58	0.46	0.28	**0.76**			
Interdependence	0.84	0.62	0.72	0.32	0.35	0.26	**0.85**		
Relationship-specific investment	0.85	0.73	0.65	0.78	0.68	0.41	0.26	**0.80**	
Trust	0.81	0.66	0.59	0.71	0.68	0.28	0.33	0.70	**0.77**

*The square root values of AVE

Discriminant validity: Square root values of AVE ≥ highest Pearson's coefficient with other constructs was tested by bootstrapping with a 5000 subsample. A *one-tail t test* was used to test the hypotheses. If $p > 0.10$, the null hypothesis that the hypothesized path is statistically significant is to be rejected. Table 9 summarizes the results of hypothesis testing using the structural model.

Figure 2 shows the results of the structural models with respective indicator loadings and path coefficients β of the structural model.

As shown in Fig. 2, all five paths are significant at the $p = 0.05$ and $p = 0.10$ levels. Strongly supported H_1 ($\beta = 0.711, t = 13.367, p < 0.001$) suggests that the alignment of objectives and trust is positively associated. Likewise, strong support for H_2 ($\beta = 0.409, t = 6.128, p < 0.001$) indicates a positive relationship between relationship-specific investment and a history of working relationships. Relations between trust and interdependence (H_3: $\beta = 0.278$, $t = 2.863$, $p < 0.05$) and that between a history of working relationships and interdependence (H_4: $\beta = 0.179$, $t = 1.625, p < 0.1$) are also supported. The proposed positive association between

Table 9 Results of hypothesis testing

Hypothesis	Path coefficient	T-value (one-tail)	Significance	Inference
H1	0.730	10.891	0.000**	Supported
H2	0.363	2.663	0.004**	Supported
H3	0.235	1.444	0.075*	Supported
H4	0.402	1.667	0.048**	Supported
H5	0.421	2.777	0.003**	Supported

*significant at 10% level
**significant at 5% level

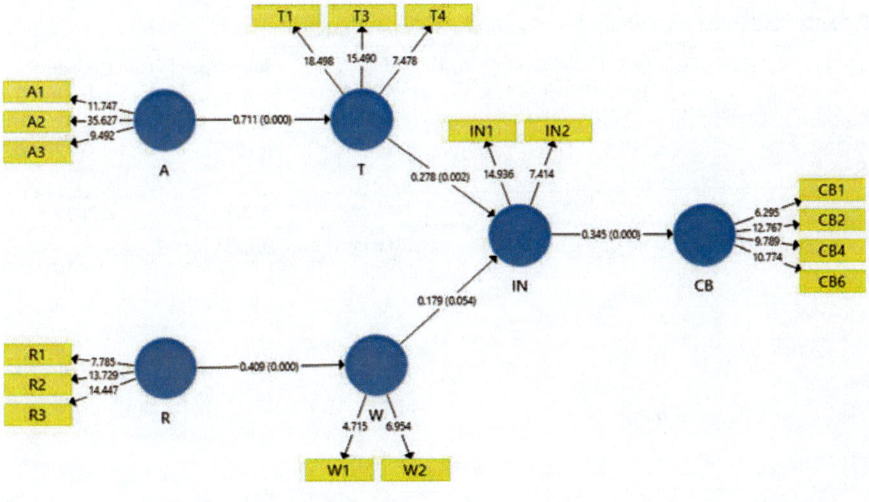

Fig. 2 The results of the structural model

interdependence and cooperative behaviour (H_5: $\beta = 0.345$, $t = 3.822$, $p < 0.001$) is also supported. Potential collinearity problems are indicated by VIF values of 5 or above in the predictor constructs [77]. No collinearity problem was found, as the VIF values range within 1.00 and 2.00.

The quality of the structural model is assessed using all 100 valid responses to determine adequacy [66]. The coefficient of determination (R^2) is an important criterion to evaluate the structural model [76]. The R^2 value measures the capacity for a model to explain the variance of endogenous constructs [69]. The R^2 value ranges from 0 to 1 with a higher value indicating more predictive accuracy. The thresholds of the R^2 value can vary according to the research discipline [69]. Falk and Mille [78] suggested that R^2 values of above 0.10 are acceptable. Adjusted R^2 values of all dependent constructs of the structural model were found to have values of above 0.10 (Table 10) and they thus exhibit acceptable predictive accuracy. However, evaluating the quality of a structural model solely based on R^2 values is inappropriate, as adding more constructs can always increase the R^2 values [79]. Therefore, other evaluation criteria, such as effect size f^2 and predictive relevance Q^2, are considered.

Q^2 values were obtained by blindfolding, which utilizes a cross-validation strategy and reports cross-validated communality and redundancy as constructs as well as indicators [62]. Q^2 values greater than 0 indicate that a model has predictive relevance for a certain endogenous construct [73]. Predictive relevance is considered low, moderate and high with respective values of 0.02, 0.15 and 0.35. The Q^2 values are above 0.15, indicating that the model has moderate predictive relevance for dependent constructs. f^2 evaluates the potential effect caused by the omission of an exogenous construct, such as the impact of removing 'Trust' from the R^2 value for 'Alignment of objectives.' f^2 values of 0.02, 0.15 and 0.35 indicate weak, moderate and strong effects, respectively [73].

Table 10 Adjusted R^2, Q^2 and f^2 values of the Structural Model

	Alignment of objectives	Trust	Relationship-specific investment	History of working relationships	Interdependence	Cooperative behaviour
Adjusted R^2	–	0.501	–	0.159	0.12	0.11
Q^2	–	–	0.482	0.249	0.202	0.269
f^2	1.023	0.083	0.201	0.034	0.135	–

Discussion and Recommendations

Mechanisms Leading to Interdependence

The PLS-SEM analysis results of the structural model shown in Fig. 1 generally support the hypotheses developed for the study. The alignment of objectives and making relationship-specific investments leverage interdependence between the contracting parties. Trust and a history of working relationships are the primary drivers. Interdependence positively contributes to the practice of cooperative behaviour. The structural model suggests that both psychological and economic perspectives leverage interdependence between contracting parties. The relationship between the alignment of objectives and interdependence via trust as interdependence can arise psychologically. Objective alignment enables the joint effort of contracting parties to address mutual objectives through a process of shared decision making. This joint decision-making process can bring about a high level of relationship investment. High levels of mutual interest among the parties are then achieved. A commitment to mutual benefits can also be expected [80]. Achieving mutual objectives would further strengthen the trust between the parties [81]. With a higher level of trust, the parties would have a more positive view of each other [3]. Whilst trust can be defined as a willingness to become vulnerable, the risk involved is only justified by the predictably of the party being trusted. Trust is thus the result of psychological interdependence.

When specific investments made in plants and humans are needed, the lock-in outcome creates a state of interdependence. For example, when targets are to be evaluated, both parties make their respective inputs and comply with the requirements [58, 59]. Setting realistic incentive targets is an efficient means to canvas joint efforts for the rewards. Economically, project-specific investments incur sunk costs and inhibit replacement [82]. As such, contracting parties are encouraged to continue their relationships [29, 51]. The likelihood of further dealings can then be enhanced. With a greater frequency of business interactions, more social interactions and affection will result [44]. Interaction over time therefore cultivates psychological attachment [27]. Accordingly, contracting parties become interdependent for both economic and sentimental reasons.

Cooperative Behaviour Arising from Interdependence

When contracting parties are highly interdependent, they appreciate the need to work together [83] and also anticipate future dealings. Thus, maintaining relationships is paramount. Cooperation is the behavioural expression of a willingness to commit to the achievement of common goals [49]. Conflict is then considered undesirable. Instead, building trust to buttress relationships is practised, with cooperation being the primary manifestation [34]. Cooperative working is also characterized by reciprocating trusting acts between parties [84]. When one party starts behaving cooperatively, cooperative responses from the counterpart are anticipated [27]. In summary, interdependence is at the very least a necessary condition for a recognition of the need for cooperation both psychologically and economically. A state of interdependence is conductive to trust building. Moreover, cooperative work can only be sustained if the parties reciprocate the positive efforts of their counterparts.

Managerial Implications

The findings of this study support the following recommendations (Schreiber et al. 2015). Project managers should realize that effective contractual strategies can be used to create a state of interdependence and enhance cooperation. Several strategies are suggested based on this conclusion:

1. Incentive schemes can be utilized to set clear mutual objectives. These objectives may have to be adjusted when necessary. Aligned objectives have the invisible effect of creating a state of interdependence;
2. A timely deployment of resources with endorsement by senior management is needed to make the necessary actions for the fulfilment of mutually agreed objectives; and
3. Communication should be streamlined. An ideal seamless information sharing system should be established. Effective and efficient communication contribute to cooperative behaviours among organizations. Furthermore, a monitoring system can be installed as a guard against back tracking or slippage. Cooperation can only be sustained as interdependence is maintained.

Summary

Construction projects require inputs from project team members from different organizations. Cooperation between team members is thus a necessary condition to accomplish the project. Moreover, team members may have different interests. Over-emphasising self-interest is a major source of dispute. When disputes arise, team members are likely to take a defensive approach instead of seeking ways to identify common goals. Lacking cooperation therefore makes problem solving more difficult. A cultural change is therefore needed to inspire cooperation among team members. It is proposed that effective contractual strategies can be used to create a state of interdependence that is considered to be an inducer of cooperation. This study (i) examines the concept of interdependence and its drivers; (ii) investigates the roles of interdependence in promoting cooperative work; and (iii) empirically tests the relationships proposed in (ii). In these connections, a conceptual framework is developed from the literature on incentives, interdependence and cooperation. A questionnaire was developed to collect data from construction professionals in Hong Kong, the PRC and Singapore. The relationships among these three constructs are tested by partial-least squares structural equation modelling (PLS-SEM). The findings suggest that cooperative behaviour can be derived from interdependence. Interdependence can have both psychological (Alignment of objectives and Trust) and economic (Relationship-specific investment and Previous work experience) bases. It is also advocated that appropriate use of incentive schemes can engender interdependence among the parties committed to the incentivisation arrangements.

Acknowledgements The empirical work of this chapter has been reported in a paper entitled "Incentivization and Interdependency in Construction Contracting" of the Journal of Management in Engineering. Special thanks to Miss Ka Wai LEE in helping with the empirical studies.

References

1. Ward D, Crane A (2003) The story so far. Partnering and collaborative working. Information Professional, London, pp 10–25
2. Latham SM (1994) Constructing the team: final report of the government/industry review of procurement and contractual arrangements in the UK construction Industry. HMSO, London
3. Cheung SO (2007) Trust in co-operative contracting in construction. City University of Hong Kong Press, Hong Kong, pp 6–11
4. CIRC (Construction Industry Review Committee) (2001) Construct for excellence. Rep. of the Construction Industry Review Committee, HKSAR Government, Hong Kong
5. Cheng EWL, Li H (2001) Development of a conceptual model of construction partnering. Eng Constr Archit Manag 8(4):292–303
6. Crowley LG, Karim M (1995) Conceptual model of partnering. J Manag Eng 11(5):33–39
7. Chan APC, Chan DWM, Chiang YH, Tang BS (2004) Exploring critical success factors for partnering in construction projects. J Constr Eng Manag 130(2):188–198
8. Larson E (1995) Project Partnering: results of study of 280 construction projects. J Manag Eng 11(2):30–35
9. Black C, Akintoye A, Fitzgerald E (2000) An analysis of success factors and benefits of partnering in construction. Int J Project Manage 8(6):424–434
10. Bresnen M, Marshall N (2000a) Motivation, commitment and the use of incentives in partnerships and alliances. Constr Manage Econ 18(5):587–598
11. Bayliss R, Cheung SO, Suen CH, Wong SP (2004) Effective partnering tools in construction: a case study on MTRC TKE contract 604 in Hong Kong. Int J Project Manage 22(3):253–263
12. Walker HTD, Hampson K, Peters R (2002) Project alliancing vs project partnering: a case study of the Australian national museum project. Supply Chain Manage Int J 7(2):83–91
13. Williamson OE (1985) The economic institutions of capitalism. Free Press, New York
14. Williamson OE (1979) Transaction-cost economics: the governance of contractual relations. J Law Econo 22:233–261
15. Henisz WJ, Levitt RE, Scott WR (2012) Toward a unified theory of project governance: economic, sociological and psychological supports for relational contracting. Eng Project Organ J 2(1):37–55
16. Dyer JH (1997) Effective interfirm collaboration: how firms minimize transaction costs and maximize transaction value. Strateg Manag J 18(7):535–556
17. Pang H-Y, Cheung S-O, Choi M-C, Chu S-Y (2015) Opportunism in construction contracting: minefield and manifestation. Int J Project Organ Manage 7(1):31–55

18. Sarkar M, Aulakh PS, Cavusgil ST (1998) The strategic role of relational bonding in inter-organizational collaborations: an empirical study of the global construction industry. J Int Manag 4(2):85–107
19. Gottschalg O, Zollo M (2007) Interest alignment and competitive advantage. Acad Manag Rev 32(2):418–437
20. Tyler TR, Blader SL (2000) Cooperation in groups: procedural justice, social identity and behavioural engagement. Psychology Press, Philadelphia
21. Lui SS, Wong Y-Y, Liu W (2009) Asset specificity roles in interfirm cooperation: reducing opportunistic behaviour or increasing cooperative behaviour? J Bus Res 62(11):1214–1219
22. Levitt B, March JG (1995) Chester I. Barnard and the intelligence of Learning. In: Williamson OE (ed) Organization theory: from Chester Barnard to the present and beyond. Oxford University Press, New York, pp 11–37
23. Cheung SO, Ng TST, Wong SP, Suen CH (2003) Behavioural aspects in construction partnering. Int J Project Manage 21(2):333–343
24. Mayer RC, Davis JH, Schoorman DF (1995) An integrative model of organizational trust. Acad Manag Rev 20(3):709–734
25. Molm LD (2009) Power and social exchange. In: Tjosvold D, Wisse B (eds) Power and Interdependence in organisations. Cambridge University Press, Cambridge
26. Anderson E, Weitz B (1992) The use of pledges to build and sustain commitment in distribution channels. J Mark Res 29(1):18–34
27. Heide JB, Miner AS (1992) The shadow of the future: effects of anticipated interaction and frequency of contact on buyer-seller cooperation. Acad Manag J 35(2):265–291
28. Axelrod R (1984) The evolution of cooperation. Basic Books, New York
29. Deakin S, Lane C, Wilkinson F (1997) contract law, trust relations, and incentives for co-operation: a comparative study. In: Deakin S, Michie J (eds) Contracts, co-operation, and competition: studies in economics, management, and law. Oxford University Press, Oxford, pp 105–139
30. Anvuur A, Kumaraswamy M (2008) Better collaboration through cooperation. In: Smyth H, Pryke S (eds) collaborative relationships in construction: developing frameworks and networks. Wiley-Blackwell, West Sussex, pp 107–128
31. Cheung SO (2001) Rationalism: construction contracting under the PRC contract law. Cost Eng AACEI 43(11):38–44
32. Ling FYY, Rahman MM, Ng TL (2006) Incorporating contractual incentives to facilitate relational contracting. J Prof Issues Eng Educ Pract 132(1):57–66
33. Cheung SO, Yiu TW, Chim PS (2006) How relational are construction contracts? J Prof Issues Eng 132(1)
34. Rahman MM, Kumaraswamy MM (2002) Joint risk management through transactionally efficient relational contracting. Constr Manag Econ 30(1):45–54
35. Yeung J, Chan A, Chan D (2012) Defining relational contracting from the Wittgenstein family-resemblance philosophy. Int J Project Manage 39(2):225–239
36. Macneil IR (2000) Relational contract theory: challenges and queries. Northwest Univ Law Rev 94(3):877–907
37. Kumaraswamy M, Anvuur A, Mahesh GM (2008) Contractual frameworks and cooperative relationships. In: Smyth H, Pryke S (eds) Collaborative relationships in construction: developing a framework and networks. Wiley-Blackwell, West Sussex, pp 78–104
38. Smith KG, Carroll SJ, Ashford SJ (1995) Intra- and inter-organizational cooperation: towards a research agenda. Acad Manag J 38(1):7–23
39. Phua FTT (2004) The antecedents of co-operative behaviour among project team members: an alternative perspective on an old issue. Constr Manag Econ 22(12):1033–1045
40. Zhang Y (2012) An investigation on cooperative behaviors in the Chinese construction industry. The University of Hong Kong
41. Suprapto M, Bakker LMH, Mooi HG, Hertogh MJCM (2015) How do contract types and incentives matter to project performance? Int J Project Manage 34(6):1071–1087

42. Turner JR, Simister SJ (2001) Project contract management and a theory of organization. Int J Project Manage 19(8):457–464
43. Schermerhorn JR (1975) Determinants of inter-organizational cooperation. Acad Manag J 18:845–856
44. Ring PS, Van de Ven AH (1994) Developmental processes of cooperative inter-organizational relationships. Acad Manag Rev 19(1):90–118
45. Campbell D (1997) The relational constitution of contract and the limits of 'Economics': Kenneth Arrow on the social background of markets. In: Deakin S, Michie J (eds) Contracts. Oxford University Press, Co-operation and Competition, Studies in Economics, Management, Oxford, pp 308–336
46. Geyskens I, Steenkamp JEM, Scheer LK, Kumar N (1996) The effects of trust and interdependence on relationship commitment: a trans-Atlantic study. Int J Res Mark 13:303–317
47. Katz M (1989) Vertical contractual relations, vol 1. Elsevier, New York
48. Young-Ybarra C, Wiersema M (1999) Strategic flexibility in information technology alliances: the influence of transaction cost economics and social exchange theory. Organ Sci 10(4):439–459
49. Lush RF, Brown JR (1996) Interdependence, contracting, and relational behaviour in marketing channels. J Mark 60(4):19–38
50. Williamson OE (1973) Markets and hierarchies: some elementary considerations. Am Econ Rev 63(2):316–325
51. Schepker DJ, Oh WY, Martynov A, Poppo L (2014) The many futures of contracts: moving beyond structure and safeguarding to coordination and adaptation. J Manag 40(1):193–225
52. Schepker DJ, Oh WY, Martynov A, Poppo L (2014) The many futures of contracts: moving beyond structure and safeguarding to coordination and adaptation. J Manage 40(1):193–225
53. Luo Y (2002) Contract, cooperation, and performance in international joint ventures. Strateg Manag J 23(10):903–919
54. Baiden BK, Price ADF, Dainty ARJ (2006) The extent of team integration within construction projects. Int J Project Manage 24(1):13–23
55. Tang W, Duffield CF, Young DM (2006) Partnering mechanisms in construction: an empirical study on the Chinese construction industry. J Constr Eng Manag 132(3):217–228
56. Zhang S, Fu YF, Gao Y, Zheng XD (2016) Influence of trust and contract on dispute negotiation behavioral strategy in construction subcontracting. J Manag Eng. https://doi.org/10.1061/(ASCE)ME.1943-5479.0000427
57. Lui SS, Ngo H-Y (2004) The role of trust and contractual safeguards on cooperation in non-equity alliances. J Manag 30(4):471–485
58. Richmond-Coggan D (2001) Construction contract incentive schemes - lessons from experience. Construction Industry Research and Information Association, London
59. Rose T, Manley K (2011) Motivation toward financial incentive goals on construction projects. J Bus Res 64(7):765–773
60. Hair JF, Black WC, Babin BJ, Anderson RE, Tatham RL (2006) Multivariate data analysis, 6th edn. Pearson Prentice Hall, New Jersey
61. Ringle CM, Wende S, Becker J-M (2015) SmartPLS 3, Boenningstedt: SmartPLS GmbH. http://www.smartpls.com
62. Hair JF, Celsi M, Money A, Samouel P, Page M (2016) Essentials of Business Research Methods, 3rd edn. CPI Group (UK) Ltd, Great Britain, pp 237–238
63. Baruch Y, Holtom B (2008) Survey response rate levels and trends in organizational research. Hum Relations 61(8):1139–1160
64. Hatcher L (1994) A step-by-step approach to using the SAS® system for factor analysis and structural equation modeling. SAS Institute Inc., Cary, NC
65. Hair JF Jr, Sarstedt M, Hopkins L, Kuppelwiser VG (2014) Partial least squares structural equation modeling (PLS-SEM): an emerging tool in business research. Eur Bus Rev 26(2):121–196
66. Hazen BT, Overstreet RE, Boone CA (2015) Suggested reporting guidelines for structural equation modelling in supply chain management

67. Podsakoff PM, MacKenzie SB, Lee JY, Podsakoff NP (2003) Common method biases in behavioural research: a critical review of the literature and recommended remedies. J Appl Psychol 88(2):879–903
68. Podsakoff NP, Organ DW (1986) Self-reporting in organizational research: problems and prospects. J Manag 12(1):531–544
69. Hair JF, Hult GTM, Ringle CM, Sarstedt M (2014) A primer on partial least squares structural equation modeling (PLS-SEM). SAGE Publications, Los Angeles
70. Li B, Tiong RLK (1999) Risk management model for international construction joint ventures. J Construct Eng Manage 125(5):377–384
71. Nunnally JC, Bernstein IH (1994) Psychometric theory. McGraw-Hill, New York
72. Dijkstra TK, Henseler J (2015) Consistent partial least squares path modelling. MIS Q 39(2):297–316
73. Chin WW (1998) Modern methods for business research. Erlbaum, Mahwah, NJ
74. Fornell C, Larcker DF (1981) Evaluating structural equation models with unobservable variables and measurement error. J Mark Res 18:39–50
75. Xiong B, Skitmore M, Xia B (2015) A critical review of structural equation modeling applications in construction research. Autom Construct 49(3):59–70
76. Henseler J, Hubona G, Ray PA (2016) Using PLS path modelling in new technology research: updated guidelines. Ind Manag Data Syst 116(1):2–20
77. Hair JF, Ringle CM, Sarstedt M (2011) PLS-SEM: indeed a silver bullet. J Mark Theor Pract 19:139–151
78. Falk RF, Miller NB (1992) A primer for soft modelling. University of Akron Press, Akron, OH
79. Hair JF Jr, Sarstedt M, Hopkins L, Kuppelwiser VG (2014) Partial least squares structural equation modeling (PLS- SEM): an emerging tool in business research. Eur Bus Rev 26(2):121–196
80. Saxton T (1997) The effects of partner and relationship characteristics on alliance outcomes. Acad Manag J 40(2):443–461
81. Das TK, Teng BS (1998) Between trust and control: developing confidence in partner cooperation in alliances. Acad Manag Rev 23(3):491–512
82. Walker A, Chau KW (1999) The relationships between construction project management theory and transaction cost economics. Eng Constr Archit Manag 6(2):166–176
83. Tjosvold D, Cho YH, Park HH, Liu WC, Sasaki (2001) Interdependence and managing conflict with sub-contractors in the construction industry in East Asia. Asia Pac J Manage 18(3):295–313
84. Cheung SO, Yiu TW, Leung AYT, Chiu OK (2008) Catastrophic transitions of construction contracting behaviour. J Constr Eng Manag 134(12):942–952

The Power of Incentivisation in Minimising Construction Disputes

Liuying Zhu and Sai On Cheung

Introduction

Construction projects often require inputs from different disciplines. The Global Construction Dispute Report [1] reports that infrastructure and public projects have the most disputes, likely because of their complexity. This high rate is because the extensive involvement of a broad range of professionals is common for these megaprojects. Some team members work interdependently without a direct contractual relationship. They tend to focus on their own interests and insist on their organisational contractual rights during the project process [2]. The insistence on an individual's interest is not conducive for amicable dispute resolution [3]. Commissioning construction projects with skewed risk and power patterns, cost problems, delayed schedules and inferior quality can lead to disputes. In addition to the loss of project efficiency, the escalation of construction disputes will also initiate adverse social effects and unproductive uses of valuable resources. In other words, avoiding disputes is one of the best ways to manage disputes.

Prevention is better than cure, and corresponding countermeasures for dispute avoidance are thus preferred over deploying extensive resources to resolve such disputes [4]. Proper contract administration at the planning stage is suggested as the most instrumental dispute strategy [1]. Similar suggestions have also been advocated by several studies [3, 5, 6]. Classical contract theory assumes it is possible to exhaust all possible risks and thereby detail the respective responsibilities. Construction activities are prone to risks and variations, and therefore providing a water-proof contract in anticipation of all potential risks is admittedly impossible. Construction contracts

L. Zhu (✉) · S. O. Cheung
Construction Dispute Resolution Research Unit, City University of Hong Kong, Hong Kong, China
e-mail: liuyinzhu3-c@my.cityu.edu.hk

S. O. Cheung
e-mail: Saion.cheung@cityu.edu.hk

S. O. Cheung (ed.), *Construction Dispute Research Expanded*, Springer Tracts in Civil Engineering, https://doi.org/10.1007/978-3-030-80256-1_13

are therefore inevitably incomplete due to uncertainties [7]. Relational contracting has been put forth as a means to maintain good interorganisational relationships and engender cooperation among project team members [8–10]. After investigating 100 construction professionals in Hong Kong, Cheung et al. [11] further found that using incentive schemes is effective at fostering interdependency, which is a key driver of cooperation. In this regard, incentivisation is proposed as a possible solution to implement relational techniques and increase the effectiveness of harnessing relationships [12]. Moreover, encouraging reports concerning the successful use of incentivisation to suppress the occurrence of opportunistic behaviour is welcomed by construction communities [13]. Incentive schemes may fundamentally suppress the occurrence of disputes [12, 14]. However, the misuse of incentives may also invite other project management issues [12]. In studying 400 construction professionals in Singapore, Yean et al. [12] found that improper/inadequate risk allocation in contractual incentives is counterproductive to soliciting collaboration. Accordingly, the ways in which incentive schemes should be designed and implemented for dispute minimisation is a very worthwhile topic. Analysing the roles of incentivisation in construction dispute management is of practical significance. To achieve this purpose, this study has the four following objectives:

(1) To identify the main underlying causes of construction disputes,
(2) To investigate the roles of incentivisation in addressing the main cause of construction disputes,
(3) To illustrate the dispute management function of incentive arrangements, and
(4) To provide recommendations for effective construction dispute minimisation.

The Underlying Causes of Construction Disputes

Construction disputes are often fuelled by the conflicting interests of the contracting parties [7]. Cheung and Pang [7] proposed that contractual and speculative disputes are the two main types of construction disputes. They further found that risks and uncertainties and collaborative conflicts would provoke contractual disputes when a construction contract is incomplete. For speculative disputes, opportunistic behaviour and the affective conflict of project team members (people factors) are the contributing subfactors. Figure 1 shows the anatomy of construction disputes adapted from Cheung and Pang [7].

In analysing the causes of construction disputes, Spittler and Jentzen [15] identified other contributing factors. These include contract ambiguity, adversarial attitude and dissimilar perceptions of fairness by project team members. Of these factors, fairness is found to be the foundation for building harmonious interorganisational relationships [16]. Indeed, conflicts are embedded together with imbalanced risk responsibility among project team members. This conceptualisation is supported by bounded rationality theory [17]. Bounded rationality suggests that people are passionate decision makers who are strongly reliant on economic analyses. Moreover, rationality is bounded to "satisfying" instead of "optimal" outcomes due to the

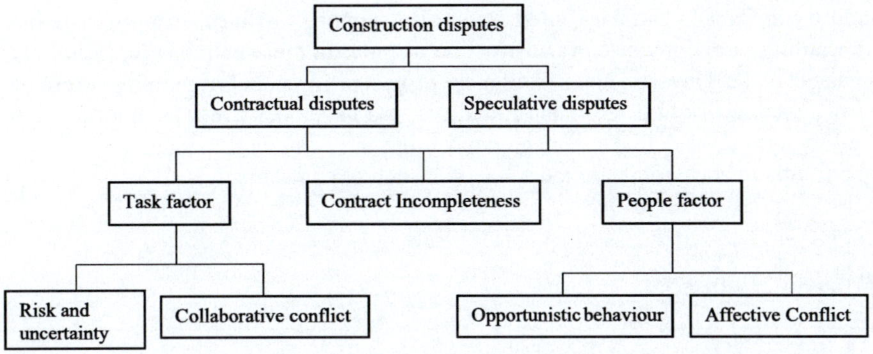

Fig. 1 The anatomy of construction disputes, adapted from Cheung and Pang [7]

boundary of decision makers' knowledge. Economically oriented decision makers affect the nature of the exchange. For example, people who have lower status or those who are more dependent on others are more sensitive to these relationship issues [18]. The perception of unfairness is expressed through emotions [16, 19]. Special care should be taken to manage the perception of unfairness, particularly distress, or adversarial attitudes may escalate the confrontation between disjointing parties. To summarise, the perception of unfairness in organisational dealings can be identified through a disproportionate distribution of risks and power [20].

Inequitable Risk Allocation

Risks can be viewed as departures from planned objectives [21]. Construction projects have enormous risks that need to be managed so that project objectives are not jeopardised. Typical risks include cost overrun, time delay, quality risks, safety risks and environmental sustainability risks [22, 23] and are hard to eliminate. In general, these risks are shared among, distributed between or transferred from one party to another [24]. Reasonable risk allocation across different construction contracting parties should be carefully articulated [4]. Moreover, unilateral risk shifting occurs in many construction projects. Risks often shift to the 'weaker' party of the contractual relationship. Data collected from 284 Chinese project professionals showed that some environmental risks caused by changes in nature, the economy, society, legality and technology are often shifted to the contractor [25]. Aggregated by the practice of contract being awarded to the lowest bid, contractors commonly find themselves in an ex ante disadvantaged position. Developers take advantage of the opportunity to avoid risk as much as possible [4]. To recover the losses or seek more compensation, opportunism is practised during the construction stage. As such, inequitable risk allocation in construction contracts tends to impede cooperative behaviour and is one of the underlying causes of disputes [4]. The investigation of the construction industry in Canada and the United States also supports this idea. Inappropriate risk allocation is a major cause of construction disputes in these nations [26]. Similarly, Zhang et al. [25] investigated 284 Chinese project professionals regarding the effects of inequitable risk allocation and found that pro-developer contractual terms often lead to significant defence damage concerning the contractor's cooperation. Thus, opportunistic behaviours are more likely in such circumstances.

Power Asymmetry

Emerson [27] defined power as pressure expressed by A to exert dominance over B in a relationship. Power is an attribute of position within a network, displayed in the participants' behaviour [28]. Under Emerson's power-dependence theory [27], the dependence of two parties deepens bilateral exchanges. The degree of dependency between two parties is determined by the motivational investment made in the relationship as well as the ability to shift to a new partner. Nevertheless, the investment of both parties may not be equal. Power asymmetry may occur in many mutually dependent relationships [29] and can be either lateral or bilateral [30]. The party with an advantageous position would have the edge in controlling or manipulating the other party, which is referred to as exerting sanction power. The power to sanction is most significant during a negotiation [30].

Sanctioning occurs as a reprisal for nonconformity in a prior act. Unilateral sanctions are often seen as punishment [31]. The intention is to influence others by making them do what is wanted through sanctioning. In construction projects, dominance and control capability indicate the strength of sanction power. Though subject to backfire due to asset specificity [32], at the beginning of a project, developers usually possess the sanction power advantage, given that the switching cost to a new partner is low. Ultimate sanction power is in the hands of the developer [33]. Additional unilateral decisions function as weapons or threats by the developer to ensure the contractor continues to conform to their requests [34]. For example, a developer can conduct payment deductions for unsatisfactory project performance. This situation may change as works progresses and with more input from the contractor. At this stage, the transaction cost of retendering and the additional costs generated by the original contractor's failure increase. Thus, a unilateral determination may now present a greater threat to the developer [35]. Coordination failure on the part of the contractor is the manifestation of differential sanction power [36]. Such failure means that the contractor is unwilling to cooperate in events that are critical to the developer. For example, the contractor may refuse to provide key project information to the employer regarding a third-party's audition, expressing their dissatisfaction with the interorganisational relationship.

Different from sanction power, bargaining power indicates the extent to which one party may yield terms that are beneficial to himself [37]. This power is the relative ability of parties in a negotiation situation to exert influence over each other, and it runs through nearly every stage of the project [28]. The bargaining power differential also originates from asset specificity [32]. The party that exercises bargaining power aims to achieve its own benefits. In construction, each party attempts to influence the behaviour of the other by either depriving them of values already possessed or by obstructing the attainment of values desired in bargaining but not yet possessed [38]. Bilateral power asymmetry is thus reflected both as domination and manipulation [38]. During the bidding stage, contractors have to compromise in the negotiation of contract price. Because of competitive tendering and lowest bid selection, developers always have bargaining power compared with contractors ex ante [16]. In the construction stage, a bargaining power reversal occurs. Delays can occur, as developers may have to compromise due to necessity [35]. Due to the sunk cost of retendering [32], developers become vulnerable if changes in the contract are needed. In this way, making variations after signing contracts may cost the developer dearly [39]. Here, an 'opportunistic margin' [40] is available to the contractor. If the opportunistic behaviour cost is less than the switching cost, the developer will compromise and accept the exploitation. Thus, it is advantageous for the contractor to operate opportunistically. This dilemma will escalate further when the project reaches its critical moments. To avoid the threat of delay, such pressure can force the developer to agree to a price for variation even if they know that the price is beyond the 'true' cost of the change [35, 41].

For both types of power asymmetry, the abusive use of power hampers interorganisational relationships [42]. During dispute negotiations, excessively high pressure will hinder concessions and reduce the chance of a settlement. Power asymmetry typically prevents both sides from moving towards a settlement [27, 43]. Agreements are thus less likely under unequal power relations [43]. In fact, power asymmetry induces retaliatory opportunistic behaviour when circumstances permit. This behaviour is the result of deteriorated interorganisational relationships [26]. The restrained use of power is therefore one of the key components of relational management [44]. Balanced power is beneficial to enhance commitments for joint benefits [27] .

The underlying causes of construction disputes and consequences are summarised in Table 1.

Table 1 The underlying causes of construction disputes and the consequences [7]

Underlying causes	Consequences	Dispute type
Risk differential	Cost wastage Adversarial attitude	Contractual
Power asymmetry	Opportunistic behaviour Impede concession Emotional conflict	Speculative

The Dispute Management Function of Incentive Arrangements

Identification and Categorisation of Incentive Arrangements

Bower et al. [45] define incentivisation as 'a process by which a provider is motivated to achieve extra 'value-added' services over those already specified and of material benefit to the user.' Incentivisation is thus defined as the process of motivating the service provider to achieve extra 'value-added' services over those originally specified and of material benefit to the user. Incentivisation in construction is commonly identified as incentive schemes. The incorporation of incentive schemes aims to align contracting parties' objectives by motivating them to adopt and work towards developers' project objectives [45, 46].

Incentivisation in construction is often objective driven. Time, cost and quality [47] are typically used as performance targets. Incentive arrangements are also linked to these project targets [48, 49]. Cost incentives include extra money in addition to contract value [45]. These incentives usually include target cost [50], an incentive fee [45], 'pain share-gain share' arrangements ([51, 52] and a guaranteed maximum price [53]. Schedule or delivery incentives are used when early completion is desired, or the project is at risk of delay [54]. Performance-based or technical incentives are tied to the qualitative performance of finished works.

Based on whether their outcome is a penalty or reward, performance driving vehicles are termed incentives or disincentives, respectively. Performance can also be motivated by penalties that aim to demotivate underperformance [55, 56]. Meng and Gallagher [57] illustrated how the combined use of incentives and disincentives had an effect on contractors' performance. The composite use of incentives is common for complex projects [48, 56] although it may cause challenges in managing the project [45].

Incentives can also be financial or non-financial. Financial incentives are the most commonly used form of incentivisation in construction contracting. Examples such as fixed-price incentive contracts, cost-plus-award fees and share-in-savings incentives all include monetary rewards. Most of the aforementioned incentive arrangements are linked to financial rewards. Contractual incentives are typically based on monetary rewards [55, 56]. On the other hand, nonfinancial incentive schemes, such as more frequent payments and letters of appreciation, are used to offer rewards in an indirect way [42]. They also have a positive motivating effect on contractors [51].

The type of reward is also related to contractual/extra-contractual incentives [56]. The differences between these two were discussed by Raham and Kumaraswamy [10]. Contractual incentives may include clear and fair risk allocation and benefit-sharing systems in contracts. Noncontractual incentives may include a revision of the risk allocation pattern [46].

Table 2 summarises the classifications and examples of incentives in construction projects.

Table 2 The classifications and examples of incentives in construction projects

Categorisation method	Types of incentive scheme	Examples in construction projects	Key references
Project objectives/outcomes	Cost (Share of saving incentives)	Fixed price incentive contract Cost reimbursable contract Cost-Plus-Incentive-Fee Contracts (CPIF) Cost-plus-fixed fee contract (CPFF)	Kwawu and Laryea [58], Bubshait [55]
	Schedule/delivery	Penalty/bonus for: Final project completion date; Intermediate milestone periods; Intermediate physical completion milestones A combination of final and milestone assessments	Abu-Hijleh and Ibbs [49]
	Performance/quality incentives	Technical Performance bonus Safety Incentive Scheme	Rose and Manley [59], Herten and Peeters [48]
	Multiple incentives	Multiple incentives	Bubshait [55]
Penalty or reward	Incentives	Gain share system for cost-saving	Meng [56], Bubshait [55]
	Disincentives	Pain share system for risks and overspending	
	Combination of both	A combination of pain & gain share system	
Type of rewards	Direct reward	Financial incentive scheme	Hughes et al. [42], Rose and Manley [59], Savio et al. (2013)
	Indirect reward	Non-financial incentive scheme	
Stage of a project	Pre-planned	Agreements before tendering	Richmond-Coggan [46]
	Responsive	Concessions in tendering for winning a project	
	Reactive	Emergency work for delay/cost overrun/technical difficulties	
Whether incorporated into contracts	Contractual incentives	Clear and fair clauses written in contracts	Meng [56], Rahman and Kumaraswamy [10]

The Role of Incentivisation in Dispute Minimisation

Incentive arrangements have been shown to be instrumental in shaping interaction and signalling trust [14]. Incorporating incentivisation is seen as an effective way to realign unfair arrangement situations and facilitate relational contracting behaviours [12, 60]. A win–win solution is expected through the use of incentivisation [61]. In analysing the functions of incentivisation, the following dispute minimising capabilities are clearly revealed.

Reallocation of Risks

Equitable risk allocation is considered to be the gateway to dispute minimisation [1]. Incentivisation is seen as an effective tool to reallocate risks. It is acknowledged that equitable risk reallocation has beneficial effects in minimising construction disputes [62]. After investigating eighty-six construction professionals in Hong Kong, it was found that the willingness to use incentives signals the willingness to trust and promote teamwork [26]. As trust is considered the necessary condition to suppress opportunism and avoid conflict, an equitable allocation of risks is beneficial for minimising construction disputes [4]. There is also a growing interest in the use of incentive contracts to efficiently balance risks between the client and contractor [58]. Two common methods of risk reallocation through incentivisation are identified by Rahman and Kumaraswamy [63].

a. Contractual incentives

For contractual management, incentivisation often includes an equitable risk-reward sharing mechanism [64]. This can be used as a risk reallocation mechanism that provides the incentives for the contractor to come to the negotiation table [63]. For example, the financial incentive scheme is the most common form of incentivisation used in construction contracting. Fixed-price incentive contracts, cost-plus-award fees and share-in-savings incentives all reduce the risk of cost overrun for contractors. For relationship management, a partnering strategy is often coupled with aims towards equal risk/benefit sharing, reflecting the sense of the equal risk/benefit-sharing arrangement [50].

b. Non-contractual incentives

It is noted that not all information required to handle future risks is available. Uncertainties are discernible in the planning stage. Many relational management strategies manifest as noncontractual incentives. Attitude changes for dealing fairly with such unforeseeable risks are also the primary manifestation of incentivisation [26]. For example, more efficient cooperation would occur when contracting parties were optimistic about their business relationship. Otherwise, an inappropriate shifting of risks would raise the bidding price in the long run. Contractors in Canada were found to add an 8–20% premium to their bids in relation to covering their perceptions of high risks in current contracts [26].

Increased Investment in Relationships

Power dependence theory [27] recommends status cognition as an effective strategy to balance power asymmetry. The party with a power advantage is advised to make more motivational and relational investments towards the party with less power [27]. Specific strategies such as goal commitments [21], shared relational attitudes [65], offering mutual support and developing mutual trust [66] are manifestations of incentivisation. Relationship investment aims to encourage the weaker party to be more involved in projects to promote activities and increase the sense of engagement within their interorganisational relationship. Status cognition also conveys the idea that the party that holds a power advantage should take the first steps towards developing a reciprocal power-dependence relationship. Status imbalance is thus mitigated through these extra moves. Incentivisation is an efficient way to embody these ideas in project management. Several manifestations were detected based on these principles.

a. Power transfer and restriction

To balance the power differential, partnering should be promoted. In contrast to the traditional principle-agent relationship, partnering emphasises the pursuit of common interests and win–win solutions when facing conflicts [50]. The transfer of power shows that the management of a project does not rely on the dominant power of the developer but is rooted in conditions of equality, cooperation and mutual assistance. Moreover, the gesture of a long-term commitment also reduces the potential threat of termination of the contract and short-term opportunistic behaviour [46]. It is suggested that sudden unreasonable withdrawal behaviour is unlikely. Some examples are similar to the performance assessment scoring system proposed by the Hong Kong Housing Authority (HKHA). Twenty-five percent of the tender evaluation score was set aside as the performance score [67]. Contractors with above-standard project performance were accorded competition advantages in future bidding evaluations. Such advantages motivate the contractor to move away from short-term financial gain and emphasise long-term mutual benefits more [46]. This system also breaks the vicious circle of opportunistic behaviour and helps achieve project targets [31].

b. Provide better conditions and more support than expected

The transfer of power also reflects how better conditions can be offered through the use of incentives. It is recognised that additional transactions, such as the offer of financial rewards, improve the bargaining position of contractors [35]. In some projects, earlier payment is recognised as an incentive for better performance. It has proven invaluable that the contractor can reduce the asymmetry between cumulative work and the payments accrued. Providing extra assistance beyond the contract also has a rather positive impact on developing interorganisational relationships. At the same time, the assistance offered by the party with a power advantage reduces hostility and makes it easier for both parties to understand each other in future events.

Increasing the Perception of Fairness

Achieving the perception of fairness is important in negotiations and project execution. The possible financial incentive goals in construction projects are summarised as (i) distributive justice, (ii) procedural justice and (iii) interactional justice through a case study in Australia [59]. The aforementioned functions of incentivisation are harnessed to make the first move towards developing cooperation and a focus on equal-footing interactions. It should also be noted that since the perception of unfairness may occur throughout the project, incentivisation should cover the entire project duration. For dispute minimisation, incentivisation can be helpful. When disputes arise, the activation of the relevant incentivizing clause will increase the possibility of achieving win–win outcomes [61, 64]. Some incentive actions, such as offering an apology, also represent the exchange of power between the offender and the offended [5].

In these situations, incentivisation focuses more on the release of emotional distress than on financial compensation. The benefits include (1) reducing the possibility of escalating emotional conflict, (2) directing win–win solutions and (3) seeking possibilities for long-term cooperation.

Figure 2 presents an incentivisation-based conceptual framework.

Generally, the framework highlights the role of incentivisation for construction disputes generated from a literature review. It is found that the functions of effective incentivisation for *ex post* contractual relations are (1) the reallocation of risks; (2) increased investment in relationships and (3) increasing perceptions of fairness. These functions are instrumental within two main construction dispute causes: inequitable risk allocation and power asymmetry. Accordingly, it is proposed that incentivisation could offer an avenue to address this imbalance and thereby reduce the occurrence of construction contract disputes.

Case Study: The Implementation of Incentivisation in a Megaproject

In this section, a case study is conducted to validate these viewpoints from the literature and the developed conceptual framework (Fig. 2). A megaproject is studied. Specific strategies are described to illustrate the dispute minimisation effect of the incentivisations used.

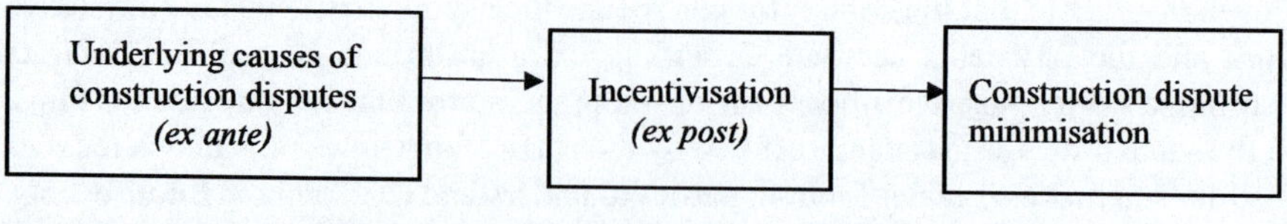

Fig. 2 An incentivisation-based conceptual framework

Background Information

As one of the largest highway projects in China, the Hong Kong-Zhuhai-Macao Bridge (HZMB hereafter) project was started in 2007. The HZMB project aims to promote economic exchanges and activate cooperation between different locations. This project contains a 29.6 km main bridge, a 6.7 km undersea tunnel, two artificial islands and ports in three cities. After 8 years of construction, the HZMB was opened for use in October 2018. During the project planning stage, it became clear that this complex project would require inputs from a wide range of disciplines. The coordination of multiple parties working together to overcome the challenges was vital for the accomplishment of the project. After 8 years of construction, the project was considered a success with all project targets neatly achieved. In particular, notable achievements were attained through the completion of time, cost and innovation targets. A total of 450 innovative patents were also collected and recognised worldwide. The outstanding records of this difficult project have set exemplary references for further megaprojects. In analysing the contributing factors for project success, project participants confirmed that incentivisation played an invaluable role.

The Incentive Behaviours of This Project

A focus group discussion was conducted to study the elements of success from the project. Ten senior managers drawn from developers and contractors participated in this discussion. A general consensus was reached between both parties that incentive behaviours were essential for minimising potential construction disputes. The developer realised that more constructive bonds between developer and contractor were needed to tackle the problems that arose during the project for improved problem solving. As summarised from the discussion, the following events are confirmed by both parties as effective non-financial incentive behaviours offered by developers during the entire project process.

Financial risks reallocation

The advance payment contained in the original contract was found to be insufficient for the contractor to start this megaproject. The contractor was facing huge financial risks when the project started. Following renegotiation, a 10% advance payment was offered by the developer to reduce the financial risks for the contractor. The contractor believed that this understanding was a good start for project cooperation. To improve operations in the middle of the project, the payment was also adjusted through split milestones into more specific targets. The earlier payment schedule ensured that the contractor maintained the project case flow. Through discussion, both parties agreed that these two methods of payment adjustments made this project more 'reasonable' and 'appropriate'. Balancing the unequal financial risks between the two parties reduced potential conflicts and possible opportunistic behaviours. Moreover, the successful negotiation experience nurtured trust and made the work smoother.

Status cognition

A spirit of partnership had been planned in the contract and was effectively implemented throughout this project. Both parties believe that this was also the key to promoting cooperation and reducing disputes. To address unforeseen challenges during the construction stage, additional assistance such as capacity, labour and time from the developer was offered in a timely fashion. All this support helped project participants face many unprecedented difficulties. The contractor was grateful for the support. A partnering strategy is believed to have substantial positive impacts in developing interorganisational relationships. The developer also acknowledged the contributions of innovations generated by the contractor. Support was offered to reduce the risks of innovations and accelerate progress. Project quality was thus improved, and the project was finished on time. When faced with problems, effective means of communication were developed.

Implications for Dispute Management

The aforementioned theoretical deliberations and practical experience support the idea that incentivisation ex ante can be an efficient way to minimise the occurrence of construction disputes. When planning incentives for this purpose, the following should be considered.

Establishment of Norms of Fairness

To minimise construction disputes, enabling behaviours that foster the perception of fairness should be an important outcome of incentivisation. Two underlying causes of disputes point to the potential damage of unequal status between the contractual parties. The preference for fairness is thus important for relational governance [36]. However, the framing effect [68] illustrates how people may have varying judgements of "fairness". Different descriptions of identical problems by individuals are indeed common. The differing view of fairness may also steer contracting behaviour towards confrontation. In this way, establishing clear norms ahead of time tends to be important [16]. To characterise a relatively fairer relation, outcome justice (all participants receive equal shares) or distributive justice (the output ratio corresponds to its inputs) should be implemented [16].

Reallocation of Risks and Investments in the Relationship

Analysing the underlying causes of construction disputes offers one pathway of using incentivisation to minimise their occurrence. Incentivisation can be an effective tool to reduce conflicts arising from inequitable risk allocation and the perception of unfairness. Although many construction incentive schemes are objective-oriented [46], the rewards attached to these schemes are secondary sources of project income for the contractors. As such, the financial return of these schemes is attractive and motivates contractors to achieve the developer's project objectives. Whenever additional targets are set and additional value is desired, the corresponding risks and responsibilities should still be taken into account [45]. Moreover, incentive schemes can be another management tool to help adjust an inequitable share of risks or power with reference to market norms or generated by changes in circumstances. In sum, when disputes result from the conflict induced by inequitable treatment, incentivisation offers a golden opportunity to adjust the inequities.

Fostering Incentivisation in a Relationship is Necessary

The purpose of incentivisation is to reach a win–win outcome rather than settle a problem with a compromise. When undertaking the suggestion above, planning and behaviour should also conform to the aforementioned norms of fairness. A mere compromise would only become an individual sacrifice without improving project outcomes. Moreover, for some megaprojects such as the HZMB, maintaining reputations may be the highest priority for the participating contracting organisations. Some non-incentive behaviour, such as saving face, is also of vital importance for them. Some competition among project team members can also be effective when performance improvement actions are taken by contracting organisations to save face.

Summary

The complexity of construction projects makes them risk-laden because of the vast uncertainties. Opportunistic behaviours may happen when aggrieved party seeks to retaliate ex post. Raising disputes has been seen as the ultimate manifestation of unaddressed tension. Incentive schemes have been used by many projects to motivate the counterpart of contracting parties to meet project goals. This study discusses the possibility of using incentivization to minimize dispute. Installing relational contracting is found to be effective through the use of incentivization. This study thus (i) identify the underlying causes of construction disputes; (ii) analyze the functions of incentivisation in minimizing disputes; (iii) examine dispute management function through case study and (iv) offer suggestions for incentivization planning to achieve dispute avoidance. Through literature review, it was found that unfair treatment of contractual parties is one of the underlying causes of construction disputes. Typically, adverse inter-organizational relationship and conflicts between contractual parties are resulted from disproportionate distribution of risks and power [20]. When incentivizations enable as *ex post* (i) reallocation of risks; (ii) investment in relation and (iii) patching perception of fairness, dispute can be minimized. Through a case study of Hong Kong-Zhuhai-Macao bridge project, it was found that reallocating financial risks through payment adjustments and promoting partnership are effective strategies in lowering dispute occurence. It is therefore suggested that incentivization can be used to provide *ex post* relational governance that moderates *ex ante* risk and power asymmetry. Furthermore, incentives should be planned to achieve: (i) establishing norms of fairness; (ii) encouraging equitable risk allocation relational investment sand (iii) fostering investments in relation.

Acknowledgements The empirical work of this chapter has been reported in a journal paper entitled "Power of Incentivization in Construction Dispute Avoidance" of the Journal of Legal Affairs and Dispute Resolution in Engineering and Construction.

References

1. ARCADIS (2018) Global construction disputes report
2. ARCADIS (2015) Global construction disputes report
3. Cheung SO, Yiu TW (2006) Are construction disputes inevitable? IEEE Trans Eng Manage 53(3):456–470
4. Cheung SO et al (2014) Construction dispute research. Springer. https://doi.org/10.1007/978-3-319-04429-3
5. Cheung SO et al (2019) How apology incentivizes construction dispute settlement. J Legal Aff Dispute Settl Constr 11(2007):2–7. https://doi.org/10.1061/(ASCE)LA.1943-4170.0000287
6. Yates JK, Epstein A (2006) Avoiding and minimizing construction delay claim disputes in relational contracting. J Prof Issues Eng Educ Pract 2:168–180
7. Cheung SO, Pang KHY (2013) Anatomy of construction disputes. J Constr Eng Manag 139(January):15–23. https://doi.org/10.1061/(ASCE)CO.1943-7862.0000532
8. Cheung SO, Yiu KTW, Chim PS (2006) How relational are construction contracts? J Prof Issues Eng Educ Pract 1(48):48–57
9. Jelodar MB, Yiu TW, Wilkinson S (2016) A conceptualisation of relationship quality in construction procurement. Int J Proj Manage 34(6):997–1011.https://doi.org/10.1016/j.ijproman.2016.03.005
10. Rahman MM, Kumaraswamy MM (2008) Relational contracting and teambuilding: assessing potential contractual and noncontractual incentives. J Manag Eng 24(1):48–63. https://doi.org/10.1061/(ASCE)0742-597X(2008)24:1(48)
11. Cheung SO, Zhu L, Lee KW (2018) Incentivization and interdependency in construction contracting. J Manag Eng 34(3):1–13. https://doi.org/10.1061/(ASCE)ME.1943-5479.0000601

12. Yean F et al (2006) Incorporating contractual incentives to facilitate relational contracting. J Prof Issues Eng Educ Pract 1(57):57–67
13. Liu J et al (2016) Incentive mechanism for inhibiting investors' opportunistic behavior in PPP projects. Int J Proj Manage 34(7):1102–1111.https://doi.org/10.1016/j.ijproman.2016.05.013
14. Dewulf G, Kadefors A (2012) Collaboration in public construction–contractual incentives, partnering schemes and trust. Eng Proj Organ J 2:240–250
15. Spittler JR, Jenztzen GH (1992) Dispute resolution: managing construction conflict with step negotiations. Trans Am Assoc Cost Eng 1
16. Kadefors A (2007) Fairness in interorganizational project relations : norms and strategies. Constr Manage Econ 6193.https://doi.org/10.1080/01446190500184238
17. Simon H (1972) Theories of bounded rationality, decision and organization. North-Holland Publishing Company. http://innovbfa.viabloga.com/files/Herbert_Simon___theories_of_bounded_rationality___1972.pdf
18. Flynn FJ, Reagans RE (2006) Helping one's way to the top: selfmonitors achieve status by helping others and knowing who helps whom. J Personality Social Psychol 91(6):1123–1137. https://doi.org/10.1037/0022-3514.91.6.1123
19. van Zomeren M, Leach CW, Spears R (2012) Protesters as "Passionate economists": a dynamic dual pathway model of approach coping with collective disadvantage. Pers Soc Psychol Rev 16(2):180–199. https://doi.org/10.1177/1088868311430835
20. Adams JS, Freedman S (1976) Equity theory revisited: Comments and annotated bibliography. Adv Exp Soc Psychol 9(C):43–90. https://doi.org/10.1016/S0065-2601(08)60058-1
21. Chapman C, Ward S (2008) Developing and implementing a balanced incentive and risk sharing contract. Constr Manag Econ 26(6):659–669. https://doi.org/10.1080/01446190802014760
22. Zou PXW, Zhang G (2009) Managing risks in construction projects: life cycle and stakeholder perspectives. Int J Constr Manag 9(1):61–77. https://doi.org/10.1080/15623599.2009.10773122
23. Zou PXW, Zhang G, Wang J (2007) Understanding the key risks in construction projects in China. Int J Project Manage 25(6):601–614. https://doi.org/10.1016/j.ijproman.2007.03.001
24. ANDI (2006) The importance and allocation of risks in Indonesian construction projects T. Constr Manage Econo 24:69–80.https://doi.org/10.1080/01446190500310338
25. Zhang S et al (2016) Contractual governance : effects of risk allocation on contractors' cooperative behavior in construction projects. J Constr Eng Manag 142(6):1–11.https://doi.org/10.1061/(ASCE)CO.1943-7862.0001111
26. Zaghloul R, Hartman F (2003) Construction contracts: the cost of mistrust. Int J Project Manage 21(6):419–424. https://doi.org/10.1016/S0263-7863(02)00082-0
27. Emerson RM (1962) Power-dependence relations. Am Sociol Rev 27(1):31–41
28. Cook KS, Emerson RM (1978) Power, equity and commitment in exchange. Am Sociol Rev 43(5):721–739. https://doi.org/10.3102/0034654311405999
29. Emerson RM (1976) Social exchange theory. Annu Rev Sociol 2(1976):335–362
30. Goldhamer H, Shils EA (1939) Types of Power and Status. Am J Sociol 45(2):171–182. https://doi.org/10.1086/218263
31. Oliver P (1980) Rewards and punishments as selective incentives for collective action: theoretical investigations. Univ Chicago Press J 85(6):1356–1375
32. Williamson OE (1979) Transaction-cost economics: the governance of contractual relations. J Law Econo 22(2)
33. Chow PT, Cheung SO, Yip KW (2005) Impact of trust and satisfaction on the commitment-withdrawal relationship. 31(Gounaris):1–8. https://doi.org/10.1061/(ASCE)ME.1943-5479.0000331
34. Back WE, Grau D, Mejia-Aguilar G (2013) Effectiveness evaluation of contract incentives on project performance. Int J Constr Educ Res 9(4):288–306. https://doi.org/10.1080/15578771.2012.729551
35. Chang C, Ive G (2007) The hold-up problem in the management of construction projects: a case study of the channel tunnel. Int J Project Manage 25:394–404. https://doi.org/10.1016/j.ijproman.2007.01.001

36. Fu Y et al (2015) Promoting cooperation in construction projects: an integrated approach of contractual incentive and trust. Constr Manage Econ 33(8):653–670.https://doi.org/10.1080/01446193.2015.1087646
37. Chang CY et al (2007) Reversal of bargaining power in construction projects: meaning, existence and implications. Constr Manag Econ 25(8):845–855. https://doi.org/10.1080/01446190601164113
38. Thibaut J, Gruder CL (1969) Formation of contractual agreements between parties of unequal power. J Pers Soc Psychol 11(1):59–65. https://doi.org/10.1037/h0027033
39. Chan APC, Yeong CM (1995) A comparison of strategies for reducing variations. Constr Manage Econ 13(6):467–473. https:// doi.org/10.1080/01446199500000054
40. Winch G (1998) Zephyrs of creative destruction: Understanding the management of innovation in construction. Build Res Inf 26(4):268–279. https://doi.org/10.1080/096132198369751
41. Masten SE, Meehan JW, Snyder EA (1991) The costs of organization. J Law Econ Organ 7(1):1–25. https://doi.org/10.1093/oxfordjournals.jleo.a037001
42. Hughes W, Yohannes I, Hillig J-B (2007) Incentives in construction contracts : should we pay for performance? CIB World Building Congress 2000, pp 2272–2283. https://www.irbnet.de/daten/iconda/CIB5038.pdf
43. Lawler EJ, Yoon J (1993) Power and the emergence of commitment behavior in negotiated exchange. Am Sociol Rev 58(4):465–481
44. Cannon JP, Achrol RC, Gundlach GT (2000) Contracts, norms, and plural form governance. J Acad Marketing Sci 28(2):180–194. https://doi.org/10.1177/0092070300282001
45. Bower D et al (2002) Incentive mechanisms for project success. J Manage Eng
46. Richmond-Coggan D (2001) Construction contract incentive schemes: lessons from experience
47. Atkinson R (1999) Project management: cost, time and quality, two best guesses and a phenomenon, its time to accept other success criteria. Int J Project Manage 17(6):337–342. https://doi.org/10.1016/S0263-7863(98)00069-6
48. Herten HJ, Peeters WAR (1986) Incentive contracting as a project management tool, pp 34–39. https://doi.org/10.1016/0263-7863(86)90060-8
49. Ibbs CW (1991) Innovative contract incentive features for construction, Constr Manage Econ 9:157–169
50. Hughes D, Williams T, Ren Z (2012) Is incentivisation significant in ensuring successful partnered projects? Eng Constr Archit Manag 19(3):306–319. https://doi.org/10.1108/09699981211219625
51. Bresnen M, Marshall N (2000a) Motivation, commitment and the use of incentives in partnerships and alliances. Constr Manage Econ 18(5):587–598. https://doi.org/10.1080/014461900407392
52. Bayliss R et al (2004) Effective partnering tools in construction: a case study on MTRC TKE contract 604 in Hong Kong. Int J Project Manage 22(3):253–263. https://doi.org/10.1016/S0263-7863(03)00069-3
53. Chan DWM et al (2011) An empirical survey of the motives and benefits of adopting guaranteed maximum price and target cost contracts in construction. Int J Project Manage 29(5):577–590. https://doi.org/10.1016/j.ijproman.2010.04.002
54. Hosseinian SM (2016) An optimal time incentive/disincentive-based compensation in contracts with multiple agents. Constr Econ Building 16(4):35–53
55. Bubshait AA (2003) Incentive/disincentive contracts and its effects on industrial projects. Int J Project Manage 21(1):63–70. https://doi.org/10.1016/S0263-7863(01)00078-3
56. Meng X (2015) Incentive mechanisms and their impact on project performance. Handbook on project management and scheduling. https://doi.org/10.1007/978-3-319-05915-0_17
57. Meng X, Gallagher B (2012) The impact of incentive mechanisms on project performance. Int J Proj Manage 30:352–362
58. Kwawu W, Laryea S (2014) Incentive contracting in construction. In: Proceedings 29th annual association of researchers in construction management conference, ARCOM 2013, pp 729–738. http://www.scopus.com/inward/record.url?eid=2-s2.0-84911379877&partnerID=tZOtx3y1

59. Rose T, Manley K (2011) Motivation toward financial incentive goals on construction projects. J Bus Res 64(7):765–773. https://doi.org/10.1016/j.jbusres.2010.07.003
60. Ashley DB, Workman BW (1986) Incentives in construction contracts. Austin (TX): CII, SD 8
61. Construction Productivity Network (1996) Incentive based contracts for win-win solutions?. In Notes of afternoon workshop jointly sponsored by ECI & CPN on 16 July. WR 96. 14L, London, CIRIA
62. Cheung SO (1997) Risk allocation: an essential tool for construction project management. J Constr Procurement 3:16–27
63. Motiar Rahman M, Kumaraswamy MM (2002) Joint risk management through transactionally efficient relational contracting. Constr Manag Econ 20(1):45–54. https://doi.org/10.1080/01446190110089682
64. Boukendour S, Hughes W (2014) Collaborative incentive contracts: Stimulating competitive behaviour without competition. Constr Manag Econ 32(3):279–289. https://doi.org/10.1080/01446193.2013.875215
65. Suprapto M et al (2015) Sorting out the essence of owner-contractor collaboration in capital project delivery. Int J Proj Manag 33(3):664–683. https://doi.org/10.1016/j.ijproman.2014.05.001
66. De Melo RSS, Granja AD, Ballard G (2013) Collaboration to extend target costing to non-multi-party contracted projects: Evidence from literature. In: 21st annual conference of the international group for lean construction 2013, IGLC 2013, pp 21–30
67. The Hong Kong Housing Authority (1998) Preferential tender award system and bonus scheme (BC 112/99)
68. Tversky A, Kahneman D (1992) Advances in prospect theory: cumulative representation of uncertainty. J Risk Uncertain 5(4):297–323
69. Bresnen M, Marshall N (2000b) Partnering in construction: a critical review of issues, problems and dilemmas. Constr Manage Econ 18(2):229–237. https://doi.org/10.1080/014461900370852

Caveats for Using Third-Party Neutrals

Sai On Cheung, Keyao Li, and Liuying Zhu

Introduction

Most employers are of the view that contracts can be drafted to deal with all eventualities. Thus, when what actually occurs is not as anticipated, disputes may arise. In fact, many disputes are related to changes that are necessitated by uncertainties. This phenomenon has been well documented [65] and explained by the concept of bounded rationality expounded by Simon [57]. Cheung and Pang [8] found that contract incompleteness is the indispensable element of all forms of construction disputes. Incomplete contracts are the minefield of opportunistic behaviours [43]. When ambiguities and gaps are found in a contract, contracting organisations may see this as an opportunity to practice opportunism [4]. Opportunistic acts of contracting party are those behaviours that pursue self-interest with deceit and at the expense of other parties. A typical example of opportunistic behaviour is withholding crucial information [62]. In construction, contractor's opportunistic behaviours significantly reduce project efficiency and are detrimental to contracting relationships [43]. For example, the withholding of key project information hinders the employer to make the necessary preparation for the project. Contractor may take advantage of any delay in provision of information by the employer that is a typical head of claim. The uncooperative attitude also cause mistrust between two parties. The willingness of contractors to enhance project efficiency will be greatly reduced. Moreover, as the

S. O. Cheung (✉) · L. Zhu
Construction Dispute Resolution Research Unit, City University of Hong Kong, Hong Kong, China
e-mail: Saion.cheung@cityu.edu.hk

L. Zhu
e-mail: liuyinzhu3-c@my.cityu.edu.hk

K. Li
Future of Work Institute, Faculty of Business & Law, Curtin University, Perth, Australia
e-mail: Keyao.li@curtin.edu.au

S. O. Cheung (ed.), *Construction Dispute Research Expanded*, Springer Tracts in Civil Engineering, https://doi.org/10.1007/978-3-030-80256-1_14

communication between the contracting parties are ineffective, it is hard for them to work together to solve problems. In this connection, [8] regarded opportunism as fuelling speculative disputes [43]. In addition, acrimonious relationships make dispute negotiation difficult. Lu et al. [44] suggested that specific strategies are needed to alleviate opportunism in construction project management. Two major counter-measures have been proposed. Formal governance like contractual management is suggested to increase the cost of opportunistic behaviours [66]. Social exchange theory offers another solution-informal governance [16] that emphasises the importance of high quality relationship management among contacting parties to combat opportunism. Cultivating trust through effective negotiation is therefore advocated. For construction dispute management, inviting third-party neutrals to help resolve disputes may help to break the bottleneck between the negotiating parties. In Hong Kong, there is a rising trend of using third-party neutrals to facilitate dispute negotiation. Dispute resolution advisors [10] and mediators [7] are notable examples of third party neutrals. In Hong Kong, mediation has been promoted by the Hong Kong Government as the mainstream alternative dispute resolution method. Mediation is a form of assisted negotiation whereby mediators are used to improve the efficiency of dispute resolution [7]. It can be said that mediators are the core force to raise the efficiency of mediation. As such, the qualities of mediators are determining factors. This study aims to raise the awareness of third-party neutrals in upholding their impartiality and neutrality. Both qualities are considered important for the proper functioning of mediators.

The Roles of Third-Party Neutrals in Construction Dispute Resolution

Third-party neutrals are professionals helping disputing parties settle their disputes through negotiation [55, 56]. There are several ways that third-party neutrals can provide their services. Dispute resolution advisors (DRAs) are appointed jointly by the employer and the contractor at the commencement of a project. The appointed DRA would then follow the project from the commencement to completion and offer advice at the earliest possible time when a problem occurs [9]. It is hoped that through early and continuous involvement in the project, a DRA can help the parties identify common ground so that a mutually satisfying settlement can be crafted [25, 39]. Mediators are appointed after a dispute has arisen, and their roles are somewhat similar to those of DRAs. Engel and Korf [17] summarised that the key functions of a mediator are to listen to the positions and interests of both parties; to make appropriate suggestions for resolution and to help them reach an agreement to which they can both commit.

The importance of mediators being impartial is recognised by many researchers [23, 46]. Impartiality is closely linked with effective functioning of mediators. For example, as the bridging agent of communication, mediators need to break the inertia against genuine exchange of views and positions. For this function, the trust of the negotiating parties is paramount. Being impartial is the first and foremost indicator of trustworthiness. Mediator must act as the neutral third party, instead of a representative of one side. Achieving this not only enhance communication, but it will also help avoiding polarization of position and escalating the problem. The opening-up for effective communication paves the path for serious attempt to resolve the dispute. It is therefore hard for a biased mediator to facilitate a better communication compared with previous two parties' negotiation.

For another angle, mediators also act as legitimizer [17]. Mediators can help both disputing parties to recognize the rights of others. Moreover, mediators should not be directly involved in arguing the case. To facilitate a mediation process, all the necessary information and assistance should also be provided to the negotiating parties so that they can resolve their differences. Any inclination to one party may lead to perception of bias which is not conducive to reconciliation of conflicts.

Apart from that, mediators are also solution explorers for the disputes. It is expected that mediators can examine the problem from different perspectives and angles [9]. As neutral third party, a mediator can consider the problem from an impartial perspective and suggests possible solutions. On the contrary, bias restricts the mediator's thinking of a limited number of aspects for one side.

As agent of reality, mediators sometimes need to perform reality testing to advise on the practicality of the parties' expectations [54]. Objective view of both party's proposals and requests would help to iron out unrealistic expectations. If one party has extreme or unrealistic goals, mediators need to let the party aware of this. An impartial mediator can objectively point out the impracticality of both sides to avoid creating hurdles against resolution.

Thus, a fair and impartial mediator would gain the trust of the disputing parties [23, 46]. Usually, mediator is jointly appointed by the disputing parties. He or she has no authority to force the parties to reach an agreement. Any suggestions made by him or her have to be agreed and accepted by the parties to move the dispute resolution forward. Although suggestions are advisory but can be pivotal if the parties have faith in the mediator and that his or her suggestions are truthful. The success or otherwise of a mediation therefore depends heavily on whether the disputing parties trust the mediator. A trusting mediator makes his or her suggestions more objective and persuasive. If the mediation process losses its efficiency, the failure caused by bias also lead to the loss of credibility of mediation and causes a vicious circle. From these aspects, impartiality is thus the very basic quality required of a mediator [25].

Most decision models are developed based on the assumption that decision-makers are rational. This may not always be the case, as humans are subject to judgement errors due to their individual limitations [2, 3, 14, 57]. Among the vast kinds of judgement flaws, biases seem to be the most notable [20, 33, 34].

It has been mentioned that impartiality is the most important attribute of a dispute resolution third-party neutral.

Moreover, being human, will the third party neutral be bias-free? In this connection, Cheung and Li [12] identified five forms of bias in construction dispute negotiation: anchoring, overconfidence, self-serving bias, hindsight and confirmation. It is further observed by third-party neutrals that anchoring and confirmation are more likely to occur [41]. The occurrence of biases may be unnoticed, and the chance of settlement silently eroded [30, 31]. This study is exploratory with the aim of investigating whether construction dispute resolution third-party neutrals are suspected of being biased.

Anchoring bias was first identified by Tversky and Kahneman [60]. The subjects of their experiment were found using arbitrary information in making their assessments. More specifically, anchoring bias can be tracked when decisions are made based on the information of the issue that first comes to the decision-maker, and the information may be irrelevant [6, 24, 63]. It is self-explanatory that unjustified use of the first available information cannot form a proper evaluation of the problem. Furthermore, this will overshadow other useful information that comes later [19, 18, 58]. It is thus undesirable for a third-party neutral to have anchoring biases. Third-party neutrals should assist the parties in using the most relevant and appropriate information to evaluate the issues at stake, and their advice must be free from pre-emptive propositions.

Confirmation bias can be identified by an unjustified purposive way to collect and interpret data [15, 35, 49]. For instance, when information was collected and analysed in such a way to substantiate preconceived positions, confirmation bias is at work [36]. In negotiations, when negotiators are searching selectively for information that supports their already formed viewpoint, they are manifesting confirmation bias [27, 29, 53, 59]. In construction dispute resolution, it is not unusual for disputants to not back away from what they have offered. Thus, in all subsequent rounds of conflict resolution, their initial positions will be insisted upon. Furthermore, they would only attend to evidence and information that reinforce these positions. Rationality and objectivity are thus compromised [1, 53, 54]. In reality, having complete information in a dispute negotiation is unlikely. A third-party neutral has to listen to the argument and positions of the disputants. The information of course has been selected to suit their version of the dispute. Third-party neutrals, therefore, have to work under such conditions. With confirmation bias in mind, third-party neutrals should also not be pinned down to their initial advice. As additional information becomes available, if they only take note of that which supports their initial advice, they have confirmation bias.

In the study of Li and Cheung [41], third-party neutrals observed that disputants have biased behaviours; this study intends to examine whether third-party neutrals also engage in biased behaviours.

The Study

Divergent views on one's rights and responsibilities under a contract are typical subject matters of construction disputes. Moreover, incomplete contracts make it more complicated when there are no specific provisions to deal with the situation [8, 47]. Inconsistency, ambiguities and incompatibilities in contracts are minefields of opportunistic behaviours [43]. The specialties of construction projects, such as long duration transactions, asset specificity and complexity also exacerbate this problem. Cheung and Yiu [11] found that opportunism is the most challenging cause of disputes because of the behavioural nature. Effective dispute resolution methods are therefore needed to alleviate opportunism. When disputes arise, two disputing parties are often holding different opinion and stick to their own version. To address this deadlock, additional relationship investment is needed to cultivate the sense of trust. The third party who is impartial and not directly involved in this particular situation is expected to participate to offer fair opinion for disputing parties [7]. Facilitation by third-party neutrals may be instrumental to managing the behavioural dimension. Moreover, when human decisions are involved, the chance of bias may not be totally avoided. Very little has been done about the bias of third-party neutrals. In construction, as practising opportunism is quite common, this study examines the bias of third-party neutrals when faced with opportunistic behaviours of the disputing parties. A simulation was used for this purpose.

The simulation involves the development of the 'Suramadu' Bridge. Practising professionals in the construction industry were invited to play the role of third-party neutrals in facilitating the resolution of disputes between an employer and a contractor. The simulation was designed to include an opportunistic chair plan used by the contractor followed by 'normal' practice. The bias of third-party neutrals was detected by comparing the responses returned for the two episodes. The simulation has three parts. In each part, the respondents were asked to indicate their level of agreement with the statements related to the case. The simulation used in the study is shown in italics.

Case Background

In 2013, the Indonesian government initiated the development of the 'Suramadu' bridge to connect the city of Madura and Surabaya, the capital city. The economic growth of Surabaya is the strongest among all cities in the province, whereas Madura, which is a small city, is the weakest. By linking the two islands together, 'Suramadu' will provide an infrastructure that offers fast and easy access for the people of Madura to expand their business to the capital city, thus supporting their economic growth. The span of 'Suramadu' is 5.4 km and will become Indonesia's longest bridge. A budget of US$6.5 billion (including US$500 million as a contingency) was approved by the government to cover both the main bridge development and the side projects. A tender was invited in the second half of 2014.

The Project

The 'Suramadu' bridge will be the first-ever bridge to be built across a strait and connects two islands with contrasting geological conditions and site topography.

'Suramadu' therefore will become the largest and most complex infrastructure project that has ever been constructed in Indonesia.

For this reason, the Indonesian government was expecting that the tender price from the contractors would take care of these complexities and challenges. A contingency of US$300 million has been included in the budget to address unforeseeable events.

Apart from the bridge, an ambitious plan to develop office towers and high-end housing estates on Madura's side of the bridge makes the project exceptionally complex. These side projects collectively shall form a satellite town for a population of 0.5 million. To achieve an early completion, a design and build procurement with a fixed price contract was adopted. Contractors C (C hereafter), with a tender price 20% lower than the second-lowest bidder, was appointed as the main contractor for this project.

Project Summary

Client: The Indonesian Government

Design and Build Contractor: C

Contract Value: US$6 billion

Project Duration: Six years, commencing on 1 January 2015.

Part A of the simulation presents the events that occurred during the first year of the project. The situations have been written to highlight the opportunistic behaviour of the contractor. The contractor raised many unjustified claims to recoup a budget shortfall due to the below-cost winning bid.

Part A: The First Year of the Project (2015)

The low tender of US$6 billion was appealing to the government, which viewed this as the best way to keep the construction costs within budget. However, land acquisition for the side projects has been slow, as only 20% of the land has been completed before 1 January 2015, the commencement date of the project. As such, the Indonesian government had not yet given possession of the site to C by 1 February 2015. C filed a claim of US$8 million to compensate for the delays due to non-possession of the site.

Three months after the project's commencement, in May 2015, C submitted a US$30 million claim for additional site investigation work. In the same month, another claim of US$18 million was raised to increase the cost of reinforcement bars due to non-possession. However, it appeared that the rise in cost was due to late orders resulting from poor communication between the supplier and domestic steel bending subcontractor, both of which are employed by C.

As the project entered the rainy season in November 2015, the work progress was greatly affected by heavy rainfall and floods. To ensure the capability of the bridge to *withstand extreme weather conditions, the Indonesian government inspected the site and opined that the site drainage system could be enhanced to mitigate the effect of floods. C provided a proposal that covered the whole contract period, but with a cost implication of US$28 million should the proposal be instructed. Furthermore, just one month after submitting claims for site drainage improvement work, in December 2015, C submitted a US$58 million claim for ground improvement works due to unforeseen ground conditions, which was revealed after the major floods in November 2015.*

Very few supporting documents were provided together with the claim submissions. In addition, during the claim negotiations, C persistently withheld crucial information and was ambivalent about what had happened. However, since the Indonesian government was keen to settle the claims quickly so that the parties could focus on the work, with the involvement of a third-party neutral, these claims for a total of US$142 million were settled for a sum of US$ 42 million. The monetary claims in the first year are listed in Table 1. Assuming you are the appointed third-party neutral, please indicate your degree of agreement with the statements about the project's dispute claims in the first year in Table 2.

Part B describes the second year of the project. In this year, several claims have also been submitted, and the claim amount progressively increases. The last claim of that year is the highest. Moreover, most of these claims do have legitimate contractual grounds. This is in notable contrast with the claim approach adopted in the first year. Part B, therefore, may resemble expected practice of a bona fide contractor.

Part B: The Second Year of the Project (2016)

In March 2016, just two months after the US$42 million settlement payment, C raised the issue that the rock is extremely hard and demanded additional investigations to

Table 1 The monetary claims in the first year

Month	Dispute issues	Amount (US$ million)
2015/03	Non-possession of site	8
2015/05	Additional site investigation	30
2015/05	Increase in reinforcement material's price	18
2015/11	Provision of extra work	28
2015/12	Additional ground improvement works	58
Total claim		142
Result		Commercial settlement 42 million

Table 2 Statements about the project's dispute claims in the first year

No	Descriptions	Strongly disagree	Disagree	Neutral	Agree	Strongly agree
		1	2	3	4	5
a	The initial budget allowance is too low					
b	C's bid is improbable					
c	Heavy claims are foreseeable in complex projects					
d	C's claims are reasonable					
e	C's claims are exaggerated					
f	The duration between each claim is too short					
g	C might have abused its right to claim					
h	Incomplete supporting documents are common in claim submission processes					
i	C's withholding of supporting documents is deliberate					
j	C behaved opportunistically					

assess the impacts of the project. C submitted yet another claim of US$8 million for additional rock quality investigations. In addition, the reinforcement bars delivered to the site have to be protected while the rock investigation is conducted. For this, an extra of US$2 million was required. Being concerned about the delay that may arise due to the quality of the rock, the Indonesian government issued instructions for the rock investigation and the protection of the reinforcement bars.

Upon receiving the instruction, C found that no relevant specification for the reinforcement protection work was included in the contract documentation. Furthermore, there were major differences in the scope of the rock investigations between C and the geotechnical division of the Indonesian government. C sought time to address the issues of specification for reinforcement protection and the scope of the investigation.

C further raised a claim of US$10 million for the idling of labour and plants for the two-week time taken to deal with these matters.

In July 2016, a claim for Extension of Time and Direct Loss and/or Expense was made by C following a labour strike that had caused a 25-day standstill of the project. C requested a sum of US$30 million for compensation.

In September 2016, C received a variation ordered by the Indonesian government to omit one high-end housing estate originally planned as residences for senior executives working in the office tower of the satellite town. This change aimed to ensure the completion time for the whole development within the 6-year contract period. C submitted a claim of US$100 million for loss of profit. The monetary claims in the second year are listed in Table 3. Assuming you are the appointed third-party neutral, please indicate your degree of agreement on the statements about the project's dispute claims in the second year in Table 4.

The contractual bases of the claims were as follows: (i) differing site conditions discovered; (ii) vague specifications in the contract documentation; (iii) inadequate information about the scope of work and (iv) order variations and project delays due to unexpected events. Prima facie, these claims appeared to have reasonable contractual grounds and could be considered legitimate. This would mark a very different approach from the claims supported in the first year. The respondents were asked to rate their level of agreement level on a 5-point scale (1 = 'Strongly Disagree' to 5 = 'Strongly Agree' on the same list of contractors' behaviours. The results were compared with the respondents' evaluations in the two parts. The aim is to identify whether the respondents had been influenced by the contractor's opportunistic moves.

In this setting, the evaluations of the respondents in Part B are expected to be lower than those in Part A if the evaluations are based on the contractual grounds raised by the contractor. If there was judgement bias due to the opportunistic behaviours of the contractor during the first year of the project, the evaluations in both parts would not show notable differences. Accordingly, no significant differences between the answers in Part A and Part B suggest that the respondents' evaluation in Part B had not duly taken into account what had occurred during the second year. The biased effect of their evaluations in Part A is demonstrated.

Part C of the simulation seeks to understand the decision process of the respondents. The representations of anchoring and confirmation bias are used. Based on the

Table 3 The monetary claims in the second year

Month	Issues in dispute	Amount (US$ million)
2016/03	Additional rock quality investigation	8
2016/03	Additional corrosion protection measure	2
2016/06	Idling of labour and plants due to inadequate information	10
2016/07	EOT and direct loss and/or expense due to labour strike	30
2016/09	Loss of profit due to omission of work	100
	Total claim	150

Table 4 Statements about the project's dispute claims in the second year

No	Descriptions	Strongly disagree	Disagree	Neutral	Agree	Strongly agree
		1	2	3	4	5
a	Initial budget allowance is too low					
b	C's bid is improbable					
c	Heavy claims are foreseeable in complex projects					
d	C's claims are reasonable					
e	C's claims are exaggerated					
f	The duration between each claim is too short					
g	C might have abused its right to claim					
h	Incomplete supporting documents are common in claim submission processes					
i	C's withholding of supporting documents is deliberate					
j	C behaved opportunistically					

relevant theoretical analysis and with appropriate operationalisation to suit the simulation context, bias representations and their categorisation are given in Table 5. The respondents were asked to rate their degree of agreement on a scale from 'Strongly Disagree = 1' to 'Strongly Agree = 5' with these manifestation statements as their decision-making approaches during the simulation.

Table 5 Bias representations and their categorisation

	Bias manifestation	References
A	*Anchoring bias*	
A1	Behaviours of either party in part A were influential	[6, 58]
A2	I referred to my previous assessments to form my final judgements	[18, 63]
A3	I compared the happenings in part B with those in part A	[6, 24]
A4	I used the reasons leading to my answers for part A in answering part B	[19, 35, 63]
A5	My opinions on the claims in part B are based on the happenings in both part A and part B	[24, 35]
B	*Confirmation bias*	
B1	I made assumptions during a reading of the case	[15, 35, 49]
B2	The repeated behaviours of C have strengthened my assessment	[37, 49, 53]
B3	My confidence in the assumptions was reinforced when supporting information was identified	[27, 29, 59]
B4	The acts of C in part B that are similar to those in part A were more notable	[22, 53]
B5	I had no intention of changing my assessments in part A after reading about the occurrences in part B	[15, 22, 35]
B6	Information supporting my assumptions are more noticeable	[53, 54]

Data Collection

Both paper-based and online forms were used to collect data. The online respondents were identified from websites of government departments, including the Housing Authority, Buildings Department, Civil Engineering and Development Department; and from websites of professional institutes, including The Hong Kong Institute of Engineers, The Hong Kong Institute of Architects, The Hong Kong Institute of Surveyors and Hong Kong Institute of Construction Managers. Paper-based data collection forms were distributed to participants of workshops and seminars of learned societies. The respondents were construction professionals. 53 valid responses were collected. The profile of the respondents are presented in Table 6.

Table 6 Profile of the respondents

Organization role	Percentage (%)	Working experience	Percentage (%)
Contractor	33	Less than 5 years	43
Employer	26	5–10 years	34
Consultant	41	More than 10 years	23
Total	100	Total	100

Findings

The Existence of Bias

A Student's t-test was used to test hypotheses about sample means [32] to identify whether there was a statistically significant difference between the mean responses in Part A and Part B. The null hypothesis was that the responses in the two parts had the same mean values. Student's t-test results are shown in Table 7. The significance value of the t-test for equality of means is 0.594 ($\geq$0.05), and the null hypothesis is therefore accepted (alpha = 0.05). As such, the respondents' answers for Part A and Part B were not significantly different. This means that, notwithstanding the built-in difference in the approach of the contractor's claims, the view of the contractor remains largely the same. In particular, opportunistic moves are first presented in Part A, which may suggest that the respondents' evaluation had been affected by their impressions of the contractor's opportunistic behaviours during the first year of the project.

Principal component factor analysis (PCFA) is a technique that explores the underlying constructs of a group of variables [11, 40]. In this study, PCFA was conducted on the data collected from Part C of the simulation. The PCFA results would indicate the suitability of the categorisation of bias representations developed for the study. The suitability of the data set for PCFA is examined with the Kaiser–Meyer–Olkin (KMO) test and Bartlett's test of sphericity [5]. The Kaiser–Meyer–Olkin (KMO) value of this study is 0.535, which is above the threshold of 0.5 [21, 64]. Bartlett's test result is also significant ($\leq$0.001). The data set is therefore sufficient to conduct PCFA [38, 51]. PCFA is useful to develop taxonomies. Similar procedures are adopted here. Table 8 presents the structure of bias. Typically, only items of factor loadings higher than 0.45 are kept within a certain extract group [42, 50, 61]. The PCFA suggests a two-factor structure: A1, A2, A3, A4 and A5 in factor 1 and B1, B2, B4, B5 in factor 2. B3 and B6 had factor loadings less than 0.45, therefore, these two statements were not included. The groupings suggested by PCFA are in line with the respective theories on bias. The overall results of PCFA indicate that the respondents' evaluations displaced the effects of anchoring and confirmation biases.

Table 7 Student's t-test results

No	Description		Score
1	T		0.535
2	Df		104
3	Sig(2-tailed)		0.594
4	Mean difference		0.040
5	Std. error difference		0.074
6	95% confidence interval	Lower	−0.107
		Upper	0.187

Table 8 The structure of bias

Bias Manifestations		Loading		SC (overall sample)	SC		
		Factor a	Factor b		Below 5 years	5–10 years	Above 10 years
A	*Anchoring bias*			*3.62*	*3.68*	*3.56*	*3.58*
A1	Behaviours of either party in part A were influential	0.529		3.8	3.9	3.6	3.8
A2	I referred to my previous assessments to form my final judgements	0.495		3.5	3.5	3.6	3.5
A3	I compared the occurrences in part B with those in part A	0.748		3.5	3.5	3.4	3.6
A4	I used the reasons leading to my answers for part A in answering part B	0.674		3.4	3.5	3.4	3.3
A5	My opinions on the claims in part B are based on the happenings in both part A and part B	0.789		3.8	4.0	3.8	3.7
B	*Confirmation bias*			*3.62*	*3.64*	*3.58*	*3.63*
B1	I made assumptions during a reading of the case		0.514	3.7	3.7	3.6	3.7

(continued)

Table 8 (continued)

Bias Manifestations		Loading		SC (overall sample)	SC		
		Factor a	Factor b		Below 5 years	5–10 years	Above 10 years
B2	Repeated behaviours of C have strengthened my assessment		0.486	3.7	3.8	3.5	3.7
B3	My confidence in the assumptions was reinforced when supporting information was identified			Variable omitted due to low factor loading			
B4	The acts of C in part B that are similar to those in part A were more notable		0.603	3.6	3.6	3.7	3.6
B5	I had no intention of changing my assessments in part A after reading about the occurrences in part B	0.562	3.5	3.5	3.5	3.6	3.6
B6	Information supporting my assumptions are more noticeable			Variable omitted due to low factor loading			

The significance score (SC) was used to examine the extent of influence in terms of the likelihood of the two types of bias [64]. The significance score is computed by the following formula:

$$SC_i = \frac{\sum_{j=1}^{n} S_{ij}}{n}$$

where SC_i is the significance score of factor i, S_{ij} is the mean score of the jth manifestation of factor i, and$n$ is the number of manifestations in factor i. A higher chance of being affected by bias will be interpreted with a high SC_i.

Applying the formula for the significance score of each factor will give the average of the mean scores of the representations of the factor. The significance scores of the two factors—anchoring bias and confirmation bias—as well as the mean scores of the representations are shown in Table 8. All representations have mean scores larger than 3 out of a maximum of 5. This may be interpreted as the occurrence of the representations being quite likely. The overall significance scores of anchoring bias and confirmation bias have the same value of 3.6. There is no notable difference in the change between these two biases. It can also be read as the respondents do not show an inherent inclination towards either form of bias. This result further suggests that the influence of bias can be quite subconscious. For anchoring bias, A1 and A5 received a rating of 3.8. These two representations, therefore, have a higher likelihood of occurrence. The overall results do suggest the evaluations in Part A have an influence on the respondents in their responses for Part B. Accordingly, the respondents might have already formed an opinion of the contractor, i.e., is practising opportunism. This impression is difficult to eliminate when assessing the behaviours of the contractor for the claims raised in the second year. The respondents' perception of the contractor remained, notwithstanding the legitimacy of the claims. For confirmation bias, B1 and B2 had the largest mean scores of 3.7. The occurrence of these two representations by the respondents is considered quite high. As such, they made certain assumptions in reading the case, and these assumptions were reinforced by further consistent information. The practice of confirmation bias means that the respondents would pay more attention to the facts in Part B that are akin to the opportunistic behaviours that occurred during the first year. In this way, the assessments for Part A were reinforced and in fact, were used as the basis for their evaluations in Part B.

Significance scores were also calculated for different groups of respondents in terms of their years of experience. The results are shown in Table 8. It can be noted that respondents with less than 5 years of experience have the highest mean scores for both anchoring bias (3.68) and confirmation bias (3.64) when compared with those of other subgroups. These scores are also higher than those of the overall sample anchoring bias (3.62) and confirmation bias (3.62). Therefore, respondents with less than 5 years of experience are more prone to the effect of biases. This suggests that training can improve the situation, especially for early-career third-party neutrals.

Discussion and Recommendations

Simulation data were collected to investigate whether third-party neutrals are subject to bias in discharging their service as dispute settlement facilitators. There are two major findings. First, it is found that third-party neutrals are also subject to bias. Second, anchoring and confirmation are the two principal forms of bias that may affect them. Anchoring can be strategically induced by a tactical disputant. The simulation results show that respondents' assessments of the contractor's behaviour in Part B did not reflect the changes in the construction game plan. It is suggested that a perception of untrustworthiness was formed towards the contractor during Part A. This perception was retained when evaluating the behaviour of the contractor in Part B. This analysis is supported by the data returned by the respondents who were also self-evaluating their behaviour in Part C. These findings inform construction dispute management. As facilitators of dispute negotiators, third-party neutrals must be free from any form of preoccupation or from having a predetermined view of the dispute. Impartiality is the most valued quality of a respected third-party neutral.

This study further examined bias in Part C by identifying the underlying constructs of biased behaviours. PCFA affirmed the existence of two forms of bias: anchoring and confirmation. In fact, this finding also pinpoints that anchoring and confirmation biases often go hand in hand, although their theoretical bases may be different. They are, in fact, instinctually linked. Regarding the particulars of the respondents, it is also noted that respondents with less experience would be more likely to be influenced by the tactics of the disputants. It is suggested that training can be an effective means to control biases. In addition, the importance of experience in the behaviour of third-party neutrals is confirmed.

For cogent construction dispute management, this study reinforces the findings of [8] that opportunistic behaviour is one of the major causes of disputes [8]. Typically, this happens with contractors being awarded projects because of their below-cost bids. With the contract confirmed, the contracting party's dependency asymmetry alters with the increase in asset-specific investments [48, 62]. With the aim of recovering project loss, some contractors take advantage of every opportunity that comes their way [13, 26, 64]. A study conducted by Ho and Liu [28] concluded that contractors' cut-throat bidding was influenced by a high chance of reclaiming profits through later opportunistic claims [52]. This study reminds us of the possibility that, under the influence of the opportunistic behaviours of project participants, third-party neutrals may make biased decisions. When third-party neutrals are biased, it is difficult for them to offer impartial advice. In fact, their credentials may be jeopardised. With the rising use of alternative dispute resolution techniques, it is expected that the demand for third-party neutrals will also increase. Less experienced third-party neutrals may be used. One of the findings of this study reminds us that this group of third-party neutrals is more vulnerable to the tactics utilised by disputants. Afterwards, third-party neutrals should regularly update their skills and knowledge so that the influence of bias can be reduced.

Summary

Construction contracts are inevitably incomplete due to the impossibility of foreseeing all future contingencies. Incomplete contracts and ambiguous terms are fertile ground for opportunistic behaviours, which were found to be major causes of construction project disputes [8, 43, 47]. This study explores the possibility of third-party neutrals having judgemental biases that are induced by project participants' opportunistic behaviours. A simulation was designed to mimic a CPDR process. The findings in this study indicate that the chance of biases in third-party neutrals' judgement is real. Two types of bias are highlighted, namely, anchoring and confirmation. As third-party neutrals are widely employed to facilitate communications between disputing parties, it is impertzive that their service be fair and impartial. Thus, relevant training to improve third-party neutrals' ability to avoid bias is imperative.

Acknowledgements The empirical works of this chapter has been reported in a paper entitled "Paradox of Bias and Impartiality in Facilitating Construction Dispute Resolution" in the Journal of Legal Affairs and Dispute Resolution in Engineering and Construction. Specials thanks for Ms. Birgitta LEVINA for her assistance in the study. The support of HKSAR GRF (11209118) is duly acknowledged.

References

1. Ask K, Granhag P (2005) Motivational sources of confirmation bias in criminal investigations: the need for cognitive closure. J Invest Psychol Off. 2(1):43–63
2. Bendle N, Cotte J (2016) Assumptions of rationality in political marketing: the case of the republican autopsy. J Nonprofit Publ Sector Mark 28(1):66–83
3. Binmore K (2009) Rational decisions. Princeton University Press, Princeton, NJ
4. Carson S, Madhok A, Wu T (2006) Uncertainty, opportunism and governance: the effects of volatility and ambiguity on formal and relational contracting. Acad Manage J 49(5):1058–1077
5. Cerny CA, Kaiser HF (1977) A study of a measure of sampling adequacy for factor-analytic correlation matrices. Multivar Behav Res 12(1):43–47
6. Chertkoff JM, Conley M (1967) Opening offer and frequency of concession as bargaining strategies. J Pers Soc Psychol 7(2):181–185
7. Cheung SO (2010) Construction mediation landscape in the civil justice system in Hong Kong. J Leg Aff Disput Resolut Eng Constr 2(3):169–174
8. Cheung SO, Pang KHY (2012) Anatomy of construction disputes. J Constr Eng Manag. https://doi.org/10.1061/(ASCE)CO.1943-7862.0000532
9. Cheung SO, Suen HC (2002) A multi-attribute utility model for dispute resolution strategy selection. Constr Manag Econ 20(7):557–568
10. Cheung SO, Yeung YW (1998) The effectiveness of the dispute resolution advisor system: a critical appraisal. Int J Project Manage 16(6):367–374
11. Cheung SO, Yiu TW (2006) Are construction disputes inevitable? IEEE Trans Eng Manage 53(3):456–470
12. Cheung SO, Li K (2019) Biases in construction project dispute resolution. Eng Constr Archit Manage
13. Crowley LG, Hancher DE (1995) Evaluation of competitive bids. Constr Eng Manag 121(2):238–245
14. Cusick W (2009) All customers are irrational. AMACOM, New York
15. Doherty ME, Mynatt CR, Tweney RD, Schiavo MD (1979) Pseudodiagnosticity. Acta Physiol (Oxf) 43(2):111–121
16. Emerson RM (1976) Social exchange theory. Ann Rev Sociol 2:335–362
17. Engel A, Korf B (2005) Negotiation and mediation techniques for natural resource management. Rome, FAO (UN Food and Agriculture Organization)

18. Englich B, Mussweiler T (2001) Sentencing under uncertainty: anchoring effects in the court room. J Appl Soc Psychol 31(7):1535–1551
19. Englich B, Mussweiler T, Strack F (2006) Playing dice with criminal sentences: the influence of irrelevant anchors on experts' judicial decision making. Pers Soc Psychol B 32(2):188–200
20. Eriksson T, Kadefors A (2017) Organisational design and development in a large rail tunnel project—influence of heuristics and mantras. Int J Proj Manag 35(3):492–503
21. Ferguson E, Cox T (1993) Exploratory factor analysis: a users' guide. Int J Sel Assess 1(2):84–94
22. Fischhoff B, Beyth-Marom R (1983) Hypothesis evaluation from a Bayesian perspective. Psychol Rev 90(3):239–260
23. Fisher RJ (1995) Pacific, impartial third-party intervention in international conflict: a review and analysis. Beyond confrontation: learning conflict resolution in the post-cold war era. University of Michigan Press, Ann Arbor, pp 33–59
24. Galinsky AD, Mussweiler T (2001) First offers as anchors: the role of perspective-taking and negotiator focus. J Pers Soc Psychol 81(4):657
25. Goldberg SB, Sander FE, Rogers NH, Cole SR (2014) Dispute resolution: negotiation, mediation and other processes. Wolters Kluwer Law & Business
26. Grogan T (1992) Low bids raise hidden costs. Eng News-Rec 228(13):30–31
27. Hendry SH, Shaffer DR (1989) On testifying in one's own behalf: interactive effects of evidential strength and defendant's testimonial demean or on jurors' decisions. J Appl Psychol 74(4):539–545
28. Ho SP, Liu LY (2004) Analytical model for analyzing construction claims and opportunistic bidding. Constr Eng Manag 130(1):94–104
29. Holstein JA (1985) Jurors' interpretations and jury decision making. Law Hum Behav 9(1):83–99
30. Honeyman C (1985) Patterns of bias in mediation. J Disp Resol 1985(9):141–149
31. Izumi C (2010) Implicit bias and the illusion of mediator neutrality. Wash UJL & Pol'y 34:71–155
32. Johnston LW (1970) Student's t-test. J Qual Technol 2(4):243–245
33. Kahneman D (2003) Maps of bounded rationality: psychology for behavioral economics. Am Econ Rev 93(5):1449–1475
34. Keren G, Teigen KH (2004) Yet another look at the heuristics and biases approach. In: Koehler DJ, Harvey N (eds) Blackwell handbook of judgment and decision making. Wiley, pp 89–109
35. Klayman J (1995) Varieties of confirmation bias. Psychol Learn Motiv 32:385–418
36. Klayman J, Ha Y (1987) Confirmation, disconfirmation, and information in hypothesis testing. Psychol Rev 94(2):211–228
37. Koriat A, Lichtenstein S, Fischhoff B (1980) Reasons for confidence. J Exp Psychol Hum Learn Mem 6(2):107–118
38. Leech NL, Barrett KC, Morgan GA (2005) SPSS for intermediate statistics, use and interpretation, 2nd edn. Lawrence Erlbaum Associates Inc., New Jersey
39. Leung RHM (2014) Hong Kong mediation handbook. Sweet & Maxwell, Hong Kong
40. Li K, On Cheung S (2018) Bias measurement scale for repeated dispute evaluations. J Manag Eng 34(4):04018016
41. Li K, Cheung SO (2019) Unveiling cognitive biases in construction project dispute resolution through the lenses of third-party neutrals. ASCE J Constr Eng Manage 145(11)
42. Llewellyn CH, van Jaarsveld CH, Johnson L, Carnell S, Wardle J (2011) Development and factor structure of the baby eating behaviour questionnaire in the Gemini birth cohort. Appetite 57(2):388–396
43. Lu W, Zhang L, Zhang L (2016) Effect of contract completeness on contractors' opportunistic behavior and the moderating role of interdependence. J Constr Eng Manag. https://doi.org/10.1061/(ASCE)CO.1943-7862.0001110
44. Lu W, Zhang L, Pan J (2015) Identification and analyses of hidden transaction costs in project dispute resolutions. Int. J Proj Manage Elsevier Ltd Int Project Manage Ass 33(3):711–718. https://doi.org/10.1016/j.ijproman.2014.08.009

45. Lui SS, Ngo HY (2004) The role of trust and contractual safeguards on cooperation in non-equity alliances. J Manage 30(4):471–485. https://doi.org/10.1016/j.jm.2004.02.002
46. McFadden D, Lim G (2015) Mediation in Singapore: a practical guide. Sweet & Maxwell
47. Mitropoulos P, Howell G (2001) Model for understanding, preventing, and resolving project disputes. J Constr Eng Manag. https://doi.org/10.1061/(ASCE)07339364(2001)127:3(223)
48. Moschandreas M (1997) The role of opportunism in transaction cost economics. J Econ Issues 31(1):39–58
49. Nickerson RS (1998) Confirmation bias: a ubiquitous phenomenon in many guises. Rev Gen Psychol 2(2):175
50. Nischal S, Bhalla GS (2015) An exploratory factor analysis of sources underlying organisational conflict-a comparative approach between public and private sector banks. Abhigyan 32(4):38–51
51. Olawale F, Garwe D (2010) Obstacles to the growth of new SMEs in South Africa: a principal component analysis approach. AJBM 4(5):729
52. Pang HY, Cheung SO, Choi MC, Chu SY (2015) Opportunism in construction contracting: minefield and manifestation. Int J Proj Org Manag 7(1):31–55
53. Peer E, Gamliel E (2013) Heuristics and biases in judicial decisions. CR, 49,114–118
54. Pennington N, Hastie R (1993) The story model for juror decision making. In: Hastie R (ed) Inside the juror: the psychology of juror decision making. Cambridge University Press, New York, pp 192–221
55. Regan PM (1996) Conditions of successful third-party intervention in intrastate conflicts. J Conflict Resolut 40(2):336–359
56. Saeb A, Mohamed OB, Danuri MSBM, Zakaria NB (2018) Critical factors for selecting a neutral to support alternative dispute resolution methods in the construction industry. Civ Eng J 4(1):11–23
57. Simon HA (1955) A behavioral model of rational choice. Q J Econ 69:99–118
58. Thompson L (2008) The truth about negotiations. Upper Saddle River, NJ, Pearson Education
59. Troutman CM, Shanteau J (1977) Inferences based on nondiagnostic information. Organ Behav Hum Perform 19(1):43–55
60. Tversky A, Kahneman D (1974) Judgment under uncertainty: heuristics and biases. Science 185(4157):1124–1131
61. Veltkamp PR, Hansen KJ, Barkley RM, Sievers RE (1996) Principal component analysis of summertime organic aerosols at Niwot Ridge, Colorado. J Geophys Res 101(D14):19495–19504
62. Williamson OE (1985) The economic institutions of capitalism: firms, markets, and relational contracting. Free Press, New York
63. Wilson T, Houston C, Etling K, Brekke N (1996) A new look at anchoring effects: Basic anchoring and its antecedents. J Exp Psychol Gen 125(4):387–402
64. Wong SP, Cheung SO (2004) Trust in construction partnering: views from parties of the partnering dance. Int J Proj Manag 22(6):437–446
65. Wong SP, Cheung SO (2004) Trust in construction partnering: views from parties of the partnering dance. Int J Proj Manag 22(6):437–446
66. Yates D (2002) Reducing the incidence of claims and disputes in construction contracts, CIB, 221–234

Intervening Decision-Making in Using Alternative Dispute Resolutions: A Parsimonious Intervention Model

Chia Kuang Lee, Tak Wing Yiu, and Sai On Cheung

Introduction

The construction industry has been plagued with various conflicts and disputes. As conflicts are common in the workplace, it can however escalate into disputes if not managed [19]. Given the fact that disputes are almost inevitable [5], disputants must rely on the available contractual avenues to resolve disputes, in addition to relying on the course of litigation. Alternative Dispute Resolution (ADR) methods have been used to successfully settle contractual disputes. However, in Malaysia, methods of ADR, especially mediation, are less applied [14].

When disputes arise, the parties may seek formal methods of dispute resolution to resolve dispute. In Malaysia, these rights are incorporated in most of the traditional forms of contracts. These rights include, but are not limited to, arbitration, mediation, adjudication, dispute review board, dispute adjudication board, and expert determination. ADR use remains largely voluntary [7]. The rights to use ADR are enshrined in most of the available standard forms produced by professional institutes, such as The Institution of Engineers, Malaysia (IEM); Pertubuhan Arkitek Malaysia (PAM); the Construction Industry Development Board (CIDB), and the Public Work Department [20]. However, in the face of disputes in any normal form of contract, it is not necessary to initiate any ADR method that is thought to be mandated. For example,

C. K. Lee (✉)
Faculty of Industrial Management, Universiti Malaysia Pahang, Pahang, Malaysia
e-mail: chia@ump.edu.my

T. W. Yiu
School of Built Environment, Massey University, Auckland, New Zealand
e-mail: tyiu@massey.ac.nz

S. O. Cheung
Construction Dispute Resolution Research Unit, City University of Hong Kong, Hong Kong, China
e-mail: Saion.cheung@cityu.edu.hk

S. O. Cheung (ed.), *Construction Dispute Research Expanded*, Springer Tracts in Civil Engineering, https://doi.org/10.1007/978-3-030-80256-1_15

Clause 7 in Model Terms of Construction contract for Subcontract Work (Published by CIDB) stipulates that parties to a dispute are encouraged to use amicable resolves.

To better understand decision making in ADR use, Lee et al. [15] have reclassified the factors underpinning ADR use into cognitive, normative, and control domains, Lee et al. [16] have proposed a Macro–Micro model in ADR use; and Lee et al. [17] have evidently shown that amongst attitude, subjective norms, and perceived behavioural control, attitude prevails as the significant predictor in explaining the intention to use ADR. Based on the empirical notion that attitudinal factor influence intention to use ADR, which is the culmination of decision making, this paper intends to revise the Technology Acceptance Model (TAM) in predicting decisions making in using alternative dispute resolution (ADR) when disputes occurred.

Literature Review

Technology Acceptance Model is a model developed based on Theory of Reasoned Action [9]. TAM model is a well-known model that deals with technology adoption [25]. TAM posits that a user's behavior to use a system is predicted by behavioural intention (INT) to use a system. The immediate determinant of behavioural intention (INT) is Attitude (ATT). Attitude (ATT) is further predicted by two major beliefs, which are perceived usefulness (PU) and perceived ease of use (PE). Perceived ease of use (PE) predicts perceived usefulness (PU).

Behavioural Intention

Intention is the "indication of how hard people are willing to try, how much of an effort willing to put in performing a behavior" [3]. Intention is depicted to be the predictor of actual behavior, and sums up the motivational factors that influence behavior [3]. In the context of ADR use, intention can be contextualized as "I plan to use mediation/negotiation to settle this project disputes", or "I will use mediation/negotiation to resolve this project disputes", or "I have the intention to use mediation/negotiation to settle this project disputes".

Attitude

Attitude refers to the generated favorable or unfavorable feelings towards a behavior [3, 4]. The decision maker eventually forms and generates attitude towards the given behavior through myriads and structures of beliefs. Attitude can be mathematically represented as $\sum_{i=1}^{n} B_i a_i$, where B_i is the sum of beliefs about the effects of performing a given behavior. It can be coined in this fashion, such as "Using mediation or negotiation will assure the quality of settlement". On the other hand, a_i refers to evaluation of the effect of performing the behavior, which can appear in the form of question such as "using mediation or negotiation and assuring the quality of settlement is good/bad". It appears that the formation of attitude aligns with Edwards' Decision Theory Model with the equation of maximizing the subjective expected utility $\sum_{i=1}^{n} SP_i U_i$ in maximizing gains or minimizing loss, where SP_i refers to the subjective probability that a certain consequence will follow with certain act, while U_i refers to maximized gains or minimized losses from the given alternatives. In TAM, attitude is influenced by two major beliefs, which are Perceived Usefulness (PU) and Perceived Ease of Use (PE).

Perceived Usefulness/Perceived Relative Advantage

Perceived usefulness (PU), which is defined as the extent to which a user believes that using a system will help them perform better in their job [8, 22]. Perceived usefulness (PU) is found to be an influential determinant of attitude [24]. Perceived usefulness (PU) has a similar notion of **"perceived relative advantage"** from the perspective of the diffusion of innovation theory.

Innovation is seen as an "idea, practice, or object" that is perceived as new by an individual [21]. In the context of ADR, it is an innovative legal practice that aid disputants to resolve disputes without the need to litigate. ADR process is seen as the result of innovation that has taken place within the judicial system [23]. According to Rogers [21], relative advantage refers to the degree to which an innovation is perceived as better than the idea it supersedes. Perceived relative advantage (PA) significantly influence innovation adoption decisions [25]. In the context of ADR use, perceived relative advantage maybe measured in time, quality, effectiveness, and financial impact, for example: "Using this ADR method would settle the dispute more quickly, or using this ADR method would improve the quality of settlement between parties, or using this ADR method would enhance the effectiveness of dispute settlement, or using the ADR method would be more economical". Compared to perceived usefulness (PU), perceived relative advantage (PA) is more apt to be conceptualized as a determinant of attitude towards ADR use.

Perceived Ease of Use

Secondly, is perceived ease of use (PE), which is defined as the degree to which a user believes that using a particular system would be free of effort [8]. It has a similar notion of "perceived complexity" in the diffusion of innovation theory. Perceived ease of use (PE) generally refers to the degree of difficulty of an innovation to be put into use and executed, while perceived complexity refers to the degree to which an innovation is perceived to be difficult to understand and use [21]. Generally, a system that is perceived to be less complex and easy to use would lead to a favorable attitude towards the system [1]. In the context of ADR use, disputants who view ADR methods to be less complex and easy to instigate would tend to have favorable feelings towards its use. Perceptions of ease of use in ADR can appear in the form of adherence, such as "instructions of using this ADR method would be easy to follow", or "it would be easy to learn/adhere to the process of using this ADR method".

Research Methodology

This study first invited 6 construction experts for an interview session. The experts were asked to provide their insights on the constructs such as Perceived Relative Advantage (PA), Perceived Ease of Use (PE), Attitude (ATT), and Intention (INT). As intention (INT) is the culmination of decision making, the experts were addressed with these main questions:

Question 1: These are the important factors that influence intention to use ADR to settle dispute. What is your opinion? Based on the model, what is the most important factor influencing Intention to use ADR?"

Question 2: "Can you propose any interventions (such as programs/strategies/policies) that can make Attitude towards using ADR to settle dispute more favorable?"

Question 3: "What are the issues that hinder the use of ADR in the Malaysian construction industry? How to deal with these issues by referring to this model? How can this model help to improve the level of ADR use in the Malaysian construction industry?"

Based on the feedback obtained in the interview sessions, thematic analysis was employed, and a parsimonious intervention model was then conceptualized. A structured questionnaire measuring the constructs of the conceptualized model were then developed, and ready to be sent out for main survey. The main survey consists of three major sections. In section 1, four different sections of dispute scenarios (Scenario A, B, C, D) were embedded in the questionnaire. These scenarios were formulated based on [2, 13] definitions of dispute. Table 1 presents the description of Scenario A, B, C and D in section 1.

Based on the chosen scenario, the respondents will be asked to reflect the types of disputed claims in the project in section 2 of the questionnaire. The Dispute Taxonomies based on [6, 13, 18] as depicted in Table 2.

With guided instructions, the respondents were then asked to make ADR selections based on the chosen Scenario. Table 3 shows the detailed Instructions for ADR Selection.

Table 1 Section 1—description of scenario A, B, C and D

Scenario category	Scenario description
A	Unable to reach settlement for a major dispute in one of the projects
B	Recently settled a major dispute in one of the projects with an ADR method
C	Settlement is in progress for a major dispute in one of the projects by using an ADR method
D	In the process of negotiating a claim/claims in one of the project

Table 2 Dispute taxonomies based on [6, 13, 18]

Category	Types of project disputes encountered
1	Change or variation order
2	Errors in drawings, specifications, and quantities
3	Differing site conditions
4	Payment (e.g., delayed progress payment or nonpayment related dispute
5	Delay (e.g., extension of time and disruption related dispute)
6	Ambiguity in contract terms or contract interpretation
7	Quality related (e.g., defects, workmanship)
8	Performance related (e.g., supply of goods, materials, execution of work, suspension issue, issue of regularly and diligently)
9	Information- and administrative-related dispute
10	Awards and decisions (e.g., dispute about adjudication or arbitration awards)
11	Professional negligence
12	Personal injuries
13	Property damages

Table 3 Detailed instructions for ADR selection

Scenario	Instructions	Choice of ADR methods
A	No further instruction given—this implies that the current ongoing dispute requires settlement	Select only one ADR method to settle the dispute
B	Imagine this dispute would reoccur in the future, and when negotiation fails	Select only one ADR methods to settle this similar dispute
C	Imagine if your current ADR method fails to settle the dispute	Select only one ADR methods to settle this similar dispute
D	Imagine if this negotiation breaks down and turns into a major dispute	Select only one ADR methods to settle this similar dispute

The choice of ADR methods include arbitration, mediation, adjudication under the Construction Industry Payment and Adjudication Act 2012 (CIPAA 2012), adjudication (contractual), expert determination, dispute review board, and dispute adjudication board (DAB). Finally, in Section 3 of the questionnaire, the respondents will be required to respond to the survey based on 7-point Likert Scale, where "1—Strongly Disagree, 2—Moderately Disagree, 3—Somewhat Disagree, 4—Neutral, 5—Somewhat Agree, 6—Moderately Agree and 7—Strongly Agree".

Administration of the Main Survey

Prior to the survey, the questionnaire was pretested by three industrial experts. After refinements, the survey was pilot tested by 20 building and civil engineering contractors. After careful modifications and improvements of the questionnaire, it was then sent out to 2000 contractor companies registered under the Construction Industry Development Board (CIDB). The questionnaires were stratified according to the contractors' tender capacities, using the stratified simple random sampling approach. Table 4 shows the contractors' Classification According to Tender Capacity, while Table 5 shows the stratified samples of the contractors according to their grade.

Results and Discussions

The results and discussions are presented in the following manner: Firstly, the demographic background of the respondents (six experts in the interview and respondents who responded to the survey) are presented, followed by project details, thematic analysis of extending the original TAM model, and assessment of the parsimonious intervention model in both measurement and structural model.

Table 4 Contractors' classification according to tender capacity

Grade	Tendering capacities USD ($)	Paid up Capital/net capital worth USD ($)	Contractor categories (Size)
G7	No limit	168,000.00	Large
G6	Not exceeding 4.5 million	111,800.00	Medium
G5	Not exceeding 1 million	55,900.00	Medium
G4	Not exceeding 680,000	33,500.00	Medium
G3	Not exceeding 220,000	11,180.00	Small
G2	Not exceeding 110,000	5,590.00	Small
G1	Not exceeding 22,000	1,120.00	Small

Table 5 Stratified samples of the contractors according to their grade

State	Grade1		Grade 2		Grade 3		Grade 4		Grade 5		Grade 6		Grade 7	
	N	S	N	S	N	S	N	S	N	S	N	S	N	S
JHR	2623	91	976	34	765	27	285	10	300	10	111	4	386	13
KDH	1959	68	432	15	214	7	103	3	108	4	54	2	181	6
KLTN	1966	68	493	17	202	7	84	3	106	3	61	2	119	4
LBN	137	4	21	1	15	1	2	1	1	0	0	0	1	0
MLK	985	34	275	10	213	7	108	4	106	4	37	1	130	5
N9	1756	61	522	18	258	9	107	4	132	4	41	1	95	3
PHG	2089	73	561	19	317	11	176	6	152	5	60	2	122	4
PRK	2415	84	592	21	397	14	191	7	226	8	83	3	156	5
PRS	864	30	105	4	46	2	15	0	27	1	6	0	31	1
PNG	1118	39	232	8	329	11	139	5	176	6	81	3	328	11
SBH	5302	184	1215	42	561	20	136	5	186	7	74	3	445	16
SRK	1971	69	777	27	358	12	132	5	169	6	78	3	436	15
SLR	3493	122	1150	40	1359	47	609	21	916	32	262	9	1252	44
TGN	2257	79	570	20	248	9	152	5	196	7	85	3	186	7
WP	1307	45	411	14	859	30	418	14	890	31	243	8	1312	46
Total		1051		290		214		93		128		44		180

Demographic Background of Respondents

Table 6 shows the interviewees background. All experts have at least 7 years of experience in the construction industry. 5 interviewees are working for the contractors, with the positions such as Assistant Project Manager, Senior Surveyor, Assistant Contract Manager, and Construction Site Manager; while another one is working as a Senior Project Executive in the consultant company. All of them are experienced in the use of ADR in dispute settlement.

Table 7 shows the background of the respondents (Scenario A and D) that responded to the survey. 37 respondents encountered on-going dispute (Scenario A), 13 respondents had recently settled disputes, and asked to imagine if such dispute would re-occur in the future (Scenario B), 14 respondents were in the stage of settling disputes (Scenario C), while 64 respondents are negotiating their claims (Scenario D). The minimum sample size for PLS path model evaluation should be at least equal to or greater than 10 times the largest number of structural paths pointing to a specific construct in the model [11, 12]. Due to sample size problems, Scenario B and Scenario C were eliminated from the study, with only Scenario A and Scenario D prevailing.

Majority of the respondents in Scenario A have high authority in decision making. The respondents hold prominent positions such as managing directors (27%), directors (16.2%), and project managers (8.1%). Respondents in Scenario D have a similar background as managing directors (21.9%), directors (21.9%), and executive directors (14.0%). 45.9% of the respondents in Scenario A have experience in ADR use, while 6% of the respondents have experience in ADR use.

Table 6 Interviewees background

Expert code	Experience in construction (years)	Designation in organisation	Type of organisation	Experience in the use of alternative dispute resolution (ADR)
E1	7	Senior Project Executive	Consultant	Yes
E2	8	Assistant Project Manager	Contractor	Yes
E3	7	Senior Surveyor	Contractor	Yes
E4	8	Senior Surveyor	Contractor	Yes
E5	7	Assistant Contract Manager	Contractor	Yes
E6	8	Construction Site Manager	Contractor	Yes

Table 7 The background of the respondents (Scenario A and D)

Designation in organization	Scenario A		Scenario D	
	Frequency	%	Frequency	%
President	2	5.4	1	1.6
Executive Director	2	5.4	9	14.0
Managing Director	10	27.0	14	21.9
Chief Executive Officer	2	5.4	4	6.3
Director	6	16.2	14	21.9
Assistant Vice President	–	–	1	1.6
Project Manager	3	8.1	2	3.1
Assistant Project Manager	–	–	1	1.6
Project Coordinator	–	–	1	1.6
Contract Manager	2	5.4	–	–
Assistant Contract Manager	–	–	1	1.6
General Manager	1	2.7	2	3.1
Site Engineer	1	2.7	–	–
Site Manager	1	2.7	–	–
Project Engineer	1	2.7	1	1.6
Project Executive	–	–	1	1.6
Executive Manager	2	5.4	–	–
Admin Manager	1	2.7	–	–
Senior Executive	1	2.7	–	–
Engineer	1	2.7	–	–
Manager	1	2.7	6	9.4
Assistant Manager	–	–	2	3.1
Contract Executive	–	–	2	3.1
Quantity Surveyor	–	–	2	3.1
Authorized to make decision for ADR use in projects				
1—strongly disagree	–	–	–	–
2—moderately disagree	–	–	–	–
3—somewhat disagree	–	–	–	–
4—neutral	–	–	–	–
5—somewhat agree	31	83.8	43	67.2
6—moderately agree	3	8.1	14	21.9
7—strongly agree	3	8.1	7	10.9
Contractors' grade				
Grade 1	–	–	3	4.7
Grade 2	12	32.4	14	21.9

(continued)

Table 7 (continued)

Designation in organization	Scenario A		Scenario D	
	Frequency	%	Frequency	%
Grade 3	5	13.5	15	23.4
Grade 4	2	5.4	13	20.3
Grade 5	3	8.1	6	9.4
Grade 6	1	2.7	2	3.1
Grade 7	14	37.9	11	17.2
Years of experience in the construction industry				
1–5 years	11	29.7	17	26.6
6–10 years	7	18.9	13	20.3
11–15 years	7	18.9	15	23.4
16–20 years	5	13.5	5	7.8
21–25 years	–	–	6	9.4
26–30 years	5	13.5	3	4.7
31–35 years	2	5.4	–	–
More than 35 years	–	–	5	7.8
Experience in ADR use	17	45.9	10	6

Project Details and Choice of ADR Methods

Table 8 shows the overall project details. Most of the projects in Scenario A include residential (32.4%), commercial (27.0%), civil & infrastructure (24.3%), and industrial projects (16.2%). The projects were located in Kuala Lumpur (16.2%), followed by Johor, Sarawak, Selangor (13.5%), Penang (10.8%), Kelantan, Pahang (8.1%), Sabah (5.4%), Kedah, Melaka, Perlis, and Terengganu, (2.7%).

On the contrary, most of the projects involved in Scenario D include civil and infrastructure (42.2%), residential (25.0%), commercial (20.3%), industrial (7.8%), healthcare (3.1%), and cultural projects (1.6%). The projects are located in Selangor (28.6%), Johor (10.9%), Sabah (9.4%,), Kuala Lumpur, Pahang (7.8%), Kelantan, Sarawak (6.3%), Negeri Sembilan, Perak, Penang, Terengganu (4.7%), Kedah and Melaka (3.1%).

Payment issues (27%) was the main encountered issue, followed by change or variation order related (21.6%), errors in drawings, specifications, and quantities (16.2%), delay issues (13.5%), quality related issues (5.4%), differing site conditions (5.4%), nomination and re-nomination issues, professional negligence, ambiguity in contract terms, and performance related issues (2.7%). Similarly, in Scenario D, payment issues (31%) was the main encountered issue, followed by change/variation order (23%), delay issues (11%,), compliance with instruction (7.8%), errors in drawings specifications and quantities, differing site conditions, information and administration related disputes (6.3%), quality related disputes (3.1%,), ambiguity

Table 8 Overall project details

Project details	A (N = 37)		D(N = 64)	
	N	%	N	%
Type of project				
1. Residential	12	32.4	16	25.0
2. Commercial	10	27.0	13	20.3
3. Cultural	–	–	1	1.6
4. Sporting	–	–	–	–
5. Healthcare	–	–	2	3.1
6. Civil and infrastructure	9	24.3	27	42.2
7. Industrial	6	16.2	5	7.8
Project location				
1. Johor	5	13.5	7	10.9
2. Kedah	1	2.7	2	3.1
3. Kelantan	3	8.1	4	6.3
4. Melaka	1	2.7	2	3.1
5. Negeri Sembilan	–	–	3	4.7
6. Pahang	3	8.1	5	7.8
7. Perak	1	2.7	3	4.7
8. Perlis	–	–	–	–
9. Penang	4	10.8	3	4.7
10. Sabah	2	5.4	6	9.4
11. Sarawak	5	13.5	4	6.3
12. Selangor	5	13.5	17	28.6
13. Terengganu	1	2.7	3	4.7
14. Kuala Lumpur	6	16.2	5	7.8
Type of project dispute/claim				
Change/variation order	8	21.6	15	23
Errors in drawings, specifications and quantities	6	16.2	4	6.3
Differing site conditions	2	5.4	4	6.3
Payment (E.G.: delayed progress payment/non–payment related dispute, etc.)	10	27.0	20	31
Delay (E.G.: extension of time and disruption related dispute, etc.)	5	13.5	7	11
Ambiguity in contract terms/contract interpretation	1	2.7	1	1.6
Quality related (E.G.: defects, workmanship, etc.)	2	5.4	2	3.1
Performance related (E.G.: supply of goods, materials, execution of work, suspension issue, issue of "regularly and diligently" etc.)	1	2.7	–	–

(continued)

Table 8 (continued)

Project details	A (N = 37)		D(N = 64)	
	N	%	N	%
Information and administrative related dispute	–	–	4	6.3
Awards and Decisions (E.G.: dispute about adjudication/arbitration awards, etc.)	–	–	–	–
Professional negligence	1	2.7	–	–
Personal injuries	–	–	1	1.6
Property damages	–	–	–	–
Nomination and re-nomination (E.G.: appointment of replacement person, etc.)	1	2.7	1	1.6
Compliance with instruction (E.G.: compliance with instruction by S.O/architect, atc.)	–	–	5	7.8
Contract sum (In USD $)				
Contract sum less than 2 million	25	67.6	59	92.2
2 million ≤ contract sum < 11 million	6	16.2	4	6.3
11 million ≤ contract sum < 22 million	2	5.4	1	1.6
22 million ≤ contract sum < 33 million	2	5.4	–	–
33 million ≤ contract sum < 44 million	–	–	–	–
44 million ≤ contract sum < 55 million	–	–	–	–
55 million ≤ contract sum	2	5.4	–	–
Intended choice of ADR to settle hypothetical dispute/claim scenario				
1. Arbitration	15	40.5	30	46.9
2. Mediation	7	18.9	9	14.1
3. Adjudication under CIPAA 2012	8	21.6	15	23.4
4. Adjudication (Contractual)	3	8.1	3	4.7
5. Expert determination	3	8.1	5	7.8
6. Dispute review board	–	–	1	1.6
7. Dispute adjudication board	–	–	1	1.6
8. Other: (facilitated negotiation)	1	2.7	–	–

in contract terms, personal injuries, nomination and re-nomination related issues (1.6%).

In Scenario A, most of the projects had contract sum less than 2 Million (67.6%), followed by 6 projects (16.2%) with value between 2 and 11 million, 2 projects (5.4%) had value between 11 and 22 Million, followed by 2 projects (5.4%) valued between 22 and 33 Million, and another 2 projects (5.4%) with more than 55 Million. On the other hand, most of the projects (92.2%) in Scenario D had contract sum less than 2 Million, followed by 4 projects (6.3%) with value between 2 and 11 million, and 1 project with contract sum value between 11 and 22 Million USD.

Majority of the respondents (40.5%) in Scenario A chose arbitration, followed by adjudication under CIPAA 2012 (21.6%); mediation (18.9%); adjudication (Contractual) and expert determination (8.1%); and facilitated negotiation (2.7%). Arbitration was the most popular method in Scenario D (46.9%), followed by adjudication under CIPAA 2012 (23.4%), mediation (14.1%), expert determination (7.8%), adjudication contractual (4.7%), dispute review board (1.6%) and dispute adjudication board (1.6%).

Interview Analysis

The significance of attitude towards intention (INT) was validated by six experts. The experts were required to address their views on the questions as follows:

"These are the important factors that influence intention to use ADR to settle dispute. What is your opinion? Based on the model, what is the most important factor influencing Intention to use ADR?"

The six interviewees (experts) are connoted with the Expert Code: E1, E2, E3, E4, E5, E6 in Table 6 for the ease of reference in the discussion. E1, E2, E3, E4, E5 and E6 claimed that attitude (ATT) is a significant important factor that influenced intention to use ADR. The corresponding feedback comments were about the relevance of attitude that influence intention (INT):

*"To me, the model is quite true. The majority important factors that affect intention would be attitude... Anyway, attitude is the most important one. When you have good **attitude** to use ADR, you would have higher intention to use ADR"* **Validation Expert 1 (E1).**

*But still, in my opinion, the most **important factor is attitude**... "* **Validation Expert 2 (E2).**

*"Basically, I will take every factor (as shown) in the model into consideration as the influential factors towards intention to use ADR. Every factor has their own strength in influencing intention, because dispute would come with different facts and cases, so it is case-by-case basis. So based on your model, in my opinion...**attitude** comes second". **Validation Expert 3 (E3).***

*ADR users would evaluate themselves and weighing the possibility against the consequences of dispute, and from there proceeds with any intention to use ADR. For example, let us say there is a dispute regarding the variation order (VO). The employer disagrees with the all the VO submitted by the main contractor, thus dispute occurs. To proceed with the use of ADR, contractor (ADR user) needs to first weigh potential loss and additional expenses and cash-flow issues affected by the dispute, while at the same time the contractor needs to collect the facts and documents involved associated with the use of ADR...**attitude** towards ADR method comes next".* **Validation Expert 4 (E4).**

"My opinion is that this model is generally ok and logical. When disputes happen, first we will try to negotiate with the client. There will be conversations through letters. If this cannot solve the dispute, we will have our opinion, they would have their opinion, and then we have no choice but to take litigation. Yes, overall these factors will influence intention to take ADR...I think attitude comes second". **Validation Expert 5 (E5).**

"Well, all these factors will influence on the intention to use the ADR. I do agree with the significance and relevance of all these factors, definitely they will influence on the intention... if you follow the sequence, ***attitude*** *comes second..."* **Validation Expert 6 (E6).**

Based on the relevance of attitude (ATT) in influencing ADR selection, the experts were further probed with this question:

"Can you propose any interventions (such as programs/strategies/policies) that can make attitude towards using ADR to settle dispute more favourable?"

The feedback included:

"Attitude towards ADR use can be enhanced if the ADR method itself is improved, or increase the ***competencies of the neutrals****, that would lead to better disposition of perceived ease of use (****PE****) and relative advantage (****PA****)..."* ***Validation Expert 1 (E1).***

*"To make attitude better, ADR users can be convinced of the ease of use (****PE****), relative advantage (****PA****) through instillations of values and propagations of knowledge by the key players in the industry..."* **Validation Expert 2 (E2).**

"Most of the construction personnel do not ***know and unaware*** *about the advantage or ease of use of ADR...Possible active interventions can begin by influencing these two factors: "perceived relative advantage", and "perceived ease of use" in the model. Education and trainings to the construction workers may be helpful".* **Validation Expert 3 (E3).**

"To foster better attitudes towards ADR use, most users need to be exposed to the characteristics and features of ADR first..." **Validation Expert 4 (E4).**

"To make people like ADR, mostly everything is concerning with money. Mediation would be ***cheaper*** *than arbitration. Very cheap, mostly more than hundred thousand if you go arbitration. The second one depends on the relationship between two parties. If very bad then arbitration, it depends, on the attitude in the ADR method. Make people more aware on ADR; make sure people think that ADR can protect their business relationship".* **Validation Expert 5 (E5).**

"As for the attitude, it depends on the person himself whether or not the ADR is more favourable to settle the dispute. It is personal ***perception on the advantage*** *of each dispute method. I think some of the factors from the* ***normative sides*** *can make attitude overall favourable.* ***I think construction industry like CIDB, and NGOs like Master Builders should encourage people to use ADR to solve dispute related***

to any projects. The interventions can be done through construction and government agencies; they should educate contractors, consultants and developers on the understanding of all these ADR. Education is important. *Even in the university, they can educate strategies to all future contractors that there is such a way to win any dispute cases by using ADR. This education, road shows, seminars might influence on the attitude more favourable.*" **Validation Expert 6 (E6).**

To suggest possible interventions, with reference to the ADR decision-making behavioural model, the interviewees were further probed with these questions:

"What are the issues that hinder the use of ADR in the Malaysian construction industry? How to deal with these issues by referring to this model? How can this model help to improve the level of ADR use in the Malaysian construction industry?"

All validation experts emphasized on the importance of propagating awareness towards improving attitude (ATT). Possible interventions could be made via road shows, industrial campaigns, educations, seminars and even training. These efforts could help intervene in the perceived advantage (PA) and ease of use (PE) of ADR methods. The qualitative interviews discussions below managed to highlight the importance of awareness:

"The biggest issue in Malaysia is still the lack of awareness. Just go back to the awareness. Because people are not trained, most contractors, developers, or consultants, they are not aware and do not know how to use and the rights. They do not know how to use, and most important thing is ***lack of positive attitude****. For example, the* ***arbitrators are not well trained****, they are lawyers, they are not from the construction, and they are not aware of the construction law and contract. To deal with this issue, we still have* ***to do more seminars****, let more personnel from construction to be involved in* **more seminars***, not only for the arbitrators or for lawyer. Your model actually can show that what are the factors that affect intention to use ADR, and through the factors we can pinpoint which are the factors which we can do something like…for example like focus on the factors, what we are lack of and which factor to improve…mostly on attitude by cultivating more awareness."* **Validation Expert 1 (E1).**

"So far the biggest issue in ADR use in the Malaysian construction industry is there is lack of protection by law. ADR does not guarantee outcomes in its use (perceived relative advantage issues) … People do not use ADR because they lack confidence, and bottleneck issues and lack of outcome of ADR. To settle this, find ways to increase perceptions of trusts on the methods and increase confidence… Interventions can be done based on ***propagating success stories about other companies through conference, trainings, brochures****, that can create more awareness for potential users. Overall, this model is easy and detail in portraying the relationship towards intention (selection behaviour), where it creates more understanding for possible interventions on the important factors."* **Validation Expert 2 (E2).**

"Lack of awareness is the main culprit behind low ADR use. Awareness refers to dos and don't's in ADR use. Well from this model, I know that all these six factors are important to look into, as you know every dispute is different, we can refer to this model and look at the problem and intervene in from ***different perspectives****. Based on your explanation, I do believe that this model and your findings can be of help to the policy maker in the Malaysian Construction industry itself. The model needs to be exposed to the industry player as this model can show them the process of decision-making in using and choosing ADR"* **Validation Expert 3 (E3)**.

"Some of the issues such as ***awareness****, and* ***lack of expertise*** *as a whole affect the use of ADR. Lack of awareness is something related to attitude, it falls under category of attitude. They do not know about the* ***ease of use and relative advantage****. Besides that, an organisation would lack of expertise, which falls under organisational competency. Construction professionals should be trained since university level to be competent with the technique. An organisation that has this competency would have more control on it. Besides, trainings and* ***propagation of knowledge improve awareness****...I agree that the model could be insightful to the key players in understanding decision-making in ADR use"* **Validation Expert 4 (E4)**.

In my opinion, the overall ADR use is greatly affected by seriousness of the case. For example, payment issue. If the client delays your payment 3 months, this is not an attitude issue; we have to look at the case. This is serious, your cash flow is tight, and you have no choice. You have to use ADR. However, depends on the status of the relationship, they will choose ADR, because some ADR has their ***own advantage****, some will destroy mutual good relationship. Good relationship means they would have future project together. Sometimes depends on lawyer advice and* ***expert suggestion, some ADR method may prove to be ease of use****. Sometimes we have our intention to take litigation. However, when some expert says that "mediation is better, not that harsh, and you can get back what you want", would actually affect our intention. Besides, most of us do not have the knowledge, expert and the team. Construction team needs the expert, they* ***do not know this right or know how to use****, and they would give up this right. To solve these issues, no knowledge, go to seminar, more aware, knowledge, and technical skill. Some of the issues can be solved by this model...Overall, at certain point, this model is easy to understand and at least can show the important factor, and if you want to influence people to use ADR, you can refer to this model. Why? Because some users* ***do not have the knowledge****, this is due to lack of awareness, so, by* ***promoting seminar*** *they will know. You focus on attitude right. For me attitude is a key role, you can convince user. How? You can use expert, because people has* ***trust on the expert****. If you go litigation, is not a play-game, it is a serious matter. You have to increase trust on the mediator (neutral)."* **Validation Expert 5 (E5)**.

"Education is one thing, probably is how the people around advise on the use of ADR. The second issue is lack of expertise, as they do not have expertise in this. They need to be educated, and made known that this ADR method will help them win any dispute. If the method itself has more advantage, it can help them. If they know

how to use, it will also increase the use of ADR. The method itself, existence and awareness of the technicality aspects of the ADR use will help. The policy maker can learn from this model, as they know how to influence the contractor's awareness in knowing that this ADR will help them in any dispute in the future. This model can help policy maker to intervene in the behaviour of ADR use." **Validation Expert 6 (E6).**

Drawing on the qualitative discussions, interventions through seminars and education by professional institutes and other professional practices, may resemble the effects of normative pressures (NP); while interventions through competencies of neutrals are operationalised as trust (TR). Normative pressures (NP) is a form of institutional pressure that derives from professionalization [10]. Trust (TR) captures attributes of ability, integrity, and benevolence of the neutrals. The qualitative results suggest that trust (TR) might not have a direct influence on attitude (ATT), but through the functions of perceived ease of use (PE) and perceived relative advantage (PA). Overall, normative pressures and favourable attributes of the neutrals are postulated to be effective in intervening the decision to use ADR through cognitive instrument process.

Development of a Parsimonious Intervention Model

Complemented by qualitative interviews, a parsimonious intervention framework can be developed as follows (Fig. 1):

The hypotheses postulated in the new conceptualized parsimonious intervention model are as follows:

H_a Attitude (ATT) relates positively with Intention (INT).
H_b Perceived Relative Advantage (PA) relates positively with Attitude (ATT).
H_c Perceived Ease of Use (PE) relates positively with Attitude (ATT).

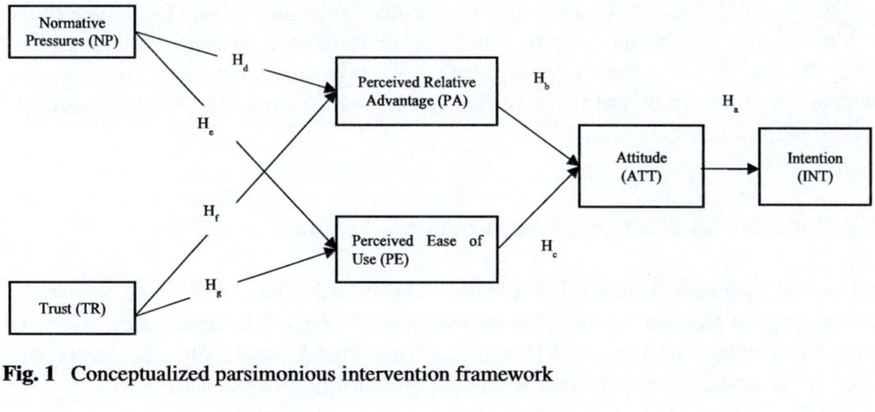

Fig. 1 Conceptualized parsimonious intervention framework

H_d Normative Pressures (NP) relates positively with Perceived Relative Advantage (PA).
H_e Normative Pressures (NP) relates positively with Perceived Ease of Use (PE).
H_f Trust (TR) relates positively with Perceived Relative Advantage (PA).
H_g Trust (TR) relates positively with Perceived Ease of Use (PE).

Evaluation of Parsimonious Intervention Framework

The procedures for assessing both measurement and structural model were presented and discussed in this section.

Measurement Model of Intervention Framework (Scenario A and D)

37 respondents were deciding ADR methods for dispute settlement (Scenario A); while 64 respondents were in the stage of negotiating claims (Scenario D). Scenario A renders decision making during dispute in route, while Scenario D portrays decision making before dispute occurs.

The reliability and validity of measurement models for both scenarios were assessed with composite reliability, indicator loadings, convergent reliability and discriminant validity. To assess the reliability of the data sets, the internal consistency reliability was assessed with composite reliability, while indicator reliability was assessed with indicator loadings. To evaluate the validity of the data sets, convergent validity and discriminant validity were assessed with Average Variance Extracted (AVE). Overall, the validity and reliability results of Measurement Model for Scenario A and D are presented in Table 9 while the cross loadings assessments and the Fornell-Larcker criterion assessment are shown in Tables 10 and 11.

The results show that all loadings for all items, and composite reliability (CR) for all constructs were above 0.7. The reliability of the datasets for both scenarios (A and D) are confirmed.

The AVE of all constructs exceeds the threshold value of 0.5, and the square roots of the AVE are greater than the constructs' correlation with other constructs respectively. No indicator loadings were higher than the opposing constructs. These results confirm the convergent and discriminant validity of the data sets (both Scenario A and D) were satisfactorily met.

Structural Model of Intervention Framework (Scenario A and D)

Following the assessment of reliability and validity of the datasets for both scenarios, the subsequent step was to evaluate the structural model of the framework. Accordingly, the result showed that all VIF values are less than 5.0 and within the acceptable range. Collinearity is not an issue in the data sets for both scenarios.

Table 9 Validity and reliability results of measurement model (intervention framework: scenario A and D)

Constructs	Items	Loadings (Scenario)		Composite reliability (Scenario)		Average variance extracted (AVE) (Scenario)	
		A	D	A	D	A	D
Attitude (ATT)	ATT_1	0.9716	0.9732	0.984	0.9801	0.9536	0.9427
	ATT_2	0.9737	0.9814				
	ATT_3	0.9843	0.958				
Intention (INT)	INT_1	0.9797	0.9716	0.9793	0.9819	0.9221	0.9312
	INT_2	0.9679	0.9743				
	INT_3	0.9405	0.9671				
	INT_4	0.9526	0.9468				
Normative pressures (NP)	NP_1	0.9244	0.9411	0.9444	0.969	0.81	0.8867
	NP_2	0.9423	0.943				
	NP_3	0.9193	0.9504				
	NP_4	0.8078	0.9321				
Perceived relative advantage (PA)	PA_1	0.9489	0.8612	0.983	0.9754	0.9061	0.8689
	PA_2	0.9463	0.9709				
	PA_3	0.9673	0.964				
	PA_4	0.9711	0.9547				
	PA_5	0.9264	0.9154				
	PA_6	0.9508	0.922				
Perceived ease of use (PE)	PE_1	0.9753	0.9933	0.9827	0.9934	0.9498	0.9804
	PE_2	0.9896	0.9892				
	PE_3	0.9586	0.988				
Trust (TR)	TR_1	0.8664	0.9381	0.9791	0.9884	0.8703	0.9243
	TR_2	0.9483	0.9638				
	TR_3	0.9395	0.9691				
	TR_4	0.9664	0.974				
	TR_5	0.9384	0.9868				
	TR_6	0.9431	0.9521				
	TR_7	0.9251	0.9452				

Both PLS Algorithm (path weighting scheme) with maximum iterations of 300 and a bootstrapping procedure of 37 cases (for the case of scenario A) and 64 cases (for the case of scenario D) and 5000 samples were undertaken to evaluate the R^2 value, structural model relationships, Effect Sizes (f^2), predictive relevance Q^2 and q^2 effect sizes.

Table 10 Cross loadings assessments (intervention framework: scenario A and D)

Scenario A							Scenario D					
	ATT	INT	NP	PA	PE	TR	ATT	INT	NP	PA	PE	TR
ATT_1	*0.9716*	0.8804	0.5806	0.6546	0.7656	0.7279	*0.9732*	0.8517	0.5851	0.7541	0.6808	0.6389
ATT_2	*0.9737*	0.8813	0.5923	0.6452	0.7318	0.6997	*0.9814*	0.8508	0.5531	0.7107	0.6311	0.6542
ATT_3	*0.9843*	0.8995	0.6122	0.6657	0.7676	0.7251	*0.958*	0.8204	0.5161	0.7224	0.6423	0.6011
INT_1	0.8919	*0.9797*	0.6835	0.7339	0.7639	0.7639	0.8761	*0.9716*	0.557	0.6587	0.5465	0.6185
INT_2	0.8536	*0.9679*	0.6816	0.6916	0.7547	0.727	0.8016	*0.9743*	0.573	0.6771	0.5058	0.6182
INT_3	0.8884	*0.9405*	0.6123	0.6544	0.7215	0.736	0.8398	*0.9671*	0.5462	0.6238	0.5125	0.6067
INT_4	0.8538	*0.9526*	0.6187	0.702	0.7124	0.7324	0.8229	*0.9468*	0.598	0.7459	0.5571	0.6337
NP_1	0.6053	0.6546	*0.9244*	0.8707	0.7883	0.8556	0.5636	0.5467	*0.9411*	0.7206	0.6003	0.6257
NP_2	0.5686	0.6439	*0.9423*	0.794	0.8008	0.8408	0.5375	0.561	*0.943*	0.7223	0.5948	0.6152
NP_3	0.5667	0.6209	*0.9193*	0.7852	0.8031	0.7525	0.513	0.5671	*0.9504*	0.6592	0.5362	0.6125
NP_4	0.4283	0.4921	*0.8078*	0.4984	0.6087	0.518	0.5238	0.5432	*0.9321*	0.6742	0.5266	0.5915
PA_1	0.6501	0.6591	0.8036	*0.9489*	0.8325	0.83	0.6819	0.6787	0.5701	*0.8612*	0.746	0.6127
PA_2	0.6407	0.6719	0.8075	*0.9463*	0.8471	0.8435	0.7392	0.6796	0.7113	*0.9709*	0.7728	0.6884
PA_3	0.6434	0.7323	0.801	*0.9673*	0.8166	0.8423	0.7135	0.6807	0.6867	*0.964*	0.7317	0.675
PA_4	0.6458	0.6919	0.801	*0.9711*	0.8318	0.8618	0.7257	0.6879	0.7392	*0.9547*	0.7523	0.6882
PA_5	0.5926	0.6569	0.7676	*0.9264*	0.8759	0.8472	0.7149	0.6286	0.7136	*0.9154*	0.7157	0.6603
PA_6	0.6584	0.7244	0.7966	*0.9508*	0.8221	0.8069	0.6186	0.5572	0.7	*0.922*	0.6336	0.5992
PE_1	0.7847	0.7545	0.8197	0.891	*0.9753*	0.8915	0.6696	0.5468	0.5931	0.7788	*0.9933*	0.7172
PE_2	0.7626	0.7575	0.8092	0.8845	*0.9896*	0.8611	0.6699	0.5598	0.6172	0.7637	*0.9892*	0.7075
PE_3	0.7115	0.7356	0.8313	0.7933	*0.9586*	0.8143	0.6541	0.527	0.5747	0.7715	*0.988*	0.695

(continued)

Table 10 (continued)

Scenario A								Scenario D				
TR_1	0.6648	0.6646	0.8074	0.8202	0.7988	*0.8664*	0.6444	0.6573	0.655	0.6666	0.7188	*0.9381*
TR_2	0.6591	0.7389	0.8275	0.8344	0.8431	*0.9483*	0.6653	0.6536	0.6358	0.6699	0.7066	*0.9638*
TR_3	0.7215	0.7452	0.7645	0.8081	0.8094	*0.9395*	0.5816	0.5833	0.6122	0.688	0.6729	*0.9691*
TR_4	0.6863	0.732	0.8039	0.8255	0.813	*0.9664*	0.6306	0.6078	0.5881	0.681	0.6801	*0.974*
TR_5	0.6843	0.7184	0.7859	0.8608	0.812	*0.9384*	0.6365	0.6408	0.6721	0.7015	0.6994	*0.9868*
TR_6	0.6947	0.731	0.7483	0.8037	0.7981	*0.9431*	0.6171	0.5836	0.6017	0.6588	0.6738	*0.9521*
TR_7	0.6881	0.7008	0.7455	0.7971	0.8605	*0.9251*	0.6003	0.5887	0.6034	0.6635	0.6484	*0.9452*

Table 11 Fornell-Larker criterion assessment

	Scenario A						Scenario D					
	ATT	INT	NP	PA	PE	TR	ATT	INT	NP	PA	PE	TR
ATT	*0.9765*						*0.9709*					
INT	0.9085	*0.9603*					0.8663	*0.965*				
NP	0.6094	0.6759	*0.9000*				0.5684	0.5888	*0.9416*			
PA	0.671	0.7243	0.8366	*0.9519*			0.7511	0.7003	0.7385	*0.9321*		
PE	0.7733	0.7689	0.8411	0.8797	*0.9746*		0.6712	0.5501	0.6011	0.779	*0.9902*	
TR	0.735	0.7707	0.8402	0.881	0.8788	*0.9329*	0.6505	0.6417	0.6496	0.7029	0.7137	*0.9614*

Table 12 R^2 values of the endogenous construct

Endogenous construct (R^2)	Scenario A (N = 37)	Scenario D (N = 64)
Intention (INT)	0.825 (Substantial)	0.751 (Substantial)
Attitude (ATT)	0.598 (Moderate)	0.583 (Moderate)
Perceived relative advantage (PA)	0.808 (Substantial)	0.632 (Moderate)
Perceived ease of use (PE)	0.808 (Substantial)	0.542 (Moderate)

Assessment of R^2 Values of the Endogenous Construct (Parsimonious Intervention Model: Scenario A and D)

The R^2 values of the endogenous construct for both scenarios were examined. Table 12 shows the R^2 values of the endogenous construct.

The intervention framework explains 82.5% (substantial) of the variance of intention to use ADR (selection behaviour) when dispute occurs (Scenario A), and 75.1% (substantial) of the variance of intention to use ADR (selection behaviour) in negotiation phase (Scenario D).

Assessment of Structural Model Path Coefficients

Empirical *t* value was assessed to determine the significance level of the path coefficient. The path significance testing results are shown in Table 13.

With reference to the path coefficient and t-statistics, the hypotheses for both scenario A and D were assessed and examined. The findings showed that the intervention framework for both scenarios were structurally good ($R^2 = 0.825$ for scenario A; $R^2 = 0.751$ for scenario D) and sufficient in predicting behavioural intention to use ADR. The findings showed that in an event of dispute (Scenario A), the perceived ease of use (PE) of the method contributes to the decision-making process. It was found that perceived relative advantage (PA) is not significant for the cognitive process. Evidently, interventions through normative pressures (NP) and trust (TR) are possible as they positively influence perceived ease of use (PE). Accordingly, perceived ease of use (PE) positively influences attitude (ATT), and a favourable attitude (ATT) contributes to intention to use ADR. In Scenario Á, all paths were significant, except H_b: Perceived Relative Advantage (PA) → Attitude (ATT).

On the contrary, when the decision makers were in the stage of negotiation claims (Scenario D), perceived ease of use (PE) does not influence attitude (ATT). Perceived relative advantage of ADR (PA) evidently forms the attitudinal cognitive structures. It was found that the path from perceived relative advantage (PA) → attitude (ATT) is significant. Normative pressures (NP) and trust (TR) would strongly influence the perceptions of utilities in ADR methods [perceived relative advantage (PA)]. In

Table 13 Path significance testing results

Scenario	Path	Hypothesis test		
		Path coefficient (β)	T-statistics (t)	Interpretation
Intention (R^2 = 0.825) Scenario A	H_a attitude → intention	0.9085	25.7468***	Supported
	H_b perceived relative advantage → attitude	−0.0414	0.0917	NS
	H_c perceived ease of use → attitude	0.8098	2.2851**	Supported
	H_d normative pressures → perceived relative advantage	0.3277	1.8236*	Supported
	H_e normative pressures → perceived ease of use	0.3496	2.2540**	Supported
	H_f trust → perceived relative advantage	0.6057	3.2053***	Supported
	H_g trust → perceived ease of use	0.5850	4.1214***	Supported
Intention (R^2 = 0.751) Scenario D	H_a attitude → intention	0.8663	22.0394***	Supported
	H_b perceived relative advantage → attitude	0.5806	3.2233***	Supported
	H_c perceived ease of use → attitude	0.2190	1.2860	NS
	H_d normative pressures → perceived relative advantage	0.4876	3.2965***	Supported
	H_e normative pressures → perceived ease of use	0.2379	1.3246	NS
	H_f trust → perceived relative advantage	0.3862	2.7414***	Supported

(continued)

Table 13 (continued)

Scenario	Path	Hypothesis test		
		Path coefficient (β)	T-statistics (t)	Interpretation
	H_g trust → perceived ease of use	0.5591	4.1733***	Supported

***Significant at $p < 0.01$, 1%, ** Significant at $p < 0.05$, 5%, *Significant at $p < 0.1$, 10%, 2 tailed test

Table 14 Effect size result for confirmed path

Scenario	Path	Effect sizes		
		f^2		Interpretation
		A	D	
Intention ($R^2 = 0.825$) Scenario A	H_a attitude → intention	–	–	–
	H_c perceived ease of use → attitude	0.368	–	Large
	H_d normative pressures→ perceived relative advantage	0.167	–	Medium
	H_e normative pressures→ perceived ease of use	0.182	–	Medium
	H_f trust→ perceived relative advantage	0.563	–	Large
	H_g trust → perceived ease of use	0.521		Large
Intention ($R^2 = 0.751$) Scenario D	H_a attitude → intention	–	–	–
	H_b perceived relative advantage → attitude	–	0.317	Medium
	H_d normative pressures→ perceived relative advantage	–	0.375	Large
	H_f trust→ perceived relative advantage		0.234	Medium
	H_g trust → perceived ease of use		0.395	Large

this scenario, all paths were significant, except H_c Perceived Ease of Use (PE) → Attitude (ATT); and H_e Normative Pressures (NP) → Perceived Ease of Use (PE).

Assessment of Effect Sizes

Subsequently all effect sizes for confirmed paths were computed. The effect sizes result for confirmed path are portrayed in Table 14.

Table 15 Predictive relevance of endogenous construct

Endogenous construct	R^2 value		Q^2 value	
	Scenario A	Scenario D	Scenario A	Scenario D
Intention (INT)	0.825 (Substantial)	0.751 (Substantial)	0.7536 (Substantial)	0.6930 (Moderate)
Attitude (ATT)	0.598 (Moderate)	0.583 (Moderate)	0.5333 (Moderate)	0.5424 (Moderate)
Perceived relative advantage (PA)	0.808 (Substantial)	0.632 (Moderate)	0.7320 (Moderate)	0.5457 (Moderate)
Perceived ease of use (PE)	0.808 (Substantial)	0.542 (Moderate)	0.7640 (Substantial)	0.5418 (Moderate)

For Scenario A (dispute phase), perceived ease of use (PE) has a large effect ($f^2 = 0.368$) on attitude (ATT). To intervene in perceived ease of use (PE) effectively, the competencies of the neutrals need to be carefully enhanced, as trust (TR) has a large effect ($f^2 = 0.521$) on perceived ease of use (PE). Meanwhile, normative pressures (NP) have a medium effect on perceived ease of use (PE).

For Scenario D (Negotiation phase), perceived relative advantage (PA) has a medium effect on attitude (ATT) ($f^2 = 0.317$). Perceived relative advantage (PA) can be influenced significantly by both normative pressures (NP) and trust (TR). Normative pressures (NP) have a larger effect on perceived relative advantage (PA), while trust (TR) has a medium effect on perceived relative advantage (PA).

Assessment of Predictive Relevance Q^2

The predictive relevance of the constructs on the confirm paths was examined with Stone-Geisser's Q^2 by using the blindfolding procedure. The omission distance (D) for both scenarios was set to 7. Cross-validated redundancy was used to examine Q^2. Table 15 shows the Predictive Relevance of Endogenous Construct. The Q^2 values are considerably above zero, providing support for the model's predictive relevance related to the endogenous construct.

The summary of predictive relevance are shown in Table 16. For Scenario A, perceived ease of use (PE) has a medium predictive relevance ($q^2 = 0.277$) for attitude (ATT); normative pressures (NP) has a medium predictive relevance ($q^2 = 0.216$) for perceived relative advantage (PA); but a small predictive relevance ($q^2 = 0.119$) for perceived ease of use (PE). Trust (TR) has both large predictive relevance for perceived relative advantage (PA) ($q^2 = 0.374$), and perceived ease of use (PE) ($q^2 = 0.385$).

For scenario D, perceived relative advantage (PA) has a medium predictive relevance ($q^2 = 0.278$) for attitude (ATT). Normative pressures (NP) has a medium predictive relevance ($q^2 = 0.276$) for perceived relative advantage (PA). Following that, trust (TR) has a medium predictive relevance ($q^2 = 0.159$) for perceived relative

Table 16 Summary of predictive relevance

Scenario	Path	Effect sizes			
		f^2 effect size		q^2 effect size	
		A	D	A	D
Intention (R^2 = 0.825) Scenario A	H_a attitude → intention	–	–		
	H_c perceived ease of use → attitude	0.368 (large)	–	0.277 (medium)	
	H_d normative pressures → perceived relative advantage	0.167(medium)	–	0.216 (medium)	
	H_e normative pressures → perceived ease of use	0.182 (medium)	–	0.119 (small)	
	H_f trust → perceived relative advantage	0.563 (large)	–	0.374 (large)	
	H_g trust → perceived ease of use	0.521 (large)		0.385 (large)	
Intention (R^2 = 0.751) Scenario D	H_a attitude → intention	–	–	–	–
	H_b perceived relative advantage → attitude	–	0.317 (medium)		0.278 (medium)
	H_d normative pressures → perceived relative advantage	–	0.375 (large)		0.276 (medium)
	H_f trust → perceived relative advantage		0.234 (medium)		0.159 (medium)
	H_g trust → perceived ease of use		0.395 (Large)		0.388 (large)

advantage (PA), but relatively a larger predictive relevance ($q^2 = 0.388$) for perceived ease of use (PE).

Discussion

Normative pressures (NP) and trust (TR) were found to influence perceived relative advantage (PA), and perceived ease of use (PE) of ADR methods. These two forces are capable to influence decision-making process in ADR use. Both perceived relative advantage (PA) and perceived ease of use (PE) were found to influence attitude (ATT). Favourable attitude (ATT) would then encourage intention (INT) to use ADR.

Consistent with Davis et al. [9] assertions, perceived ease of use (PE) will affect attitude (ATT) on the self-efficacy basis, where it is perceived to be easier to deal with, one will have more sense of control and sense of efficacy in using the system. Perceived ease of use (PE) in ADR significantly affects attitude (ATT) when disputes occur (Scenario A). One's affection and attitude towards the use of ADR may include feelings about the process of using ADR methods (Perceived ease of use, PE). The intrinsic motivation elements of ease of use in ADR lead to the beneficial emotions in ADR use to circumvent more problems in dispute situations if the ADR instructions are easy to obey, learn and stick to.

However, the empirical finding shows no proof of perceived relative advantage (PA) to attitude (ATT) when a dispute emerges. Overall, this means that the decision to use ADR must be directed solely by the direct effects of feelings (positive or negative) on the use of ADR, guided by a sense of effectiveness by ease of use of the methods. To the point before negotiations failed (Scenario D), the decision-making process through the use of ADR was driven by instrumental cognitive assessments. Attitudinal cognitive assessments of the effects of the use of ADR strongly rely on the relative benefit of ADR methods. Although perceived ease of use (PE) has lost its influence on attitude (ATT), perceived relative advantage (PA) retains its influence on attitude (ATT). Intention (INT) to use ADR strongly influenced by affective feelings (attitudinal dimensions) cantered on the instrumental results of ADR methods.

Overall, when there is a dispute (Scenario A) that compels decision makers to prefer ADR methods, the perceived ease of use (PE) is calculated to affect only the attitude. Favourable perceptions and evaluations of behaviour (ATT) will then affect the decision to use the ADR. The perceived relative advantage (PA) in the decision-making process is not important. Trust (TR) in neutrals has a greater impact on perceived ease of use, whereas normative stresses (NP) have a medium influence on perceived ease of use (PE).

In the phase of negotiation claims (Scenario D), only perceived relative advantage (PA) is found to influence attitude (ATT). Favourable attitudes (ATT) would then influence intention (INT) to use ADR. Perceived ease of use (PE) is not significant in the decision-making process. Normative pressures (NP) on the other hand have a larger influence on perceived relative advantage (PA), while trust (TR) has a medium effect on relative advantage (PA).

Summary

This study presents the development of a parsimonious ADR intervention model based on qualitative interview and structural equation modelling. Six expert interviews were used to extend the existing TAM Model. Another 101 decision makers responded to the main survey and tested the conceptualized parsimonious intervention model. Drawing on the interviews, normative pressures (NP) and trust (TR) were found to be influencing force on the attitudinal elements of a decision. In general, the intervention model was shown to be structurally sound and relevant, in two distinctive scenarios: the phase when an actual dispute occurs and the phase of negotiating claims prior to the occurrence of a dispute. Both normative pressures (NP) and trust (TR) were clearly shown to be influential towards perceived ease of use (PE) and perceived relative advantage (PA) in ADR use. Effective interventions for ADR use could include ADR campaigns and propagation of ADR values that dispel any myths about ADR and foster acceptance of ADR (fostering higher values of normative pressures), and training that enhances competencies of neutrals (fostering higher values of trust in neutrals). Favourable normative pressures (NP) and trusts (TR) would then lead to a favourable attitude (ATT) and increase the overall intention (INT) to use ADR through the interactions of perceived ease of use (PE) and relative advantage (PA) with attitude (ATT).

Acknowledgements The empirical work of this chapter has been reported in "Understanding Intention to Use Alternative Dispute Resolution in Construction Projects: Framework Based on Technology Acceptance Model." of the Journal of Legal Affairs and Dispute Resolution in Engineering and Construction. 10(1):04517021. The texts have been substantially re-written.

References

1 Abroud A, Choong Y, Muthaiyah S, Fie D (2013) Adopting e-finance: decomposing the technology acceptance model for investors. Service Business 1–22
2. Aibinu AA, Ofori G, Ling FY (2008) Explaining cooperative behavior in building and civil engineering projects' claims process: interactive effects of outcome favorability and procedural fairness. J Constr Eng Manag 134:681–691
3. Ajzen I (1991) The theory of planned behavior. Org Beh Human Decis Proces 50:179–291
4. Ajzen I, Fishbein M (2000) Attitudes and the attitude-behavior relation: reasoned and automatic processes. Eur Rev Soc Psychol 11:1–33
5. Al-Humaidi H (2014) Arbitration in Kuwait: study of current practices and suggestions for improvements. J Legal Affairs Dispute Resol Eng Constr 6:03013001
6. Charoenngam C, Mahavarakorn W (2011) Collaborative negotiation behaviors in Thai construction projects. J Leg Aff Disput Resolut Eng Constr 3:109–115
7. Cheung S (2006) Mandatory use of ADR in construction—a fundamental change from voluntary participation. J Prof Issues Eng Educ Pract 132:224–224
8. Davis FD (1989) Perceived usefulness, perceived ease of use, and user acceptance of information technology. MIS Quart 13:319–340
9. Davis FD, Bagozzi RP, Warshaw PR (1989) User acceptance of computer technology: a comparison of two theoretical models. Manage Sci 35:982–1003

10. Dimaggio PJP, Walter W (1983) The iron cage revisited: institutional isomorphism and collective rationality in organizational fields. Am Sociol Rev 48:147–160
11. Hair JF, Hult GTM, Ringle CM, Sarstdet M (2014) A primer on partial least squares structural equation modeling (PLS-SEM). California, SAGE Publications Inc., Thousand Oaks
12. Hair JF, Sarstedt M, Pieper TM, Ringle CM (2012) The use of partial least squares structural equation modeling in strategic management research: a review of past practices and recommendations for future applications. Long Range Plan 45:320–340
13. Kumaraswamy M (1997) Conflicts, claims, and disputes in construction. Eng Constr Arch Manage 4:95–111
14. Lee CK, Lee MS, Thurasamy R (2020) Using mediation in project disputes based on theory of planned behavior and technology acceptance model. J Legal Affairs Dispute Resol Eng Constr 12:04519044(1)–04519044(10)
15. Lee CK, Yiu TW, Cheung SO (2016) Selection and use of alternative dispute resolution (ADR) in construction projects-past and future research. Int J Project Manage 34:494–507
16. Lee CK, Yiu TW, Cheung SO (2018) A macro-micro framework of ADR Use in the Malaysian construction industry. Springer, pp 97–106
17. Lee CK, Yiu TW, Cheung SO (2018) Predicting intention to use alternative dispute resolution (ADR): an empirical test of theory of planned behaviour (TPB) model. Int J Constr Manage 1–14
18. Love P, Davis P, Ellis J, Cheung SO (2010) Dispute causation: identification of pathogenic influences in construction. Eng Constr Archit Manag 17:404–423
19. Lu W, Zhang L, Pan J (2015) Identification and analyses of hidden transaction costs in project dispute resolutions. Int J Project Manage 33:711–718
20. Oon CK (2002) Standard construction contracts in Malaysia—issues and challenges. In: Innovations in construction contracts, 31 May 2002 Melaka
21. Rogers EM (2003) Diffusion of innovations, Free Press, New York
22. Son H, Lee S, Kim C (2015) What drives the adoption of building information modeling in design organizations? An empirical investigation of the antecedents affecting architects' behavioral intentions. Autom Constr 49, Part A:92–99
23. Sourdin T (2013) Innovation and alternative dispute resolution. The Arbitrator and Mediator, December 2013, pp 111–126
24. Taylor S, Todd PA (1995) Understanding information technology usage: a test of competing models. Inf Syst Res 6:144–176
25 Tornatzky L, Klein K (1982) Innovation characteristics and innovation adoptio-implementation: a meta-analysis of findings. IEEE Trans Eng Manage 29:28–45
26. Zhang Y, Sun J, Yang Z, Wang Y (2018) Mobile social media in inter-organizational projects: aligning tool, task and team for virtual collaboration effectiveness. Int J Project Manage 36:1096–1108